Robot Mechanisms

International Series on
INTELLIGENT SYSTEMS, CONTROL, AND AUTOMATION: SCIENCE AND ENGINEERING

VOLUME 60

For further volumes:
www.springer.com/series/6259

Jadran Lenarčič · Tadej Bajd · Michael M. Stanišić

Robot Mechanisms

Springer

Jadran Lenarčič
J. Stefan Institute
Ljubljana-Vic-Rudnik, Slovenia

Tadej Bajd
Faculty of Electrical Engineering
University of Ljubljana
Ljubljana-Vic-Rudnik, Slovenia

Michael M. Stanišić
Aerospace and Mechanical Engineering
Notre Dame University
Notre Dame, IN, USA

ISBN 978-94-007-9291-3 ISBN 978-94-007-4522-3 (eBook)
DOI 10.1007/978-94-007-4522-3
Springer Dordrecht Heidelberg New York London

Printed on acid-free paper

Springer is part of Springer Science+Business Media (www.springer.com)

Foreword

The objective of this book is to provide a comprehensive introduction to the area of robot mechanisms, primarily considering industrial manipulators and humanoid arms. The book is intended for both teaching and self-study. Emphasis is given to the fundamentals of kinematic analysis and the design of robot mechanisms. The coverage of topics is untypical. Our focus is on robot kinematics, leaving out other important areas in robot mechanics and control. It is not our intention to cover a wide spectrum of areas or to consider one specific area in all its details. We strive to create a balance between theoretical and practical aspects in the development and application of robot mechanisms, and to include the latest achievements and trends in robot science and technology.

Clearly, subjects of both theory and practice of robot mechanisms are too extensive to be covered in a single book. We include topics which in our opinion are essential for robot mechanism analysis and design and which also open new research perspectives. Where possible, we present results and instructive examples of our own investigations and studies.

This book is primarily intended for students of graduate or senior undergraduate studies who are familiar with the basics of mechanism analysis and design. Because of the contemporary topics considered, we expect this book will be useful in advanced education and research activities, especially at the postgraduate level.

In the last decade robotics has undergone a major metamorphosis. Emerging from a past of largely industrial focus, robotics has extended itself into our daily lives. Robots today are intended for use in uncertain, dynamic, unstructured environments, and for intelligent interaction with humans. This requires robots to posses not only more advanced and efficient sensor-based control strategies and intelligence, but also in some cases, human-like or animal-like morphologies of their mechanisms.

This book began as a collection of personal notes for course lectures and was partially published in the Slovenian language in 2003, by the University of Ljubljana (Slovenia). The second edition came to light in 2009. A warm acceptance of the two editions by our robotics students, who mostly have an electrical engineering background, has exceeded our expectations. This encouraged us to rewrite the book in English, add additional material and offer it to a wider international audience.

We are grateful to our young colleagues Mitja Veber, Gregorij Kurillo, Jan Babič, and others, who allowed us to use some of their numerical results. We are indebted to the staff of Springer who brought this book together in a most efficient manner.

Ljubljana, Slovenia

J. Lenarčič
T. Bajd
M.M. Stanišić

Preface

The first robots, as they are understood today, were developed in the sixties of the previous century. Their mechanisms were almost without exceptions simple serial mechanical arms possessing six or preferably fewer degrees of freedom, and used almost exclusively in different industrial applications. Fifty years later, robot mechanisms are still extensively being investigated for a variety of robot applications in everyday life. A significant impulse to the development of robot mechanisms was given in the nineties by a dramatic appearance of parallel mechanisms and later by human-like or animal-like mechanisms. New designs are mainly due to many new discoveries and better understanding of human or animal motion. They also result from new technological advancements such as new materials, more efficient or conceptually different actuators and even artificial muscles.

Although the theory of robot mechanisms largely relies on methods of classical mechanics, in recent decades it has made huge steps forward. Authors, such as B. Roth, K. Hunt and J. Duffy and L.-W. Tsai, J. Angeles, J.M. McCarthy and many others, have contributed to genuine advances in science and in industrial practice. Today, for instance, all major manufacturers of industrial robots offer robots with high precision parallel kinematic structures and all major research laboratories develop humanoid robots or body parts possessing the morphology and other characteristics of humans.

The book consists of ten chapters. In the first chapter we introduce the mathematical tools which are used throughout the book, making the presentation self-contained. Translation and rotation of the body are treated as a transformation between vector spaces. Special attention is focused on the rotation matrix and its characteristic properties, as well as on the homogeneous transformation matrix which describes the position and orientation of a body, together with its velocities and accelerations. In the second chapter, we review the basic characteristics of robot mechanisms, calculate the number of degrees of freedom and present some typical arrangements of links and joints used in robot mechanisms. We present the parameters and variables of a kinematic pair and then develop a step by step procedure which enables us to model an entire mechanism.

In the third chapter we develop equations for the kinematic analysis of serial mechanisms. These equations represent the position and orientation, the linear and angular velocity and the linear and angular acceleration. We introduce the Jacobian matrix and the Hessian matrix of the mechanism. Both are of crucial importance in the kinematic analysis of robots. In the second part of the chapter we define the direct and the inverse kinematics problem. The inverse kinematics problem for serial mechanisms of general geometry is difficult to solve. In general, a closed-form solution does not exist and special-purpose numerical iterative methods are to be applied. Their main drawbacks are associated with the convergence and with multiple solutions which are pertinent to the inverse kinematics problem. Difficulties are also due to the fact that the existence of a real solution is not always guaranteed. In the fourth chapter, we define the criteria to examine different functional properties of robot mechanisms from a perspective of a user or a designer robots. Among these we study the reachable and the dexterous workspace, the kinematic flexibility and singularity, as well as the manipulability and the kinematic index. The fifth chapter deals with describing kinematic singularities in industrial robots. The description is in terms of singular planes. The singular planes show that the origin of kinematic singularities in industrial robots are pointing singularities. This leads to a discussion singularity free pointing systems and how they can be incorporated into singularity free robots.

A redundant mechanism is referred to as a mechanism which contains more degrees of freedom than is needed to perform a given task. Redundant mechanisms can solve a given primary task in an infinite number of ways. This feature allows to the robot to simultaneously solve additional secondary tasks. The kinematic redundancy of mechanisms is the theme of the sixth chapter of this book. Mathematically speaking, the system of differential equations defining the kinematics of a redundant mechanism is underdetermined and the related Jacobian matrix is rectangular. This requires special mathematical approaches to solve the inverse kinematics problem. Humans or animals are using kinematic redundancy as a tool to optimize their motion. Some simplified examples are described in the end of the section. The seventh chapter is devoted to parallel mechanisms. These mechanisms contain one or more closed kinematic chains. Different types of parallel mechanisms are described and the computation of direct and inverse kinematics of parallel mechanisms is discussed. Attention is given also to the difficulties that arise when a parallel mechanism enters in a singular position.

In the eight and the ninth chapter we describe the mathematical basis and learn about the effects of robotic touch and grip. Robotic grip is understood as a set of contacts between the fingers and the object. The problem is how to determine the conditions to restrict an object's movement. In the tenth and the last chapter of the book we present the direct and the inverse kinematic model of the thumb and fingers of the human hand. The introduced kinematic models were obtained based on a series of optical measurements of the human hand. The introduced models enable us to analyze the motion of the human hand depending on the length and width of the palm.

The book deals primarily with the analytical study of kinematics of robot mechanisms which includes the geometry of motion of mechanisms without regards to the forces and moments that cause or result from the motion. In addition to recognized areas, this book also presents examples of recent advances in emerging areas such as the design and control of humanoids and humanoid subsystems, and the analysis, modeling and simulation of human body motions.

Contents

Chapter 1
Kinematics of Rigid Bodies

Abstract The motion of rigid bodies is presented using standard vector and matrix algebra. Combinations of translations and rotations, as well as linear and angular velocities and linear and angular translations, are studied. The characteristic properties of the rotation matrix and of the homogeneous transformation matrix are described. Different ways to represent the orientation of the body are introduced, such as the Euler angles, the YPR angles and the invariants of the rotation matrix.

In this chapter the basic mathematical tools used in the study of rigid body kinematics will be introduced. We describe relevant kinematic variables such as a body's position and orientation, and its translational and angular velocities and accelerations. Kinematics will be considered from the perspective of linear algebra, in such a way that translation and rotation will be considered as transformations between vector spaces. Special attention will be focused on the rotation matrix and its characteristic properties. Also introduced will be homogeneous transformations which describe the position and orientation of a body, together with its velocities and accelerations.

1.1 Position and Displacement of a Point

The position of a point in space is completely determined by three parameters called coordinates. The coordinates can be either Cartesian, cylindrical, or spherical [67, 77]. In this chapter we will show how the position, velocity, and acceleration of a point can be defined in these different types of coordinates. In the very beginning we will introduce the function $\arctan_2$, which we shall find helpful in symbolic calculations.

1.1.1 The Function $\arctan_2$

The function $\arctan_2$ is used in computer programs and is usually denoted as `ATAN2`. It is an inverse trigonometric function, which takes into account the quad-

J. Lenarčič et al., *Robot Mechanisms*,
Intelligent Systems, Control and Automation: Science and Engineering 60,
DOI 10.1007/978-94-007-4522-3_1, © Springer Science+Business Media Dordrecht 2013

Table 1.1 Definition of function $\arctan_2 a/b$

$a > 0$ in $b > 0$	$\arctan_2 a/b = \arctan a/b$
$a > 0$ in $b < 0$	$\arctan_2 a/b = \pi + \arctan a/b$
$a < 0$ in $b > 0$	$\arctan_2 a/b = \arctan a/b$
$a < 0$ in $b < 0$	$\arctan_2 a/b = \arctan a/b - \pi$

Table 1.2 Values of the function $\arctan_2 a/b$ at the boundaries of the numerator and denominator

$a = 0$ in $b > 0$	$\arctan_2 a/b = 0$
$a = 0$ in $b < 0$	$\arctan_2 a/b = \pm\pi$
$a > 0$ in $b = 0$	$\arctan_2 a/b = \pi/2$
$a < 0$ in $b = 0$	$\arctan_2 a/b = -\pi/2$
$a = 0$ in $b = 0$	not defined

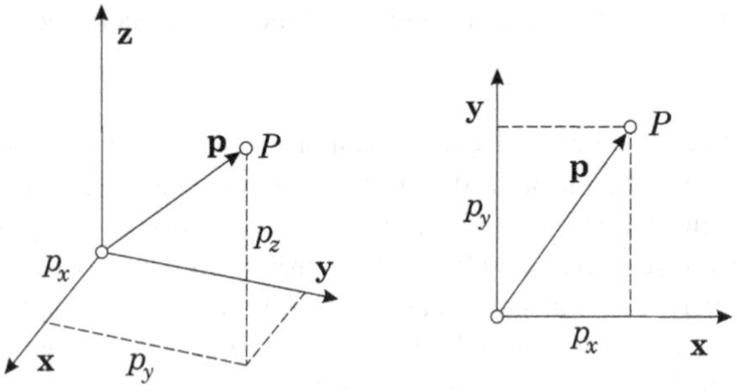

Fig. 1.1 Position of a point P in Cartesian coordinates

rant of the solution. It is defined as a function of a fraction a/b, while taking into consideration the sign of the numerator a and the denominator b. The function is given in Table 1.1.

The values of the function $\arctan_2$ at the boundaries of the numerator and the denominator are given in Table 1.2. The value of $\arctan_2$ is not defined, when numerator and denominator equal zero at the same time ($a = 0$ in $b = 0$).

1.1.2 Points in Cartesian Coordinates

The position of a point P in Cartesian coordinates is determined by a vector $\mathbf{p}$ with the components p_x, p_y, and p_z as shown in Fig. 1.1. In the plane only two coordinates are necessary to describe the position of the point P. In Fig. 1.1 these are the coordinates p_x and p_y.

When a point P on a moving body displaces, its position with respect to the reference frame is changing, which defines the point's velocity and acceleration. The

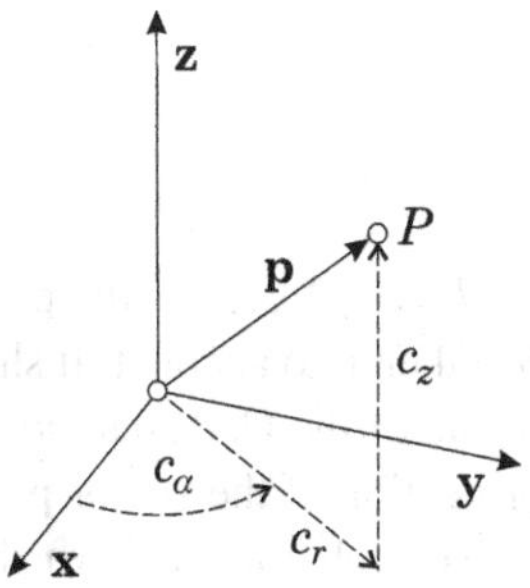

Fig. 1.2 Position of a point P in the cylindrical coordinate frame

velocity is directed tangential to the trajectory of the moving point and is represented by the time derivative of the position vector **p**

$$\mathbf{v} = \frac{d}{dt}\mathbf{p},$$

and the acceleration is the time derivative of the velocity vector

$$\mathbf{a} = \frac{d}{dt}\mathbf{v} = \frac{d^2}{dt^2}\mathbf{p}.$$

The acceleration vector has both tangential and radial components, the latter is perpendicular to the trajectory of the moving point and pointing towards the center of curvature.

As we shall see later in the text, there are singularities in the transformations between the different types of coordinates. These must be avoided since at these singularities a finite velocity expressed in Cartesian coordinates transforms into an infinite velocity in either cylindrical or spherical coordinates.

1.1.3 Points in Cylindrical Coordinates

The position of a point P in cylindrical coordinates is determined by the parameters c_r, c_α, and c_z, as shown in Fig. 1.2. These are related to the Cartesian coordinates as follows

$$\begin{aligned} p_\mathrm{x} &= c_\mathrm{r} \cos c_\alpha, \\ p_\mathrm{y} &= c_\mathrm{r} \sin c_\alpha, \\ p_\mathrm{z} &= c_\mathrm{z}. \end{aligned} \tag{1.1}$$

Equations (1.1) present the Cartesian coordinates in terms of the cylindrical coordinates. More caution should be given to the inverse relation,

$$c_\mathrm{r} = \sqrt{p_\mathrm{x}^2 + p_\mathrm{y}^2}, \tag{1.2}$$

$$c_\alpha = \arctan_2 \frac{p_y}{p_x} \pm 2k\pi, \tag{1.3}$$

$$c_z = p_z, \tag{1.4}$$

for $k = 1, 2, \ldots$, as the periodicity of the trigonometrical function arctangent must be taken into account. It should be recognized that angle c_α can be determined only if one of the Cartesian coordinates p_x or p_y is nonzero. This is the case, when the projection of the vector **p** onto the **x**–**y** plane is nonzero, i.e. when $c_r \neq 0$.

The relation between the velocities in Cartesian and cylindrical coordinates is linear. By differentiating (1.1), we obtain

$$\begin{aligned}
\dot{p}_x &= \dot{c}_r \cos c_\alpha - \dot{c}_\alpha c_r \sin c_\alpha, \\
\dot{p}_y &= \dot{c}_r \sin c_\alpha + \dot{c}_\alpha c_r \cos c_\alpha, \\
\dot{p}_z &= \dot{c}_z.
\end{aligned}$$

The above equations can be written in a vector and matrix form

$$\begin{bmatrix} \dot{p}_x \\ \dot{p}_y \\ \dot{p}_z \end{bmatrix} = \mathbf{C} \begin{bmatrix} \dot{c}_r \\ \dot{c}_\alpha \\ \dot{c}_z \end{bmatrix}, \tag{1.5}$$

where

$$\mathbf{C} = \begin{bmatrix} \cos c_\alpha & -c_r \sin c_\alpha & 0 \\ \sin c_\alpha & c_r \cos c_\alpha & 0 \\ 0 & 0 & 1 \end{bmatrix}. \tag{1.6}$$

The inverse relation is obtained by inverting the matrix **C**

$$\begin{bmatrix} \dot{c}_r \\ \dot{c}_\alpha \\ \dot{c}_z \end{bmatrix} = \mathbf{C}^{-1} \begin{bmatrix} \dot{p}_x \\ \dot{p}_y \\ \dot{p}_z \end{bmatrix}. \tag{1.7}$$

This is possible when **C** is nonsingular, i.e. when the determinant of **C**, which is c_r, is nonzero.

The time derivative of the velocity (1.5), yields the acceleration,

$$\begin{bmatrix} \ddot{p}_x \\ \ddot{p}_y \\ \ddot{p}_z \end{bmatrix} = \dot{\mathbf{C}} \begin{bmatrix} \dot{c}_r \\ \dot{c}_\alpha \\ \dot{c}_z \end{bmatrix} + \mathbf{C} \begin{bmatrix} \ddot{c}_r \\ \ddot{c}_\alpha \\ \ddot{c}_z \end{bmatrix}. \tag{1.8}$$

The time derivative, $\dot{\mathbf{C}}$, has the following form

$$\dot{\mathbf{C}} = \frac{\partial \mathbf{C}}{\partial c_r} \dot{c}_r + \frac{\partial \mathbf{C}}{\partial c_\alpha} \dot{c}_\alpha, \tag{1.9}$$

where

$$\frac{\partial \mathbf{C}}{\partial c_{\mathrm{r}}} = \begin{bmatrix} 0 & -\sin c_\alpha & 0 \\ 0 & \cos c_\alpha & 0 \\ 0 & 0 & 0 \end{bmatrix} \tag{1.10}$$

and

$$\frac{\partial \mathbf{C}}{\partial c_\alpha} = \begin{bmatrix} -\sin c_\alpha & -c_{\mathrm{r}} \cos c_\alpha & 0 \\ \cos c_\alpha & -c_{\mathrm{r}} \sin c_\alpha & 0 \\ 0 & 0 & 0 \end{bmatrix}. \tag{1.11}$$

The inverse relation between the accelerations in Cartesian and cylindrical coordinates is found as follows

$$\begin{bmatrix} \ddot{c}_{\mathrm{r}} \\ \ddot{c}_\alpha \\ \ddot{c}_{\mathrm{z}} \end{bmatrix} = \mathbf{C}^{-1} \left(\begin{bmatrix} \ddot{p}_{\mathrm{x}} \\ \ddot{p}_{\mathrm{y}} \\ \ddot{p}_{\mathrm{z}} \end{bmatrix} - \dot{\mathbf{C}} \begin{bmatrix} \dot{c}_{\mathrm{r}} \\ \dot{c}_\alpha \\ \dot{c}_{\mathrm{z}} \end{bmatrix} \right),$$

which requires that $\mathbf{C}$ be invertible. Thus the transformations of position, velocity and acceleration, from Cartesian coordinates to cylindrical coordinates, are all singular when $c_{\mathrm{r}} = 0$. The above equations can also be written in the following form

$$\begin{bmatrix} \ddot{c}_{\mathrm{r}} \\ \ddot{c}_\alpha \\ \ddot{c}_{\mathrm{z}} \end{bmatrix} = \mathbf{C}^{-1} \left(\begin{bmatrix} \ddot{p}_{\mathrm{x}} \\ \ddot{p}_{\mathrm{y}} \\ \ddot{p}_{\mathrm{z}} \end{bmatrix} - \dot{\mathbf{C}}\mathbf{C}^{-1} \begin{bmatrix} \dot{p}_{\mathrm{x}} \\ \dot{p}_{\mathrm{y}} \\ \dot{p}_{\mathrm{z}} \end{bmatrix} \right). \tag{1.12}$$

1.1.4 Points in Spherical Coordinates

The position of a point P in spherical coordinates is determined by the parameters s_{r}, s_α and s_β, shown in Fig. 1.3. These are related to the Cartesian coordinates through the following equations

$$\begin{aligned} p_{\mathrm{x}} &= s_{\mathrm{r}} \cos s_\alpha \cos s_\beta, \\ p_{\mathrm{y}} &= s_{\mathrm{r}} \sin s_\alpha \cos s_\beta, \\ p_{\mathrm{z}} &= s_{\mathrm{r}} \sin s_\beta. \end{aligned} \tag{1.13}$$

Here the Cartesian coordinates are functions of the spherical coordinates. We must be careful when considering the inverse relation. The first spherical coordinate represents the length of the vector $\mathbf{p}$

$$s_{\mathrm{r}} = \sqrt{p_{\mathrm{x}}^2 + p_{\mathrm{y}}^2 + p_{\mathrm{z}}^2}. \tag{1.14}$$

The following two spherical coordinates, which may be considered azimuth and elevation angles, are problematic. These are given in terms of the Cartesian coordinates

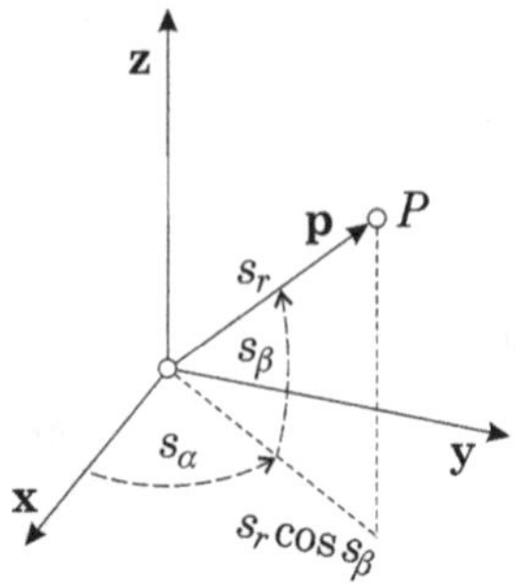

Fig. 1.3 Position of a point P in the spherical coordinate frame

by

$$s_\alpha = \arctan_2 \frac{p_y}{p_x} \pm 2k\pi, \tag{1.15}$$

$$s_\beta = \arcsin \frac{p_z}{\sqrt{p_x^2 + p_y^2 + p_z^2}} \pm 2k\pi, \tag{1.16}$$

where $k = 0, 1, \ldots$, accounts for the periodical behavior of the functions arctangent and arcsine. The angle s_α can be determined only when at least one of the Cartesian coordinates p_x or p_y is nonzero, while angle s_β can be determined only when at least one of the Cartesian coordinates p_x, p_y or p_z differs from zero. This is the case, when the projection of vector **p** onto the **x**–**y** plane is nonzero, i.e. $s_r \cos s_\beta \neq 0$.

The relation between velocities expressed in Cartesian and spherical coordinates is also linear. The time derivative of (1.13) results in

$$\begin{aligned}
\dot{p}_x &= \dot{s}_r \cos s_\alpha \cos s_\beta - \dot{s}_\alpha s_r \sin s_\alpha \cos s_\beta - \dot{s}_\beta s_r \cos s_\alpha \sin s_\beta, \\
\dot{p}_y &= \dot{s}_r \sin s_\alpha \cos s_\beta + \dot{s}_\alpha s_r \cos s_\alpha \cos s_\beta - \dot{s}_\beta s_r \sin s_\alpha \sin s_\beta, \\
\dot{p}_z &= \dot{s}_r \sin s_\beta + \dot{s}_\beta s_r \cos s_\beta
\end{aligned}$$

which can be rewritten in matrix form as,

$$\begin{bmatrix} \dot{p}_x \\ \dot{p}_y \\ \dot{p}_z \end{bmatrix} = \mathbf{S} \begin{bmatrix} \dot{s}_r \\ \dot{s}_\alpha \\ \dot{s}_\beta \end{bmatrix}, \tag{1.17}$$

where

$$\mathbf{S} = \begin{bmatrix} \cos s_\alpha \cos s_\beta & -s_r \sin s_\alpha \cos s_\beta & -s_r \cos s_\alpha \sin s_\beta \\ \sin s_\alpha \cos s_\beta & s_r \cos s_\alpha \cos s_\beta & -s_r \sin s_\alpha \sin s_\beta \\ \sin s_\beta & 0 & s_r \cos s_\beta \end{bmatrix}. \tag{1.18}$$

Transformation of velocity from Cartesian to spherical coordinates is singular when the determinant of **S** is zero. This determinant is given by $s_r \cos s_\beta$. Thus the singularity in transformation of velocity from Cartesian to spherical coordinates occurs when the singularity in transformation of Cartesian coordinates into spherical coordinates occurs, i.e. when $s_r \cos s_\beta = 0$. When $s_r \cos s_\beta \neq 0$, $\mathbf{S}$ is invertible and

$$\begin{bmatrix} \dot{s}_r \\ \dot{s}_\alpha \\ \dot{s}_\beta \end{bmatrix} = \mathbf{S}^{-1} \begin{bmatrix} \dot{p}_x \\ \dot{p}_y \\ \dot{p}_z \end{bmatrix}. \tag{1.19}$$

The relation between acceleration in Cartesian and spherical coordinates is obtained by differentiating the velocities

$$\begin{bmatrix} \ddot{p}_x \\ \ddot{p}_y \\ \ddot{p}_z \end{bmatrix} = \dot{\mathbf{S}} \begin{bmatrix} \dot{s}_r \\ \dot{s}_\alpha \\ \dot{s}_\beta \end{bmatrix} + \mathbf{S} \begin{bmatrix} \ddot{s}_r \\ \ddot{s}_\alpha \\ \ddot{s}_\beta \end{bmatrix}. \tag{1.20}$$

Here, we must consider

$$\dot{\mathbf{S}} = \frac{\partial \mathbf{S}}{\partial s_r}\dot{s}_r + \frac{\partial \mathbf{S}}{\partial s_\alpha}\dot{s}_\alpha + \frac{\partial \mathbf{S}}{\partial s_\beta}\dot{s}_\beta, \tag{1.21}$$

where

$$\frac{\partial \mathbf{S}}{\partial s_r} = \begin{bmatrix} 0 & -\sin s_\alpha \cos s_\beta & -\cos s_\alpha \sin s_\beta \\ 0 & \cos s_\alpha \cos s_\beta & -\sin s_\alpha \sin s_\beta \\ 0 & 0 & \cos s_\beta \end{bmatrix}, \tag{1.22}$$

$$\frac{\partial \mathbf{S}}{\partial s_\alpha} = \begin{bmatrix} -\sin s_\alpha \cos s_\beta & -s_r \cos s_\alpha \cos s_\beta & s_r \sin s_\alpha \sin s_\beta \\ \cos s_\alpha \cos s_\beta & -s_r \sin s_\alpha \cos s_\beta & -s_r \cos s_\alpha \sin s_\beta \\ 0 & 0 & 0 \end{bmatrix}, \tag{1.23}$$

$$\frac{\partial \mathbf{S}}{\partial s_\beta} = \begin{bmatrix} -\cos s_\alpha \sin s_\beta & s_r \sin s_\alpha \sin s_\beta & -s_r \cos s_\alpha \cos s_\beta \\ -\sin s_\alpha \sin s_\beta & -s_r \cos s_\alpha \sin s_\beta & -s_r \sin s_\alpha \cos s_\beta \\ \cos s_\beta & 0 & -s_r \sin s_\beta \end{bmatrix}. \tag{1.24}$$

The inverse relation can be expressed in the following form

$$\begin{bmatrix} \ddot{s}_r \\ \ddot{s}_\alpha \\ \ddot{s}_\beta \end{bmatrix} = \mathbf{S}^{-1} \left(\begin{bmatrix} \ddot{p}_x \\ \ddot{p}_y \\ \ddot{p}_z \end{bmatrix} - \dot{\mathbf{S}} \begin{bmatrix} \dot{s}_r \\ \dot{s}_\alpha \\ \dot{s}_\beta \end{bmatrix} \right),$$

which is valid only when $s_r \cos s_\beta \neq 0$ (as with position and velocity). The above equations can also be written in the following form

$$\begin{bmatrix} \ddot{s}_r \\ \ddot{s}_\alpha \\ \ddot{s}_\beta \end{bmatrix} = \mathbf{S}^{-1} \left(\begin{bmatrix} \ddot{p}_x \\ \ddot{p}_y \\ \ddot{p}_z \end{bmatrix} - \dot{\mathbf{S}}\mathbf{S}^{-1} \begin{bmatrix} \dot{p}_x \\ \dot{p}_y \\ \dot{p}_z \end{bmatrix} \right). \tag{1.25}$$

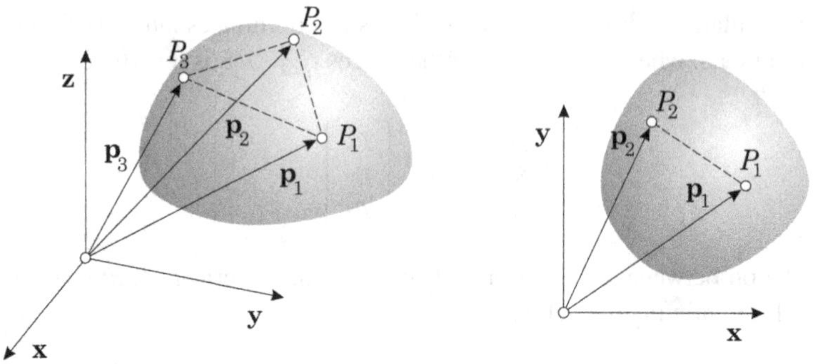

Fig. 1.4 The pose of a body in space and plane

1.2 Pose and Displacement of a Body

The pose of a body in space is determined by six independent parameters, which can be expressed in various ways. Usually they are split into position and orientation of a body.

1.2.1 Pose of a Body

The pose of a rigid body in space can be determined by use of so called natural coordinates. Natural coordinates are given by the position of three non-collinear points on the body P_1, P_2 and P_3. The three points are in Fig. 1.4 and are denoted by Cartesian vectors $\mathbf{p}_1$, $\mathbf{p}_2$ and $\mathbf{p}_3$. As every spatial vector has three coordinates, we are dealing with a total of nine coordinates. Only six of them are associated with the spatial pose of a body and are independent. The other three coordinates are dependent on the first six, and are determined from the known geometry of the triangle created by the points P_1, P_2 and P_3.

The pose of a body lying in a plane is determined by only two points. These are points P_1 in P_2 shown in Fig. 1.4. They are represented by vectors $\mathbf{p}_1$ and $\mathbf{p}_2$, each having two coordinates. All together we are dealing with four coordinates, however only three are necessary to determine the pose of a body in a plane and are independent. The fourth coordinate is dependent on the first three, and is determined from the known distance between the points.

The pose of a body can also be described as a geometrical relation between a local coordinate frame attached to the body and a reference coordinate frame, as shown in Fig. 1.5. The pose of the body is given as the position and orientation of a local coordinate frame $\mathbf{x}_1$, $\mathbf{y}_1$, $\mathbf{z}_1$, with respect to a reference coordinate frame $\mathbf{x}_0$, $\mathbf{y}_0$, $\mathbf{z}_0$.

The spatial position of a body is represented by a Cartesian vector $\mathbf{p}$, connecting the origin of the local coordinate frame with the origin of the reference frame. In the plane, a vector $\mathbf{p}$ is described with only two coordinates.

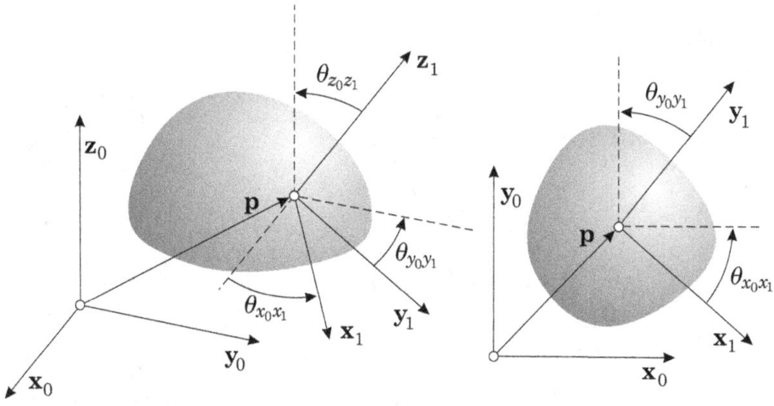

Fig. 1.5 Position and orientation of a body in space and plane

The orientation of a body in space is given by three angles, which can be measured between the axes of the local and the reference coordinate frame. One possibility of describing the orientation of a body is to select the angle between the corresponding pairs of axes of the local and reference frame, as shown in Fig. 1.5, where the angles are denoted as $\theta_{\mathrm{x_0x_1}}$, $\theta_{\mathrm{y_0y_1}}$ and $\theta_{\mathrm{z_0z_1}}$. In the plane, the orientation of a body can be described by a single angle, because of the equivalence $\theta_{\mathrm{x_0x_1}} = \theta_{\mathrm{y_0y_1}}$ and the fact that $\theta_{\mathrm{z_0z_1}} = 0$.

1.2.2 Finite Displacement of a Body

A displacement of a body in space or in a plane is a combination of a translation and a rotation [1, 10, 58].

Figure 1.6 shows a body which undergoes a planar displacement. The motion is viewed in a direction perpendicular to the plane of motion. The initial pose is denoted by i while the final pose is k. The translation is in the direction of a vector laying in the plane of motion while the rotation takes place around an axis (a line) which is perpendicular to the plane of motion. The center of rotation is an important point of planar motion. It is represented by the intersection of the rotational axis with the plane of motion.

From a given initial pose, the body can be displaced into a given final pose in an infinite number of ways, depending on the selected position of the center of rotation, and the sequence of the translation and rotation. Figure 1.6 shows two possible combinations of translation and rotation for a given initial and final pose of a body. In the upper example, the body rotates around a center S_1 and is thus displaced from the initial pose i into an intermediate pose j_1. It is afterwards translated into the final pose k. In the second case, the body is first rotated around a different center S_2 into an intermediate pose j_2 and is from there displaced along a line into the final

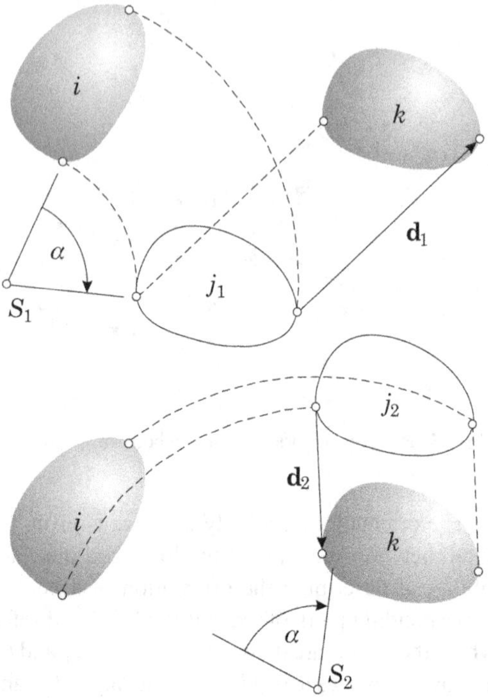

Fig. 1.6 Finite displacement of a body in plane

position k. It can be seen that the angle of rotation α is the same in both examples, regardless of the selected position for the center of rotation. However, there are differences in the directions and lengths of the translational displacement ($\mathbf{d}_1$ and $\mathbf{d}_2$), as can be seen.

There exists a unique center of rotation S, shown in the upper example of Fig. 1.7, where the movement of the body from the initial pose i into the final pose k involves only a rotation and no translation. This unique center of rotation lies at the intersection of the perpendicular bisectors of the line segments AB and CD, connecting the initial and final positions of two arbitrary points on the body.

In the pure translational displacement shown in the lower example of the figure, the line segments AB and CD are parallel and their perpendicular bisectors intersect at infinity, in the direction of the perpendicular bisectors. This translational displacement can be described as an infinitesimal rotation around a center of rotation infinitely distant from the body in a direction perpendicular to the pure translation.

We have observed that planar displacement of a body from a given initial pose to a final pose consists of a rotation followed by a translation. The angle of rotation can be uniquely determined from the known initial and final pose of the body. However, the direction and the length of the translation are dependent on the choice of the center of rotation. We have also observed that there exists a unique center of rotation whereby the body can be displaced from an initial pose to a final pose by a pure rotation.

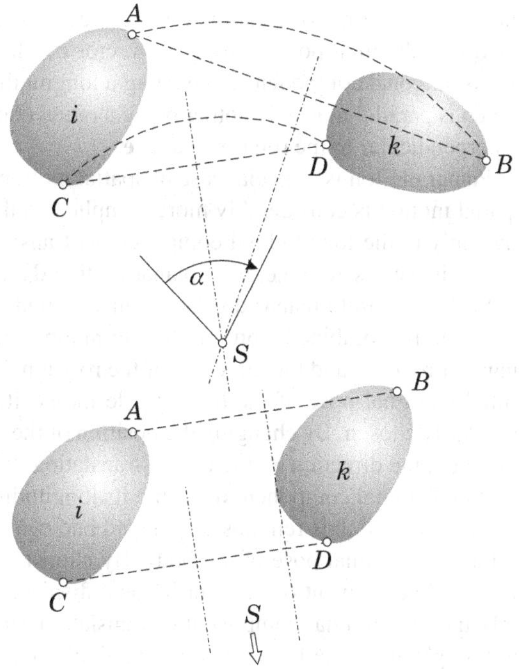

Fig. 1.7 Finite displacement of a body in a plane achieved either by rotation only or translation only

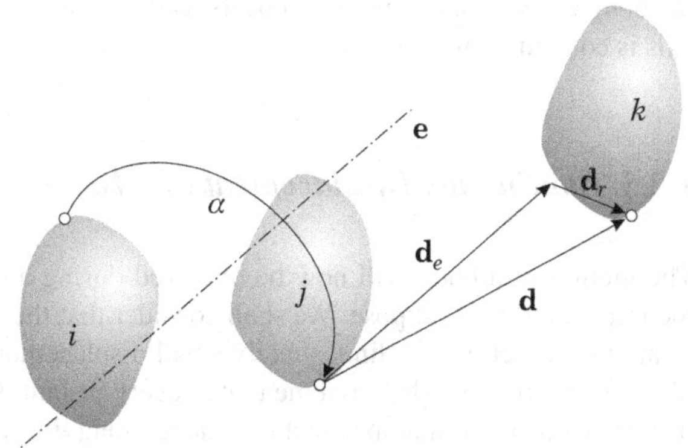

Fig. 1.8 Finite displacement of a body in space

The finite displacement of a body in a 3-dimensional space (spatial motion) can also be described by a rotation and a translation, similar to planar motion. In spatial motion, the axes of translation and rotation in general are not perpendicular and the center of rotation is replaced by the axis of rotation. In Fig. 1.8 a body is displaced from an initial pose i to a final pose k in such a way that it is first rotated by angle α around the **e** axis. The direction of axis **e** is uniquely defined by the body's orientation in the initial and final pose, although technically one could also use the vector -**e** and a rotation angle -α as a second solution which is actually equivalent to the first solution. After the rotation, the body is in an intermediate pose j which has the correct final orientation, but incorrect position. From here the body is translated into the final pose k along the vector $\mathbf{d}$. This translation can be divided into two orthogonal components, one being a longitudinal component of translation, $\mathbf{d}_e$, which is parallel to $\mathbf{e}$ and another being a radial component of translation, $\mathbf{d}_r$, which is perpendicular to the rotational axis $\mathbf{e}$.

Planar motion is a special case of spatial motion. Although it appears at first that spatial motion is considerably more complicated than planar motion, the difference lays only in the longitudinal component of translation. In planar motion this component is always zero, i.e. for a planar motion $\mathbf{d}_e$ in Fig. 1.8 is always zero.

Similar to finite planar displacement of a body, finite spatial displacement of a body can be combined from an infinite number of translations and rotations. The angle of rotation and the direction of the rotational axis are uniquely defined by the initial and final pose of the body, while the position of the rotational axis can be arbitrarily chosen. By changing the position of the rotational axis, but maintaining it direction, the direction and length of translation $\mathbf{d}$ changes. The change in $\mathbf{d}$ occurs only in its radial component $\mathbf{d}_r$, while its longitudinal component $\mathbf{d}_e$, being parallel to the rotational axis remains unchanged and can be uniquely determined from the initial and the final pose of the body. By choosing the location of axis $\mathbf{e}$ correctly, the radial component of the translational displacement can be brought to zero and only the longitudinal component of translation remains. In this case, the body can be brought from the initial pose to the final pose by a rotation about axis $\mathbf{e}$ and a translation along axis $\mathbf{e}$. This is known as a screw motion, since the movement of the body from pose i to pose k appears as if it where rotating about a screw, whose axis is coincident with $\mathbf{e}$.

1.2.3 Continuous Displacement of a Body

The motion of a body will now be observed during a continuous motion between the initial and the final pose. We shall consider that the motion of the body consists of an infinite series of infinitesimally small displacements [10]. In this observation of spatial motion we shall assume that at every instant the body is rotated about the instantaneous rotational axis and translated along the same axis. The translation in the radial direction is now replaced by the displacement of the rotational axis in a spatial motion or by the displacement of the center of rotation in a planar motion.

The rigid body in Fig. 1.9 is continuously displaced from the known initial pose into the known final pose. The motion of the body is illustrated by a series of snapshots taken at discrete time intervals $k = 0, 1, \ldots, 5$. During the motion each point of the body travels along its own trajectory. The point A travels along the trajectory from A_0 to A_5 and the point B from B_0 to B_5.

The instantaneous center of rotation of the body also changes its position and travels along a curve in the plane of motion. This curve is obtained by determining the instantaneous position of the center of rotation S_{ij} between two neighboring body poses i and j, as was explained in the preceding section. The trajectory of

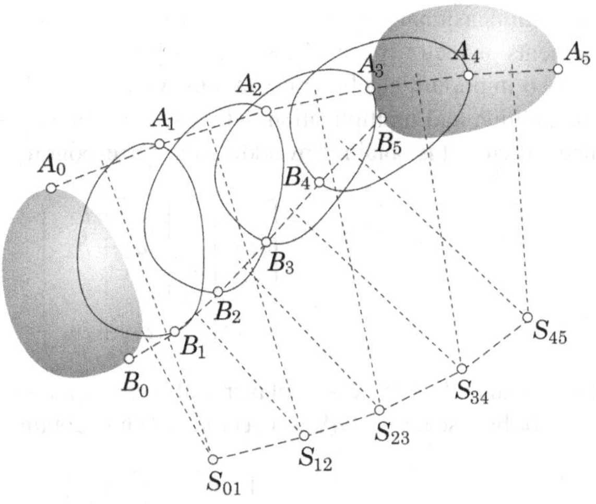

Fig. 1.9 Continuous displacement of a body in plane

the center of rotation is defined as a series of points between S_{01} and S_{45}. This trajectory can be unstable and in the case when the body's motion is predominantly translational, it can move to infinity.

When studying the continuous motion of a body, besides the trajectories of points, the velocities and accelerations of a body must also be considered. We should know that translational velocity is a property of a point in the body and represents the time rate of change of its position. All points in the body experience the same angular velocity, which is given as the rate of change of body orientation or rate of change in the angle around the rotational axis. Angular velocity is a property of the body whereas linear velocity is the property of a particular point belonging to the body.

Translational acceleration is the time rate of change of the translational velocity of a selected point in the body. The angular acceleration is the same for all points in a body and is represented by the time rate of change in angular velocity. This time rate of change in angular velocity accounts for both the time rate of change of its magnitude and its direction. During steady motion of a body, both accelerations are zero. In robot control, jerk, which is the time derivative of acceleration, is also used.

1.3 Operations in Vector Space

In general, a vector is represented by an m-tuple of numbers

$$\mathbf{a} = \begin{bmatrix} a_1 \\ a_2 \\ \vdots \\ a_m \end{bmatrix},$$

where m describes the dimension of the vector. An infinite set of m-dimensional real vectors $\mathbf{a}$, $\mathbf{b}$, …, including also the vector $\mathbf{0}$ with the coordinates 0, represents the m-dimensional real vector or linear space $\Re^m$. We say that vectors $\mathbf{a}, \mathbf{b}, \ldots$ are elements of vector space $\Re^m$, $\mathbf{a}, \mathbf{b}, \ldots \in \Re^m$.

Two important algebraic operations with vectors belonging to vector space $\Re^m$ are addition and multiplication with a scalar. The sum of two vectors $\mathbf{c} = \mathbf{a} + \mathbf{b}$ is also a vector. It is obtained by adding the corresponding components of both vectors

$$\begin{bmatrix} c_1 \\ c_2 \\ \vdots \\ c_m \end{bmatrix} = \begin{bmatrix} a_1 + b_1 \\ a_2 + b_2 \\ \vdots \\ a_m + b_m \end{bmatrix}.$$

The calculated vector $\mathbf{c}$ is coplanar with vectors $\mathbf{a}$ and $\mathbf{b}$. The result of multiplying a vector by a scalar $\mathbf{c} = a\mathbf{b}$ is a vector, which is obtained in the following way

$$\begin{bmatrix} c_1 \\ c_2 \\ \vdots \\ c_m \end{bmatrix} = \begin{bmatrix} ab_1 \\ ab_2 \\ \vdots \\ ab_m \end{bmatrix}.$$

The resulting vector has the same direction as the vector $\mathbf{b}$. However, it can have different length and if $a < 0$, then $\mathbf{c}$ points in the opposite direction of $\mathbf{b}$. It is not difficult to realize that $0\mathbf{a} = \mathbf{0}$, $1\mathbf{a} = \mathbf{a}$, and $\mathbf{0} + \mathbf{a} = \mathbf{a}$. When adding vectors, or multiplying them by a scalar, the laws of commutativity and associativity apply. Both are well known from the scalar algebra $\mathbf{a}+\mathbf{b} = \mathbf{b}+\mathbf{a}$, $(\mathbf{a}+\mathbf{b})+\mathbf{c} = \mathbf{a}+(\mathbf{b}+\mathbf{c})$ and $a\mathbf{b} = \mathbf{b}a$, $a(\mathbf{b}+\mathbf{c}) = a\mathbf{b} + a\mathbf{c}$.

Operation of vector transposing is of formal character only. It transforms a column vector into a row vector and vice versa

$$\mathbf{a}^{\mathrm{T}} = (a_1, a_2, \ldots, a_m) \quad \text{and} \quad \left(\mathbf{a}^{\mathrm{T}}\right)^{\mathrm{T}} = \mathbf{a}.$$

The scalar product (dot product) of two vectors can be defined by the use of a transposed vector

$$c = \mathbf{a}^{\mathrm{T}}\mathbf{b} = a_1 b_1 + a_2 b_2 + \cdots + a_m b_m.$$

The resulting vector is a scalar. The scalar product is commutative $\mathbf{a}^{\mathrm{T}}\mathbf{b} = \mathbf{b}^{\mathrm{T}}\mathbf{a}$ and distributive $(\mathbf{a}+\mathbf{b})^{\mathrm{T}}\mathbf{c} = \mathbf{a}^{\mathrm{T}}\mathbf{c} + \mathbf{b}^{\mathrm{T}}\mathbf{c}$. It is not difficult to understand that the scalar product of the vector with itself results in a scalar which is greater than or equal to zero, $\mathbf{a}^{\mathrm{T}}\mathbf{a} \geq 0$.

The scalar product can be useful when discussing metrics in a vector space. As an example let us consider the relations in the space $\Re^3$, where vector $\mathbf{c}$ represents the position of the point C and vector $\mathbf{b}$ the position of the point B. The vector $\mathbf{a} = \mathbf{c} - \mathbf{b}$ is connecting the points C and B. The Euclidean norm of vector $\mathbf{a}$ is defined by the scalar product

$$a = \sqrt{\mathbf{a}^{\mathrm{T}}\mathbf{a}} = |\mathbf{a}|,$$

and represents the distance between the points C and B, i.e. the length of vector **a**.

The length of the unit vector **c** equals 1. The unit vector, having the same direction and pointing orientation as the vector **a**, can be calculated with a simple formula $\mathbf{c} = \mathbf{a}/a$. By the use of the scalar product, and with some algebraic dexterity, the angle θ between two vectors of the vector space can be calculated

$$\mathbf{a}^{\mathrm{T}}\mathbf{b} = ab\cos\theta.$$

When vectors **a** and **b** are mutually orthogonal, the scalar product equals $\mathbf{a}^{\mathrm{T}}\mathbf{b} = 0$. When they are either parallel or collinear, their scalar product is $\mathbf{a}^{\mathrm{T}}\mathbf{b} = ab$. The real vector space in which the metrics or scalar product with the properties described above is defined, is called Euclidean space.

The basis of a vector space $\Re^m$ is represented by the m-tuple of the normalized and orthogonal linearly independent vectors $\mathbf{v}_1, \mathbf{v}_2, \ldots, \mathbf{v}_m$. By linear combination of the basis vectors, any element of the vector space can be obtained in the following way

$$\mathbf{a} = a_1\mathbf{v}_1 + a_2\mathbf{v}_2 + \cdots + a_m\mathbf{v}_m,$$

where the coefficients $a_1, a_2, \ldots, a_m$ represent the coordinates of vector **a**.

Three-dimensional Cartesian space with coordinate system **x**, **y**, **z** and Cartesian vectors **a**, **b**, …, is just an example of a vector space $\Re^3$. The Cartesian vector **a** is geometrically considered as a directed line segment arising from the origin of coordinate frame **x**, **y**, **z**. It can be considered as an element of a three-dimensional real vector space and will be represented by a column vector

$$\mathbf{a} = \begin{bmatrix} a_1 \\ a_2 \\ a_3 \end{bmatrix},$$

$\mathbf{a} \in \Re^3$. The basis of this vector space is represented by vectors

$$\mathbf{x} = \begin{bmatrix} 1 \\ 0 \\ 0 \end{bmatrix}, \qquad \mathbf{y} = \begin{bmatrix} 0 \\ 1 \\ 0 \end{bmatrix}, \qquad \mathbf{z} = \begin{bmatrix} 0 \\ 0 \\ 1 \end{bmatrix},$$

which are mutually perpendicular and normalized, therefore

$$\mathbf{x}^{\mathrm{T}}\mathbf{y} = \mathbf{x}^{\mathrm{T}}\mathbf{z} = \mathbf{y}^{\mathrm{T}}\mathbf{z} = 0$$

and

$$\mathbf{x}^{\mathrm{T}}\mathbf{x} = \mathbf{y}^{\mathrm{T}}\mathbf{y} = \mathbf{z}^{\mathrm{T}}\mathbf{z} = 1.$$

The cross product (also called the vector product) **c** of vectors **a** and **b** is defined as

$$\mathbf{c} = \mathbf{a} \times \mathbf{b} = \begin{bmatrix} a_2b_3 - a_3b_2 \\ a_3b_1 - a_1b_3 \\ a_1b_2 - a_2b_1 \end{bmatrix}.$$

The cross product is useful in handling Cartesian vectors. It is not difficult to show that the resulting vector $\mathbf{c}$ is directed perpendicular to both vectors $\mathbf{a}$ and $\mathbf{b}$. It is perpendicular to the plane in which both vectors lie, thus $\mathbf{a}^T\mathbf{c} = \mathbf{b}^T\mathbf{c} = 0$. Also valid is the relation $\mathbf{a} \times \mathbf{b} = -\mathbf{b} \times \mathbf{a}$. The length of vector $\mathbf{c}$ is equal to the area of the parallelogram with sides a in b

$$c = ab \sin\theta,$$

where θ stands by the angle between vectors $\mathbf{a}$ and $\mathbf{b}$. The cross product of two parallel vectors is therefore zero. The double cross product $\mathbf{a} \times (\mathbf{b} \times \mathbf{c}) = (\mathbf{a}^T\mathbf{c})\mathbf{b} - (\mathbf{a}^T\mathbf{b})\mathbf{c}$ is coplanar to vectors $\mathbf{b}$ and $\mathbf{c}$.

Linear transformation from vector space $\Re^n$ into vector space $\Re^m$ is represented by a matrix of dimension $m \times n$. A matrix is $m \times n$-tuple of numbers, which are organized into m rows and n columns

$$\mathbf{A} = \begin{bmatrix} a_{11} & a_{12} & \dots & a_{1n} \\ a_{21} & a_{22} & \dots & a_{2n} \\ \vdots & \vdots & \vdots & \vdots \\ a_{m1} & a_{m2} & \dots & a_{mn} \end{bmatrix}.$$

The product of a vector $\mathbf{b}$ of dimension n and a matrix $\mathbf{A}$ of dimension $m \times n$ is a vector $\mathbf{c}$ of dimension m

$$\mathbf{Ab} = \mathbf{c}.$$

Vector $\mathbf{c}$ is a transformation of the vector $\mathbf{b}$ into the space $\Re^m$. In general, vector $\mathbf{c}$ can differ from vector $\mathbf{b}$ in its length, direction and dimension. The above equation is a compact form of a system of linear algebraic equations, which can be written, considering the components of the matrix and both vectors, in the following way

$$\begin{aligned} a_{11}b_1 + a_{12}b_2 + \dots + a_{1n}b_n &= c_1, \\ a_{21}b_1 + a_{22}b_2 + \dots + a_{2n}b_n &= c_2, \\ &\vdots \\ a_{m1}b_1 + a_{m2}b_2 + \dots + a_{mn}b_n &= c_m. \end{aligned}$$

The elements of matrix $\mathbf{A}$ are the coefficients of the linear equations, while the components of vectors $\mathbf{b}$ and $\mathbf{c}$ are either dependent or independent variables. When the vector $\mathbf{b}$ is known, it is not difficult to calculate the vector $\mathbf{c}$. It is a different situation when we know the components of vector $\mathbf{c}$ and it is our aim to calculate vector $\mathbf{b}$. When $m > n$, the number of equations exceeds the number of unknowns. The system of equations is over-determined. The exact solution exists only when the equations of the system are not in contradiction. Otherwise only an approximate solution of the vector $\mathbf{b}$ can be found. When $m < n$, the system of equations is under-determined, because there are more unknowns than equations. Such a system has an infinite number of solutions for vector $\mathbf{b}$. Typically, the number of equations equals the number of unknowns, i.e. $m = n$. In this case we are dealing with a square matrix $\mathbf{A}$. Such a matrix can be inverted if it is of full rank, meaning it is not singular and its determinant is not equal to zero. In this case there is only a single solution for vector $\mathbf{b}$

$$\mathbf{b} = \mathbf{A}^{-1}\mathbf{c}.$$

Matrix $\mathbf{A}^{-1}$ operates in the opposite direction from matrix $\mathbf{A}$, as it transforms vector $\mathbf{c}$ into vector $\mathbf{b}$.

Similar to the vector operations, we can also introduce matrix operations, such as the transpose of a matrix, multiplication of a matrix with scalar, multiplication of two matrices, inverting of a square matrix. These operations are well described in linear algebra textbooks. Here, we shall introduce two less usual operations with vectors and matrices, which will be defined in the $\Re^3$ space.

First, based on the cross product of two vectors, define a product $\mathbf{a} \otimes \mathbf{B}$ of a 3-dimensional vector $\mathbf{a}$ and 3×3-dimensional matrix $\mathbf{B}$ which can be written by columns

$$\mathbf{B} = \begin{bmatrix} \mathbf{b}_1 & \mathbf{b}_2 & \mathbf{b}_3 \end{bmatrix},$$

in the following way

$$\mathbf{a} \otimes \mathbf{B} = \begin{bmatrix} \mathbf{a} \times \mathbf{b}_1 & \mathbf{a} \times \mathbf{b}_2 & \mathbf{a} \times \mathbf{b}_3 \end{bmatrix}. \tag{1.26}$$

By use of the above definition, the following rules can be proven

$$\begin{aligned} \mathbf{B}(\mathbf{a} \otimes \mathbf{A}) &= (\mathbf{B}\mathbf{a}) \otimes (\mathbf{B}\mathbf{A}), \\ (\mathbf{a} \otimes \mathbf{A})\mathbf{B} &= \mathbf{a} \otimes (\mathbf{A}\mathbf{B}), \\ (\mathbf{a} \otimes \mathbf{A})\mathbf{b} &= \mathbf{a} \times (\mathbf{A}\mathbf{b}). \end{aligned} \tag{1.27}$$

When multiplying with scalar k, we have

$$k(\mathbf{a} \otimes \mathbf{B}) = (k\mathbf{a}) \otimes \mathbf{B} = \mathbf{a} \otimes (k\mathbf{B}), \tag{1.28}$$

and

$$\begin{aligned} (\mathbf{a} + \mathbf{b}) \otimes \mathbf{C} &= \mathbf{a} \otimes \mathbf{C} + \mathbf{b} \otimes \mathbf{C}, \\ \mathbf{a} \otimes (\mathbf{B} + \mathbf{C}) &= \mathbf{a} \otimes \mathbf{B} + \mathbf{a} \otimes \mathbf{C}. \end{aligned} \tag{1.29}$$

The operation $\mathbf{a} \otimes \mathbf{I}$, where $\mathbf{I}$ is the identity matrix, results in a skew-symmetric matrix with the components of vector $\mathbf{a}$

$$\mathbf{a} \otimes \mathbf{I} = \begin{bmatrix} 0 & -a_3 & a_2 \\ a_3 & 0 & -a_1 \\ -a_2 & a_1 & 0 \end{bmatrix}. \tag{1.30}$$

Let vector $\mathbf{b}$ be defined in a 3-dimensional Cartesian space as the product of matrix $\mathbf{B}$ and vector $\mathbf{c}$, that is $\mathbf{b} = \mathbf{B}\mathbf{c}$. Let us examine what is the cross product of vector $\mathbf{a}$

and vector $\mathbf{b}$. Applying the rules in (1.27), the following relation can be shown

$$\mathbf{a} \times \mathbf{b} = \mathbf{a} \times (\mathbf{B}\mathbf{c}) = (\mathbf{a} \otimes \mathbf{B})\mathbf{c}, \tag{1.31}$$

resulting in

$$(\mathbf{a} \otimes \mathbf{B})\mathbf{c} = (\mathbf{a} \otimes \mathbf{I})\mathbf{B}\mathbf{c} \tag{1.32}$$

and finally

$$\mathbf{a} \times \mathbf{b} = (\mathbf{a} \otimes \mathbf{I})\mathbf{b} = \mathbf{A}\mathbf{b}, \tag{1.33}$$

where $\mathbf{A}$ is a skew-symmetric matrix with components of a vector $\mathbf{a}$, such as written in (1.30). This expression is not a surprising find. However, it is often used in robot kinematics because it is transforming a nonstandard operation, such as the cross product of two Cartesian vectors, into the multiplication of a matrix and a vector, which is a standard operation in linear algebra. Nevertheless, the cross product of two vectors should not be overlooked, as it has important geometrical meaning in the Cartesian space, where it is defined.

In theoretical studies two other concepts are of importance: linear vector and linear scalar invariants [1], which will be defined for a 3×3-dimensional matrix $\mathbf{A}$ with the elements a_{uv}. The vector invariants $\mathrm{vect}(\mathbf{A})$ is a 3-dimensional vector $\mathbf{s}$

$$\mathbf{s} = \mathrm{vect}(\mathbf{A}) = \frac{1}{2}\begin{bmatrix} a_{32} - a_{23} \\ a_{13} - a_{31} \\ a_{21} - a_{12} \end{bmatrix}. \tag{1.34}$$

The invariant $\mathrm{tr}(\mathbf{A})$ is a scalar s, defined as a sum of diagonal elements

$$s = \mathrm{tr}(\mathbf{A}) = a_{11} + a_{22} + a_{33}. \tag{1.35}$$

When a matrix is symmetric ($\mathbf{A} = \mathbf{A}^{\mathrm{T}}$), then $\mathrm{vect}(\mathbf{A}) = \mathbf{0}$. When a matrix is skew-symmetric ($\mathbf{A} = -\mathbf{A}^{\mathrm{T}}$), then $\mathrm{tr}(\mathbf{A}) = 0$. The following relation can be added

$$\mathrm{vect}(k\mathbf{A}) = k\,\mathrm{vect}(\mathbf{A}), \qquad \mathrm{tr}(k\mathbf{A}) = k\,\mathrm{tr}(\mathbf{A}), \tag{1.36}$$

$$\mathrm{vect}(\mathbf{A} + \mathbf{B}) = \mathrm{vect}(\mathbf{A}) + \mathrm{vect}(\mathbf{B}), \qquad \mathrm{tr}(\mathbf{A} + \mathbf{B}) = \mathrm{tr}(\mathbf{A}) + \mathrm{tr}(\mathbf{B}). \tag{1.37}$$

An important relation, which will be used several times in the remaining text, is the following

$$\mathrm{vect}(\mathbf{s} \otimes \mathbf{I}) = \mathbf{s}, \tag{1.38}$$

where $\mathbf{I}$ is the identity matrix.

1.3.1 Rotation Matrix

Rotation of a body can be regarded as a linear transformation between two vector spaces. In our case we are dealing with two 3-dimensional Cartesian spaces, whose

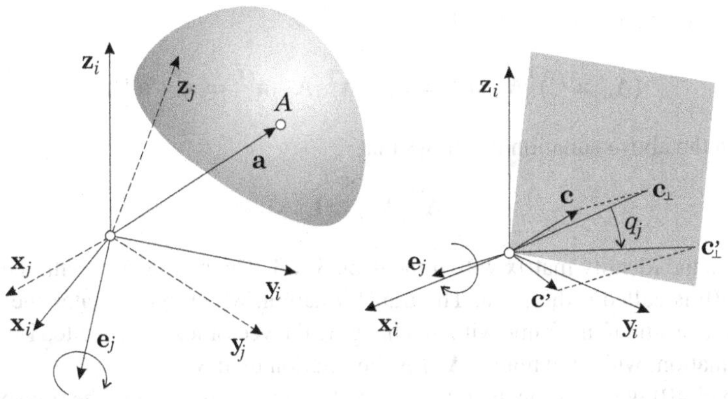

Fig. 1.10 Rotation of a coordinate frame

bases are a local coordinate frame attached to the body and a fixed reference coordinate frame [1, 10, 46]. This transformation has several properties. Most importantly, the rotation does not alter vectors lengths.

The pose of a body shown in Fig. 1.10 can be observed either with respect to frame $\mathbf{x}_i$, $\mathbf{y}_i$, $\mathbf{z}_i$ or with respect to frame $\mathbf{x}_j$, $\mathbf{y}_j$, $\mathbf{z}_j$. The coordinate frames have the same origin, while they are rotated around vector $\mathbf{e}_j$. The position of a point A, represented by Cartesian vector $\mathbf{a}$, can be expressed either with respect to the basis $\mathbf{x}_i, \mathbf{y}_i, \mathbf{z}_i$, where its position is represented by vector $\mathbf{a}^{(i)} \in \Re_i^3$, or with respect to the basis $\mathbf{x}_j, \mathbf{y}_j, \mathbf{z}_j$, where it is represented by vector $\mathbf{a}^{(j)} \in \Re_j^3$.

The operator transforming an element of space $\Re_j^3$ into space $\Re_i^3$ is represented by a matrix of dimension 3×3, which we shall denote as $\mathbf{A}_{i,j}$. The following relation exists

$$\mathbf{a}^{(i)} = \mathbf{A}_{i,j}\mathbf{a}^{(j)}. \tag{1.39}$$

The vector spaces $\Re_i^3$ and $\Re_j^3$ are isomorphic, as they have the same dimension. The matrix $\mathbf{A}_{i,j}$, describing the transformation between them, is called isomorphism. The usual rules for calculations with matrices can be applied

$$\mathbf{A}_{i,j}\left(k\mathbf{a}^{(j)}\right) = k\mathbf{A}_{i,j}\mathbf{a}^{(j)} = k\mathbf{a}^{(i)},$$

$$\mathbf{A}_{i,j}\left(\mathbf{a}^{(j)} + \mathbf{b}^{(j)}\right) = \mathbf{A}_{i,j}\mathbf{a}^{(j)} + \mathbf{A}_{i,j}\mathbf{b}^{(j)} = \mathbf{a}^{(i)} + \mathbf{b}^{(i)},$$

where k is an arbitrary scalar.

It is obvious that the length of the vector $\mathbf{a}$ is the same regardless of the frame in which it was measured. As the rotation of the coordinate frame does not alter the length of the vector, we can write

$$\mathbf{a}^{\mathrm{T}(i)}\mathbf{a}^{(i)} = \mathbf{a}^{\mathrm{T}(j)}\mathbf{a}^{(j)}.$$

By the use of relation (1.39) we obtain

$$\left(\mathbf{A}_{i,j}\mathbf{a}^{(j)}\right)^{\mathrm{T}}\mathbf{A}_{i,j}\mathbf{a}^{(j)} = \mathbf{a}^{\mathrm{T}(j)}\mathbf{A}_{i,j}^{\mathrm{T}}\mathbf{A}_{i,j}\mathbf{a}^{(j)} = \mathbf{a}^{\mathrm{T}(i)}\mathbf{a}^{(i)},$$

and from the above equation it follows that

$$\mathbf{A}_{i,j}^{\mathrm{T}}\mathbf{A}_{i,j} = \mathbf{I}, \tag{1.40}$$

where $\mathbf{I}$ is the identity matrix with dimension 3×3. The matrix $\mathbf{A}_{i,j}$ with the property (1.40) is called orthogonal. The transformation, which only rotates the vector around the origin of the frame without changing the vector length, is called isometric transformation, while the matrix $\mathbf{A}_{i,j}$ is the rotation matrix.

From (1.40) we can show that inverse rotation matrix is equal to the transpose of the rotation matrix

$$\mathbf{A}_{i,j}^{-1} = \mathbf{A}_{i,j}^{\mathrm{T}}, \tag{1.41}$$

from here the relation between the determinants can also be obtained

$$\det(\mathbf{A}_{i,j}) = \det\left(\mathbf{A}_{i,j}^{-1}\right) = 1.$$

The scalar products of individual rows or columns of an orthogonal matrix with themselves are equal one, while their mutual scalar products are zero

$$\sum_{j=1}^{3} a_{i,j}^2 = \sum_{j=1}^{3} a_{j,i}^2 = 1 \tag{1.42}$$

and

$$\sum_{j=1}^{3} a_{i,j}a_{k,j} = \sum_{j=1}^{3} a_{j,i}a_{j,k} = 0, \tag{1.43}$$

for every $i = 1, 2, 3$, $k = 1, 2, 3$ in $k \neq i$, where the scalars a_{uv} are the elements of the orthogonal matrix.

Since the rotation matrix $\mathbf{A}_{i,j}$ does not alter the length of a vector, the rotation matrix is distributive with regards to the cross product

$$\mathbf{A}_{i,j}\left(\mathbf{a}^{(j)} \times \mathbf{b}^{(j)}\right) = \left(\mathbf{A}_{i,j}\mathbf{a}^{(j)}\right) \times \left(\mathbf{A}_{i,j}\mathbf{b}^{(j)}\right) = \mathbf{a}^{(i)} \times \mathbf{b}^{(i)}.$$

In the same way as we transformed the vector $\mathbf{a}$ from the space $\Re_j^3$ into the space $\Re_i^3$, we can also transform the base vectors $\mathbf{x}_j, \mathbf{y}_j, \mathbf{z}_j$

$$\begin{aligned} \mathbf{x}_j^{(i)} &= \mathbf{A}_{i,j}\mathbf{x}_j^{(j)} \\ \mathbf{y}_j^{(i)} &= \mathbf{A}_{i,j}\mathbf{y}_j^{(j)} \\ \mathbf{z}_j^{(i)} &= \mathbf{A}_{i,j}\mathbf{z}_j^{(j)} \end{aligned}.$$

The above equations can be written in a matrix form

$$\begin{bmatrix}\mathbf{x}_j^{(i)} & \mathbf{y}_j^{(i)} & \mathbf{z}_j^{(i)}\end{bmatrix} = \mathbf{A}_{i,j}\begin{bmatrix}\mathbf{x}_j^{(j)} & \mathbf{y}_j^{(j)} & \mathbf{z}_j^{(j)}\end{bmatrix}.$$

As we have the identity matrix on the right side of the above equation

$$\begin{bmatrix}\mathbf{x}_j^{(j)} & \mathbf{y}_j^{(j)} & \mathbf{z}_j^{(j)}\end{bmatrix} = \mathbf{I}$$

it follows

$$\mathbf{A}_{i,j} = \begin{bmatrix}\mathbf{x}_j^{(i)} & \mathbf{y}_j^{(i)} & \mathbf{z}_j^{(i)}\end{bmatrix}. \tag{1.44}$$

The rotation matrix $\mathbf{A}_{i,j}$ therefore consists of the transformations of the vectors representing the basis of the vector space $\Re_j^3$ in the space $\Re_i^3$.

As the following matrix

$$\begin{bmatrix}\mathbf{x}_i^{(i)} & \mathbf{y}_i^{(i)} & \mathbf{z}_i^{(i)}\end{bmatrix} = \mathbf{I}$$

is an identity matrix, its transpose can be used to premultiply $\mathbf{A}_{i,j}$ in (1.44)

$$\mathbf{A}_{i,j} = \begin{bmatrix}\mathbf{x}_i^{(i)} & \mathbf{y}_i^{(i)} & \mathbf{z}_i^{(i)}\end{bmatrix}^{\mathrm{T}}\begin{bmatrix}\mathbf{x}_j^{(i)} & \mathbf{y}_j^{(i)} & \mathbf{z}_j^{(i)}\end{bmatrix},$$

which gives

$$\mathbf{A}_{i,j} = \begin{bmatrix}\mathbf{x}_i^{\mathrm{T}(i)}\mathbf{x}_j^{(i)} & \mathbf{x}_i^{\mathrm{T}(i)}\mathbf{y}_j^{(i)} & \mathbf{x}_i^{\mathrm{T}(i)}\mathbf{z}_j^{(i)} \\ \mathbf{y}_i^{\mathrm{T}(i)}\mathbf{x}_j^{(i)} & \mathbf{y}_i^{\mathrm{T}(i)}\mathbf{y}_j^{(i)} & \mathbf{y}_i^{\mathrm{T}(i)}\mathbf{z}_j^{(i)} \\ \mathbf{z}_i^{\mathrm{T}(i)}\mathbf{x}_j^{(i)} & \mathbf{z}_i^{\mathrm{T}(i)}\mathbf{y}_j^{(i)} & \mathbf{z}_i^{\mathrm{T}(i)}\mathbf{z}_j^{(i)}\end{bmatrix}. \tag{1.45}$$

As the lengths of the base vectors are 1, their scalar vectors can be replaced by the cosines of the angles between the vectors

$$\mathbf{A}_{i,j} = \begin{bmatrix}\cos\theta_{\mathrm{x}_i\mathrm{x}_j} & \cos\theta_{\mathrm{x}_i\mathrm{y}_j} & \cos\theta_{\mathrm{x}_i\mathrm{z}_j} \\ \cos\theta_{\mathrm{y}_i\mathrm{x}_j} & \cos\theta_{\mathrm{y}_i\mathrm{y}_j} & \cos\theta_{\mathrm{y}_i\mathrm{z}_j} \\ \cos\theta_{\mathrm{z}_i\mathrm{x}_j} & \cos\theta_{\mathrm{z}_i\mathrm{y}_j} & \cos\theta_{\mathrm{z}_i\mathrm{z}_j}\end{bmatrix}. \tag{1.46}$$

The components of the rotation matrix $\mathbf{A}_{i,j}$ are the cosines of the angles between the axes of the coordinate frames representing the bases of the vector spaces $\Re_j^3$ and $\Re_i^3$. As the elements of the orthogonal matrix are related to each other ((1.42)–(1.43)), we only have to know three angles between the axes of the coordinate frames in order to determine the matrix $\mathbf{A}_{i,j}$. Here, at most two of the angles can correspond to the same axis of one of the frames. The rest of the angles can then be calculated from (1.42)–(1.43).

The rotation matrix can be derived by the use of Rodrigues formula of finite rotations. It gives the components of the vector $\mathbf{c}'$, which results from rotation of the vector $\mathbf{c}$ around the axis $\mathbf{e}_j$ by the angle q_j (Fig. 1.10)

$$\mathbf{c}' = \mathbf{c}\cos q_j + \mathbf{e}_j\mathbf{e}_j^{\mathrm{T}}\mathbf{c}(1-\cos q_j) + \mathbf{e}_j \times \mathbf{c}\sin q_j. \tag{1.47}$$

All the vectors must be expressed in the same basis. The axis of rotation is represented by a unit vector, thus $\mathbf{e}_j^{\mathrm{T}}\mathbf{e}_j = 1$. The angle q_j is measured between the projections of the vectors $\mathbf{c}$ and $\mathbf{c}'$ in the plane perpendicular to the vector $\mathbf{e}_j$. This is the angle between the vectors $\mathbf{c}_\perp$ and $\mathbf{c}'_\perp$. The coordinate frames $\mathbf{x}_i$, $\mathbf{y}_i$, $\mathbf{z}_i$ and $\mathbf{x}_j$, $\mathbf{y}_j$, $\mathbf{z}_j$ coincide before performing the rotation. Afterwards the coordinate frame $\mathbf{x}_j$, $\mathbf{y}_j$, $\mathbf{z}_j$ is rotated around the unit vector $\mathbf{e}_j$ by the angle q_j relative to the reference frame $\mathbf{x}_i$, $\mathbf{y}_i$, $\mathbf{z}_i$. The Rodrigues formula (1.47) applied to the base vectors gives

$$\begin{aligned}
\mathbf{x}_j^{(i)} &= \mathbf{x}_i^{(i)}\cos q_j + \mathbf{e}_j^{(i)}\mathbf{e}_j^{\mathrm{T}(i)}\mathbf{x}_i^{(i)}(1-\cos q_j) + \mathbf{e}_j^{(i)} \times \mathbf{x}_i^{(i)}\sin q_j,\\
\mathbf{y}_j^{(i)} &= \mathbf{y}_i^{(i)}\cos q_j + \mathbf{e}_j^{(i)}\mathbf{e}_j^{\mathrm{T}(i)}\mathbf{y}_i^{(i)}(1-\cos q_j) + \mathbf{e}_j^{(i)} \times \mathbf{y}_i^{(i)}\sin q_j,\\
\mathbf{z}_j^{(i)} &= \mathbf{z}_i^{(i)}\cos q_j + \mathbf{e}_j^{(i)}\mathbf{e}_j^{\mathrm{T}(i)}\mathbf{z}_i^{(i)}(1-\cos q_j) + \mathbf{e}_j^{(i)} \times \mathbf{z}_i^{(i)}\sin q_j.
\end{aligned}$$

The above equations can be rewritten in matrix form

$$\begin{aligned}
\begin{bmatrix}\mathbf{x}_j^{(i)} & \mathbf{y}_j^{(i)} & \mathbf{z}_j^{(i)}\end{bmatrix} = \mathbf{e}_j^{(i)} \otimes \begin{bmatrix}\mathbf{x}_i^{(i)} & \mathbf{y}_i^{(i)} & \mathbf{z}_i^{(i)}\end{bmatrix}\sin q_j \\
+ \mathbf{e}_j^{(i)}\mathbf{e}_j^{\mathrm{T}(i)}\begin{bmatrix}\mathbf{x}_i^{(i)} & \mathbf{y}_i^{(i)} & \mathbf{z}_i^{(i)}\end{bmatrix}(1-\cos q_j) \\
+ \begin{bmatrix}\mathbf{x}_i^{(i)} & \mathbf{y}_i^{(i)} & \mathbf{z}_i^{(i)}\end{bmatrix}\cos q_j.
\end{aligned}$$

After rearranging, the formula describing the rotation matrix is obtained in the following general form [46]

$$\mathbf{A}_{i,j} = \boldsymbol{\Delta}_{i,j}\sin q_j + (\mathbf{I} - \boldsymbol{\Lambda}_{i,j})\cos q_j + \boldsymbol{\Lambda}_{i,j}. \tag{1.48}$$

Here $\mathbf{I}$ is the identity matrix, while the matrices $\boldsymbol{\Delta}_{i,j}$ and $\boldsymbol{\Lambda}_{i,j}$ depend on the rotation axis vector $\mathbf{e}_j^{(i)} = (e_{j1}, e_{j2}, e_{j3})^{\mathrm{T}}$

$$\boldsymbol{\Delta}_{i,j} = \mathbf{e}_j^{(i)} \otimes \mathbf{I} = \begin{bmatrix} 0 & -e_{j3} & e_{j2} \\ e_{j3} & 0 & -e_{j1} \\ -e_{j2} & e_{j1} & 0 \end{bmatrix} \tag{1.49}$$

and

$$\boldsymbol{\Lambda}_{i,j} = \mathbf{e}_j^{(i)}\mathbf{e}_j^{\mathrm{T}(i)} = \begin{bmatrix} e_{j1}e_{j1} & e_{j1}e_{j2} & e_{j1}e_{j3} \\ e_{j1}e_{j2} & e_{j2}e_{j2} & e_{j2}e_{j3} \\ e_{j1}e_{j3} & e_{j2}e_{j3} & e_{j3}e_{j3} \end{bmatrix}. \tag{1.50}$$

Equation (1.48) defines the rotation matrix describing the transformation between the vector spaces $\Re_i^3$ and $\Re_j^3$, which have the same origin and are rotated by angle q_j about the axis passing through the origin in the direction of unit vector $\mathbf{e}_j$. It is important to realize, that in the Rodrigues formula the angle q_j appears as argument of the trigonometric functions sine and cosine.

In special cases the rotation matrix is considerably simplified. When $q_j = 0$ we have the identity matrix $\mathbf{A}_{i,j} = \mathbf{I}$. When the vector of rotation $\mathbf{e}_j$ is parallel to $\mathbf{x}_i$, we have

$$\mathbf{A}_{i,j} = \begin{bmatrix} 1 & 0 & 0 \\ 0 & \cos q_j & -\sin q_j \\ 0 & \sin q_j & \cos q_j \end{bmatrix}, \tag{1.51}$$

when the rotation vector $\mathbf{e}_j$ is parallel to $\mathbf{y}_i$, it follows

$$\mathbf{A}_{i,j} = \begin{bmatrix} \cos q_j & 0 & \sin q_j \\ 0 & 1 & 0 \\ -\sin q_j & 0 & \cos q_j \end{bmatrix} \tag{1.52}$$

and when the rotation vector $\mathbf{e}_j$ is parallel to $\mathbf{z}_i$, we have

$$\mathbf{A}_{i,j} = \begin{bmatrix} \cos q_j & -\sin q_j & 0 \\ \sin q_j & \cos q_j & 0 \\ 0 & 0 & 1 \end{bmatrix}. \tag{1.53}$$

The rotational matrix is orthogonal, while the matrices $\boldsymbol{\Delta}_{i,j}$ and $\boldsymbol{\Lambda}_{i,j}$ have several other interesting properties. The matrix $\boldsymbol{\Delta}_{i,j}$ is skew-symmetric, while the matrix $\boldsymbol{\Lambda}_{i,j}$ is symmetric

$$\begin{aligned} \boldsymbol{\Delta}_{i,j} &= -\boldsymbol{\Delta}_{i,j}^{\mathrm{T}}, \\ \boldsymbol{\Lambda}_{i,j} &= \boldsymbol{\Lambda}_{i,j}^{\mathrm{T}}. \end{aligned} \tag{1.54}$$

It can be easily demonstrated that $\boldsymbol{\Lambda}_{i,j}$ is an idempotent matrix and that the product of $\boldsymbol{\Delta}_{i,j}$ and $\boldsymbol{\Lambda}_{i,j}$ is always zero

$$\begin{aligned} \boldsymbol{\Delta}_{i,j}\boldsymbol{\Delta}_{i,j} &= \boldsymbol{\Delta}_{i,j}^2 = \boldsymbol{\Lambda}_{i,j} - \mathbf{I}, \\ \boldsymbol{\Lambda}_{i,j}\boldsymbol{\Lambda}_{i,j} &= \boldsymbol{\Lambda}_{i,j}^2 = \boldsymbol{\Lambda}_{i,j}, \\ \boldsymbol{\Delta}_{i,j}\boldsymbol{\Lambda}_{i,j} &= \boldsymbol{\Lambda}_{i,j}\boldsymbol{\Delta}_{i,j} = \mathbf{0}. \end{aligned} \tag{1.55}$$

Useful expressions are also obtained from products with the unit vector of the rotation axis

$$\begin{aligned} \boldsymbol{\Delta}_{i,j}\mathbf{e}_j^{(i)} &= \mathbf{0}, \\ \boldsymbol{\Lambda}_{i,j}\mathbf{e}_j^{(i)} &= \mathbf{e}_j^{(i)}, \\ \mathbf{e}_j^{(i)} \otimes \boldsymbol{\Delta}_{i,j} &= \boldsymbol{\Lambda}_{i,j} - \mathbf{I}, \\ \mathbf{e}_j^{(i)} \otimes \boldsymbol{\Lambda}_{i,j} &= \mathbf{0}, \end{aligned} \tag{1.56}$$

where the product of a vector and matrix is calculated according to (1.26).

By the use of the properties of the matrices $\mathbf{\Delta}_{i,j}$ and $\mathbf{\Lambda}_{i,j}$ we can prove

$$\mathbf{A}_{i,j}\mathbf{A}_{i,j}^{\mathrm{T}} = \mathbf{I},$$

the product of the rotation vector and the rotation matrix can also be derived

$$\mathbf{e}_j^{(i)} \otimes \mathbf{A}_{i,j} = \mathbf{e}_j^{(i)} \otimes \mathbf{\Delta}_{i,j} \sin q_j + \left(\mathbf{e}_j^{(i)} \otimes \mathbf{I} - \mathbf{e}_j^{(i)} \otimes \mathbf{\Lambda}_{i,j}\right) \cos q_j + \mathbf{e}_j^{(i)} \otimes \mathbf{\Lambda}_{i,j},$$

which gives

$$\mathbf{e}_j^{(i)} \otimes \mathbf{A}_{i,j} = \mathbf{\Delta}_{i,j} \cos q_j - (\mathbf{I} - \mathbf{\Lambda}_{i,j}) \sin q_j.$$

It follows, that the product of the rotation vector and the rotation matrix is equal to the derivative of the rotation matrix with respect to the angle of rotation

$$\mathbf{e}_j^{(i)} \otimes \mathbf{A}_{i,j} = \frac{\partial \mathbf{A}_{i,j}}{\partial q_j}. \tag{1.57}$$

This observation can also be generalized for the higher derivatives by introducing multiple successive multiplication with the rotation vector

$$\mathbf{e}_j^{(i)} \otimes \frac{\partial^{k-1} \mathbf{A}_{i,j}}{\partial (q_j)^{k-1}} = \frac{\partial^k \mathbf{A}_{i,j}}{\partial (q_j)^k}, \tag{1.58}$$

where $k = 1, 2, \ldots$ is the order of the derivative.

Let us also consider the vector and scalar invariants of a matrix (1.34), (1.35) and relate them to the matrices $\mathbf{\Delta}_{i,j}$ and $\mathbf{\Lambda}_{i,j}$. It is obvious, that

$$\begin{aligned} \operatorname{vect}(\mathbf{\Delta}_{i,j}) &= \mathbf{e}_j^{(i)}, \\ \operatorname{vect}(\mathbf{\Lambda}_{i,j}) &= 0, \\ \operatorname{tr}(\mathbf{\Delta}_{i,j}) &= 0, \\ \operatorname{tr}(\mathbf{\Lambda}_{i,j}) &= \mathbf{e}_j^{\mathrm{T}(i)} \mathbf{e}_j^{(i)} = 1. \end{aligned} \tag{1.59}$$

From the above equations we obtain

$$\operatorname{vect}(\mathbf{A}_{i,j}) = \mathbf{e}_j^{(i)} \sin q_j \tag{1.60}$$

and

$$\operatorname{tr}(\mathbf{A}_{i,j}) = 2\cos q_j + 1. \tag{1.61}$$

Assume that $\mathbf{X}$ is an arbitrary 3×3 dimensional matrix. Then we have

$$\mathbf{\Delta}_{i,j}\mathbf{X} = \left(\mathbf{e}_j^{(i)} \otimes \mathbf{I}\right)\mathbf{X} = \mathbf{e}_j^{(i)} \otimes \mathbf{X}. \tag{1.62}$$

When $\mathbf{Y}$ is an orthogonal 3×3 matrix with the property $\mathbf{Y}^{\mathrm{T}}\mathbf{Y} = \mathbf{I}$, the following equation is also valid

$$\mathbf{Y}\mathbf{\Delta}_{i,j} = \mathbf{Y}\left(\mathbf{e}_j^{(i)} \otimes \mathbf{I}\right) = \left(\mathbf{Y}\mathbf{e}_j^{(i)}\right) \otimes \mathbf{Y}. \tag{1.63}$$

Fig. 1.11 Translation of coordinate frame

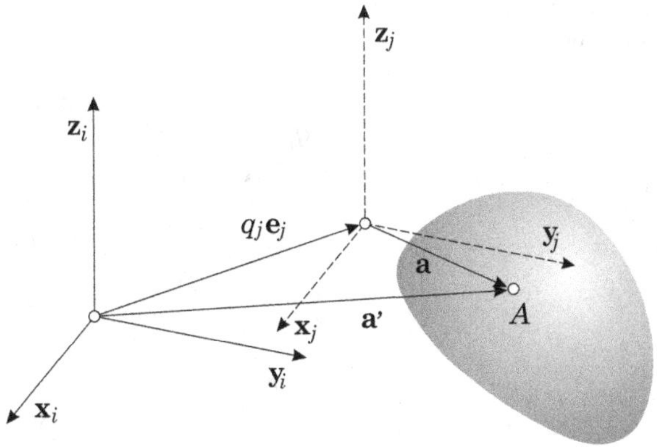

In the literature a somewhat different form of the rotation matrix can also be found. It is obtained by replacing the matrix $\mathbf{\Lambda}_{i,j}$ in (1.48) by the expression derived from (1.55), as follows

$$\mathbf{\Lambda}_{i,j} = \mathbf{\Delta}_{i,j}^2 + \mathbf{I}.$$

The result of this replacement is

$$\mathbf{A}_{i,j} = \mathbf{I} + \mathbf{\Delta}_{i,j} \sin q_j + \mathbf{\Delta}_{i,j}^2 (1 - \cos q_j). \tag{1.64}$$

This form of Rodrigues formula has no significant advantage.

1.3.2 Translation Vector

In Fig. 1.11, coordinate frame $\mathbf{x}_j$, $\mathbf{y}_j$, $\mathbf{z}_j$ is displaced from frame $\mathbf{x}_i$, $\mathbf{y}_i$, $\mathbf{z}_i$ in the direction of the unit vector $\mathbf{e}_j$ the distance q_j. The corresponding pairs of axes of both frames are parallel. The body is referred to as translated and all its points are displaced in the same direction by the same distance. The translation between the frames is determined by the translation vector $q_j\mathbf{e}_j$. The position of the point A is given by the vector $\mathbf{a}$ and is not changing with respect to the frame $\mathbf{x}_j$, $\mathbf{y}_j$, $\mathbf{z}_j$, which is attached to the body. The position of the point A expressed in the frame $\mathbf{x}_i$, $\mathbf{y}_i$, $\mathbf{z}_i$ is dependent on the length and direction of the vector $\mathbf{a}'$.

Let $\mathbf{a}^{(i)}$ denote the vector $\mathbf{a}$ as expressed in the vector space $\Re_i^3$, while $\mathbf{a}^{(j)}$ denotes the vector as expressed in the space $\Re_j^3$. The vectors are identical, as the translation of a vector in space does not change its coordinates with respect to the reference frame

$$\mathbf{a}^{(i)} = \mathbf{a}^{(j)}.$$

The position of the point A is expressed in the space $\Re_i^3$ by the following sum

$$\mathbf{a}'^{(i)} = \mathbf{a}^{(i)} + q_j\mathbf{e}_j^{(i)}. \tag{1.65}$$

Fig. 1.12 Rotation before translation

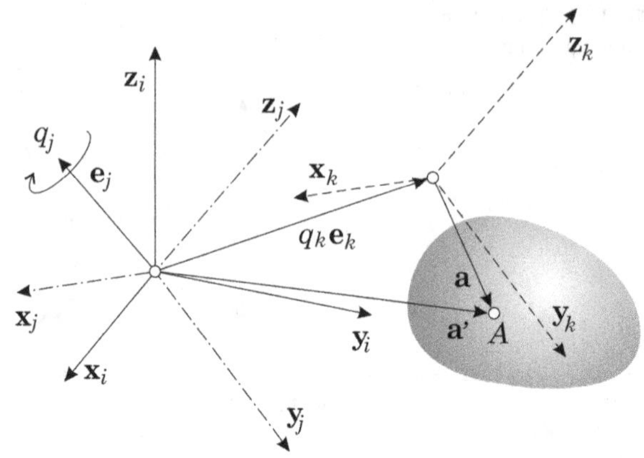

The translation is therefore mathematically described as a sum of vectors, which all have to be expressed in the same coordinate frame.

1.4 Sequences of Translations and Rotations

Let us examine a few examples demonstrating how sequences of rotations and translations are presented. For this purpose we shall observe the coordinate frames $\mathbf{x}_i$, $\mathbf{y}_i$, $\mathbf{z}_i$, $\mathbf{x}_j$, $\mathbf{y}_j$, $\mathbf{z}_j$, and $\mathbf{x}_k$, $\mathbf{y}_k$, $\mathbf{z}_k$.

1.4.1 Rotation Before Translation

In the example shown in Fig. 1.12, the coordinate frame $\mathbf{x}_k$, $\mathbf{y}_k$, $\mathbf{z}_k$ is translated by the distance q_k in the direction of the vector $\mathbf{e}_k$ relative to the frame $\mathbf{x}_j$, $\mathbf{y}_j$, $\mathbf{z}_j$, which is rotated by the angle q_j around the axis $\mathbf{e}_j$ relative to the reference coordinate frame $\mathbf{x}_i$, $\mathbf{y}_i$, $\mathbf{z}_i$. The sequence of rotation and translation can be mathematically described by the rotation matrix $\mathbf{A}_{i,j}$ and translation vector $q_k\mathbf{e}_k$. The bases of the spaces $\Re^3_k$ and $\Re^3_j$ are parallel. Therefore, the representation of the vector $\mathbf{a}$ in the space $\Re^3_j$ is to the same as the representation in the space $\Re^3_k$

$$\mathbf{a}^{(j)} = \mathbf{a}^{(k)}.$$

The position of the point A, i.e. vector $\mathbf{a}'$, is calculated in the space $\Re^3_j$

$$\mathbf{a}'^{(j)} = \mathbf{a}^{(j)} + q_k\mathbf{e}_k^{(j)}.$$

Let us now express the vector $\mathbf{a}'$ in the reference coordinate frame

$$\mathbf{a}'^{(i)} = \mathbf{A}_{i,j}\mathbf{a}'^{(j)},$$

Fig. 1.13 Translation before rotation

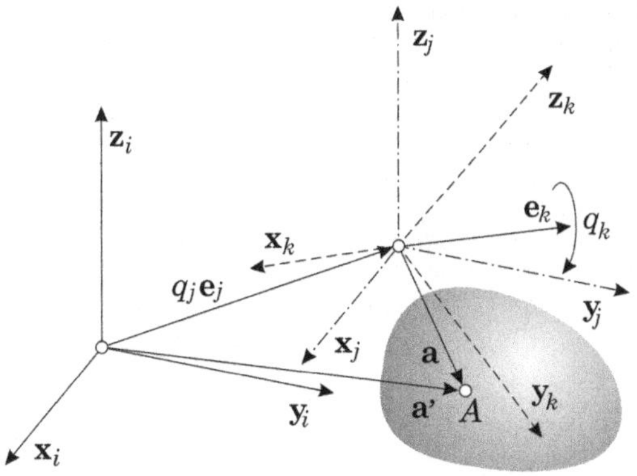

from where we have

$$\mathbf{a}'^{(i)} = \mathbf{A}_{i,j}\big(\mathbf{a}^{(j)} + q_k\mathbf{e}_k^{(j)}\big). \tag{1.66}$$

1.4.2 Translation Before Rotation

Figure 1.13 shows the coordinate frame $\mathbf{x}_k$, $\mathbf{y}_k$, $\mathbf{z}_k$, which is rotated by the angle q_k around the axis $\mathbf{e}_k$ relative to the frame $\mathbf{x}_j$, $\mathbf{y}_j$, $\mathbf{z}_j$. The latter is displaced by the distance q_j in the direction of the vector $\mathbf{e}_j$ relative to the frame $\mathbf{x}_i$, $\mathbf{y}_i$, $\mathbf{z}_i$. The sequence of translation and rotation is mathematically described by the translation vector $q_j\mathbf{e}_j$ and the rotation matrix $\mathbf{A}_{j,k}$.

First we must express the vector **a**, whose representation is in the vector space $\Re_k^3$ denoted by the vector $\mathbf{a}^{(k)}$, in the vector space $\Re_j^3$

$$\mathbf{a}^{(j)} = \mathbf{A}_{j,k}\mathbf{a}^{(k)}.$$

As the bases of the spaces $\Re_j^3$ and $\Re_i^3$ are parallel, the representation of the vector **a** in the space $\Re_i^3$ equals

$$\mathbf{a}^{(i)} = \mathbf{a}^{(j)}.$$

The vector $\mathbf{a}'$, representing the position of the point A in the reference coordinate frame $\Re_i^3$,

$$\mathbf{a}'^{(i)} = \mathbf{a}^{(i)} + q_j\mathbf{e}_j^{(i)},$$

can be written in the following form

$$\mathbf{a}'^{(i)} = \mathbf{A}_{j,k}\mathbf{a}^{(k)} + q_j\mathbf{e}_j^{(i)}. \tag{1.67}$$

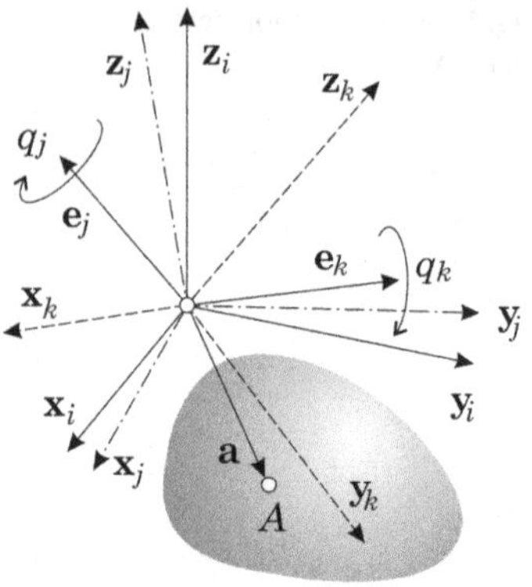

Fig. 1.14 Two consecutive rotations

1.4.3 Two Rotations

In Fig. 1.14 the coordinate frame $\mathbf{x}_k, \mathbf{y}_k, \mathbf{z}_k$ is rotated by the angle q_k around the axis $\mathbf{e}_k$ relative to the frame $\mathbf{x}_j, \mathbf{y}_j, \mathbf{z}_j$, which itself is rotated by the angle q_j around the axis $\mathbf{e}_j$ relative to the frame $\mathbf{x}_i, \mathbf{y}_i, \mathbf{z}_i$. The sequence of rotations is mathematically described by the rotation matrices $\mathbf{A}_{i,j}$ and $\mathbf{A}_{j,k}$. They can be obtained by the use of formula (1.48). First we shall transform the vector $\mathbf{a}$, whose components in the vector space $\Re_k^3$ is represented by $\mathbf{a}^{(k)}$, into the vector space $\Re_j^3$

$$\mathbf{a}^{(j)} = \mathbf{A}_{j,k}\mathbf{a}^{(k)},$$

and then we shall transform $\mathbf{a}^{(j)}$, into the vector space $\Re_i^3$

$$\mathbf{a}^{(i)} = \mathbf{A}_{i,j}\mathbf{a}^{(j)}.$$

By combining the last two equations

$$\mathbf{a}^{(i)} = \mathbf{A}_{i,j}\mathbf{A}_{j,k}\mathbf{a}^{(k)}$$

it follows

$$\mathbf{A}_{i,k} = \mathbf{A}_{i,j}\mathbf{A}_{j,k}. \tag{1.68}$$

The rotation matrix describing two consecutive rotations is the product of two rotation matrices representing each individual rotation. It is therefore a function of both angles q_j and q_k and both rotation vectors $\mathbf{e}_j$ and $\mathbf{e}_k$. The matrix $\mathbf{A}_{i,k}$ is orthogonal

$$\mathbf{A}_{i,k}^{\mathrm{T}}\mathbf{A}_{i,k} = \mathbf{A}_{j,k}^{\mathrm{T}}\mathbf{A}_{i,j}^{\mathrm{T}}\mathbf{A}_{i,j}\mathbf{A}_{j,k} = \mathbf{A}_{j,k}^{\mathrm{T}}\mathbf{A}_{j,k} = \mathbf{I}.$$

Any rotation matrix can be seen as a product of several intermediate rotation matrices. Likewise, a rotation matrix resulting from several consecutive rotations, can be seen as a single rotation.

A special case occurs, when two consecutive rotations take place around two parallel axes, thus $\mathbf{e}_j = \mathbf{e}_k$. The rotation matrices $\mathbf{A}_{i,j}$ and $\mathbf{A}_{j,k}$ have the same form.

Also the product matrix $\mathbf{A}_{i,k}$ has the same form, while the angle of the total rotation is represented by the sum of the two individual rotation angles. This can be demonstrated with a simple example. Let us consider the following two parallel rotation vectors

$$\mathbf{e}_j^{(i)} = \mathbf{e}_k^{(j)} = \begin{bmatrix} 0 \\ 0 \\ 1 \end{bmatrix}.$$

The corresponding matrices are calculated by the formula (1.48)

$$\mathbf{A}_{i,j} = \begin{bmatrix} \cos q_j & -\sin q_j & 0 \\ \sin q_j & \cos q_j & 0 \\ 0 & 0 & 1 \end{bmatrix},$$

$$\mathbf{A}_{j,k} = \begin{bmatrix} \cos q_k & -\sin q_k & 0 \\ \sin q_k & \cos q_k & 0 \\ 0 & 0 & 1 \end{bmatrix}.$$

By the use of trigonometric rules we finally obtain

$$\mathbf{A}_{i,k} = \mathbf{A}_{i,j}\mathbf{A}_{j,k} = \begin{bmatrix} \cos(q_j + q_k) & -\sin(q_j + q_k) & 0 \\ \sin(q_j + q_k) & \cos(q_j + q_k) & 0 \\ 0 & 0 & 1 \end{bmatrix}. \tag{1.69}$$

1.5 Position and Orientation of a Body

The pose of a body in space is determined by a vector which defines the position of the origin of the coordinate frame attached to the body relative to the reference frame and by a rotation matrix which defines the orientation of the frame attached to the body relative to the reference frame.

1.5.1 Rotation Angle and Vector

The orientation of the local coordinate frame $\mathbf{x}_j$, $\mathbf{y}_j$, $\mathbf{z}_j$ relative to the reference coordinate frame $\mathbf{x}_i$, $\mathbf{y}_i$, $\mathbf{z}_i$ is defined by the rotation matrix $\mathbf{A}_{i,j}$. In this way the orientation can be regarded as a rotation around the axis $\mathbf{e}_j$ by the angle q_j, as shown in Fig. 1.15. We understand that there are two possible combinations of the same orientation, $\mathbf{e}_j$ and q_j and/or the equivalent combination $-\mathbf{e}_j$ in $-q_j$. Also, periodicity must be included, $q_j \pm 2k\pi$ and $-q_j \pm 2k\pi$, $k = 0, 1, \ldots$.

Knowing the axis of rotation $\mathbf{e}_j$ (its representation in the vector space $\Re_i$ is denoted by the vector $\mathbf{e}_j^{(i)}$) and the angle of rotation q_j, we find the elements of the

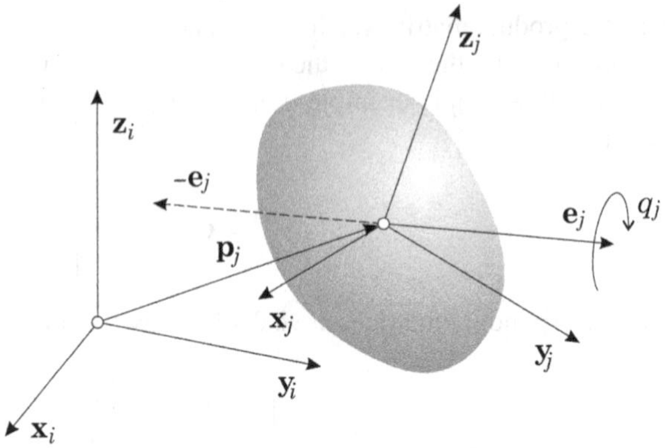

Fig. 1.15 Position and orientation of body

rotation matrix by the use of formula (1.48). From there, the inverse relations are also derived. As there is

$$2\cos q_j = a_{11} + a_{22} + a_{33} - 1,$$

where a_{uv} are the elements of the $\mathbf{A}_{i,j}$ matrix. Both possible solutions for the rotation angle can be found in the following form

$$q_j = \pm\arccos\frac{a_{11} + a_{22} + a_{33} - 1}{2}. \tag{1.70}$$

The sign of the rotation angle q_j is transferred through the sine function to the signs of the elements of the rotation vector $\mathbf{e}_j$

$$e_{j1} = \frac{a_{32} - a_{23}}{2\sin q_j}, \tag{1.71}$$

$$e_{j2} = \frac{a_{13} - a_{31}}{2\sin q_j}, \tag{1.72}$$

$$e_{j3} = \frac{a_{21} - a_{12}}{2\sin q_j}. \tag{1.73}$$

Here, e_{ju} is the corresponding element of the rotation vector $\mathbf{e}_j^{(i)}$. The above equations are valid when $\sin q_j \neq 0$. When $\sin q_j = 0$, there is the trivial case of $q_j = 0 \pm k\pi$, $k = 0, 1, \ldots$, and the rotation vector $\mathbf{e}_j^{(i)}$ cannot be uniquely determined.

At first it appears that we need four parameters for representing the spatial orientation of a body, the rotation angle q_j and three elements of the rotation vector $\mathbf{e}_j$. Only three of these four parameters are independent, as there is the following constraint among the elements of the rotation vector

$$e_{j1}^2 + e_{j2}^2 + e_{j3}^2 = 1.$$

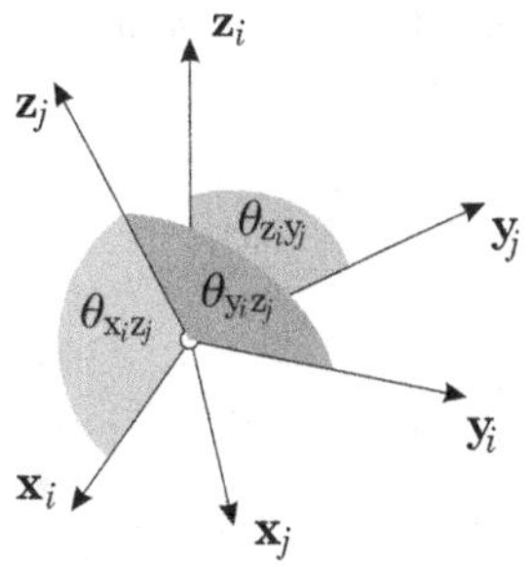

Fig. 1.16 Angles between the pairs of coordinate axes

From this, if when we know two components of the rotation vector, the third can be computed. Be aware however, that information about the sign of the third component is lost.

1.5.2 The Angles Between Pairs of Coordinate Axes

One method of defining the spatial orientation of a body is by measuring the angles between the corresponding axes of the local $\mathbf{x}_j, \mathbf{y}_j, \mathbf{z}_j$ and the reference coordinate frame $\mathbf{x}_i, \mathbf{y}_i, \mathbf{z}_i$ [84]. The cosines of these angles are the elements of the matrix $\mathbf{A}_{i,j}$ (1.46). To define the orientation of a body uniquely, three appropriately selected angles are sufficient, the rest of the angles are dependent on the three selected angles.

The three selected orientation angles must not correspond to only one axis of either the local or reference coordinate frame. In other words, the cosines of the three selected angles must not belong to the same row, or the same column, of the rotation matrix. For example, suppose the three angles between vector $\mathbf{z}_j$ and three vectors $\mathbf{x}_i$, $\mathbf{y}_i$, $\mathbf{z}_i$ were chosen to define the relative orientation of frames i and j. These are the angles, $\theta_{\mathrm{x_iz_j}}$, $\theta_{\mathrm{y_iz_j}}$ and $\theta_{\mathrm{z_iz_j}}$, all belong to the third column of the rotation matrix in (1.46). In this case the body could rotate around the axis $\mathbf{z}_j$ without changing the values of the three angles, hence these three angles are not sufficient to describe the relative orientation of the two frames. In fact, an infinite number of relative orientations correspond to these three angles. There are 78 triples of angles, which uniquely define the orientation of a body. An example is the triple $\theta_{\mathrm{x_iz_j}}, \theta_{\mathrm{y_iz_j}}, \theta_{\mathrm{z_iy_j}}$ shown in Fig. 1.16.

In this case the orientation angles are obtained from the rotation matrix $\mathbf{A}_{i,j}$ by considering the matrix (1.46) in the following way

$$\theta_{\mathrm{x_iz_j}} = \pm \arccos a_{13} \pm 2k\pi, \quad k = 0, 1, \ldots, \tag{1.74}$$

$$\theta_{\mathrm{y_iz_j}} = \pm \arccos a_{23} \pm 2k\pi, \quad k = 0, 1, \ldots, \tag{1.75}$$

$$\theta_{\mathrm{z_iy_j}} = \pm \arccos a_{32} \pm 2k\pi, \quad k = 0, 1, \ldots. \tag{1.76}$$

Because of the properties of the function arccosine, the orientation of the body does not depend on the sign of individual angles.

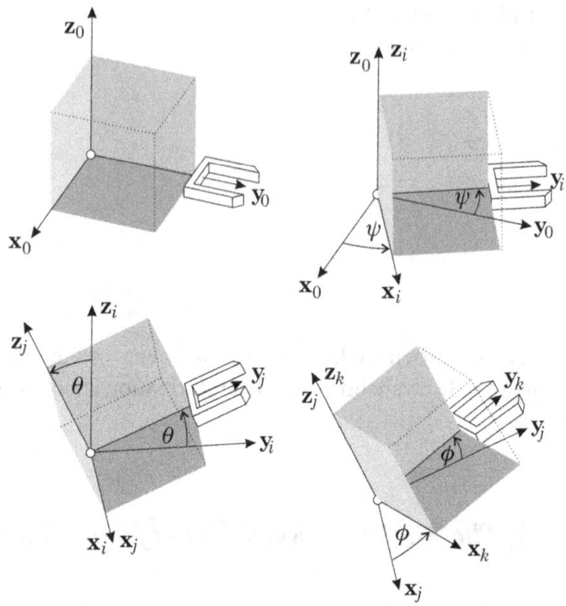

Fig. 1.17 Euler orientation angles

Description of the orientation of a body by the relative angles between the pairs of corresponding axes of the local and reference coordinate frame is mathematically correct and appealing. Nevertheless, it has little practical value, as such a description does not provide clear visualization of the spatial pose of a body.

1.5.3 Euler Orientation Angles

Euler orientation angles, usually denoted as ψ, θ, and ϕ, are shown in Fig. 1.17. This is the most widely used method for determining the spatial orientation of a body in mechanics, aeronautics, navigation, and robotics [67, 86]. The Euler angles are treated as the coordinates of three consecutive rotations about the axes of a coordinate frame.

Figure 1.17 shows only one of the six possibilities. We shall consider four coordinate frames. The reference coordinate frame is $\mathbf{x}_0, \mathbf{y}_0, \mathbf{z}_0$. It is followed by the frame $\mathbf{x}_i, \mathbf{y}_i, \mathbf{z}_i$, which is rotated by the angle ψ around the axis $\mathbf{z}_0$, and the frame $\mathbf{x}_j, \mathbf{y}_j, \mathbf{z}_j$, which is rotated by the angle θ around the axis $\mathbf{x}_i$. The coordinate frame $\mathbf{x}_k, \mathbf{y}_k, \mathbf{z}_k$ is attached to the body and is rotated by the angle ϕ around the $\mathbf{z}_j$ axis of the preceding coordinate frame. The rotation matrix, describing the orientation of the body, is a product of three rotations

$$\mathbf{A}_{0,k} = \mathbf{A}_{0,i}\mathbf{A}_{i,j}\mathbf{A}_{j,k}, \tag{1.77}$$

where we are dealing with the following separate rotation matrices

$$\mathbf{A}_{0,i} = \begin{bmatrix} \cos\psi & -\sin\psi & 0 \\ \sin\psi & \cos\psi & 0 \\ 0 & 0 & 1 \end{bmatrix},$$

$$\mathbf{A}_{i,j} = \begin{bmatrix} 1 & 0 & 0 \\ 0 & \cos\theta & -\sin\theta \\ 0 & \sin\theta & \cos\theta \end{bmatrix},$$

$$\mathbf{A}_{j,k} = \begin{bmatrix} \cos\phi & -\sin\phi & 0 \\ \sin\phi & \cos\phi & 0 \\ 0 & 0 & 1 \end{bmatrix}.$$

Let us multiply the three matrices (1.77), while writing only the third column and the third row of the resulting matrix

$$\mathbf{A}_{0,k} = \begin{bmatrix} . & . & \sin\psi\sin\theta \\ . & . & -\cos\psi\sin\theta \\ \sin\theta\sin\phi & \sin\theta\cos\phi & \cos\theta \end{bmatrix}.$$

From here the Euler orientation angles can be derived. The angle θ has two solutions

$$\theta = \pm \arccos a_{33} \pm 2k\pi, \quad k = 0, 1, \ldots. \tag{1.78}$$

By choosing the positive solution by the angle θ, we have

$$\psi = \arctan_2 \frac{a_{13}}{-a_{23}} \pm 2k\pi, \quad k = 0, 1, \ldots, \tag{1.79}$$

$$\phi = \arctan_2 \frac{a_{31}}{a_{32}} \pm 2k\pi, \quad k = 0, 1, \ldots, \tag{1.80}$$

when selecting the negative solution by the angle θ, it follows

$$\psi = \arctan_2 \frac{-a_{13}}{a_{23}} \pm 2k\pi, \quad k = 0, 1, \ldots, \tag{1.81}$$

$$\phi = \arctan_2 \frac{-a_{31}}{-a_{32}} \pm 2k\pi, \quad k = 0, 1, \ldots. \tag{1.82}$$

Both triples of the Euler angles are equivalent and belong to the same orientation of a body.

The above equations are valid when at least one of the elements a_{13} and a_{23} and one of the elements a_{31} and a_{32} is nonzero. Otherwise the rotations ψ and ϕ are parallel and the Euler angles ψ and ϕ cannot be uniquely determined. When $a_{33} = 1$ and $\theta = 0 \pm 2k\pi$, $k = 0, 1, \ldots$, we can only calculate the sum

$$\psi + \phi = \arctan_2 \frac{a_{21}}{a_{11}} \pm 2k\pi, \quad k = 0, 1, \ldots,$$

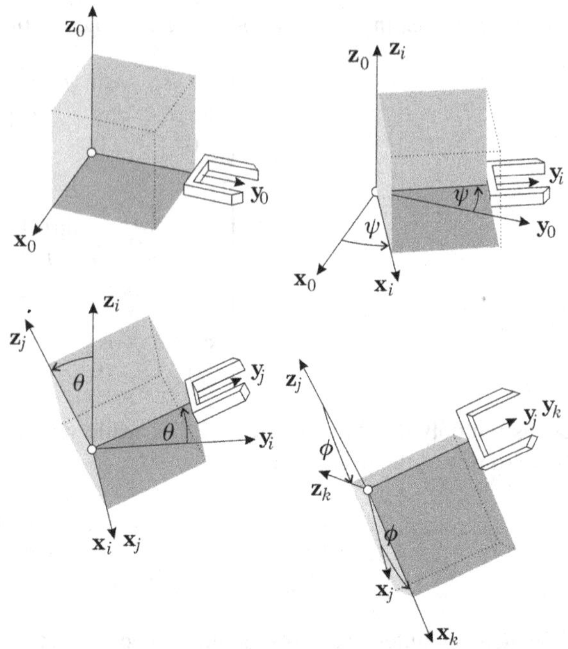

Fig. 1.18 YPR orientation angles

and when $a_{33} = -1$ and $\theta = \pi \pm 2k\pi$, $k = 0, 1, \ldots$, we can only calculate the difference

$$\psi - \phi = \arctan_2 \frac{a_{21}}{a_{11}} \pm 2k\pi, \quad k = 0, 1, \ldots.$$

1.5.4 Orientation Angles YPR

The abbreviation YPR comes from the technical terms describing the direction of a motion of a vessel. The term *yaw* means an instantaneous rotation that displaces the leading end of the vehicle to the left or the right, *pitch* means an instantaneous rotation that causes a rising or descent of the leading end of the vehicle and *roll* stands for an instantaneous rotation around the axis defined by the direction of motion of the vehicle. This notation is well established in robotics, as it clearly describes the orientation of a gripper, welding gun etc. [67].

The YPR angles will be denoted by ψ (*yaw*), θ (*pitch*), and ϕ (*roll*). In the same way as the Euler angles, also the YPR angles are treated as the coordinates of three consecutive rotations with the mutually perpendicular axes of rotation. There are six possibilities and one of them is shown in Fig. 1.18.

The reference coordinate frame is denoted as $\mathbf{x}_0$, $\mathbf{y}_0$, $\mathbf{z}_0$. It is followed by the frame $\mathbf{x}_i$, $\mathbf{y}_i$, $\mathbf{z}_i$, which is rotated by the angle ψ around the axis $\mathbf{z}_0$. The next coordinate frame is $\mathbf{x}_j$, $\mathbf{y}_j$, $\mathbf{z}_j$, which is rotated by the angle θ around the $\mathbf{x}_i$ axis with respect to the preceding frame. The coordinate frame $\mathbf{x}_k$, $\mathbf{y}_k$, $\mathbf{z}_k$ is attached to the body. It is rotated by the angle ϕ around the $\mathbf{y}_j$ axis. The collective rotation matrix is

$$\mathbf{A}_{0,k} = \mathbf{A}_{0,i}\mathbf{A}_{i,j}\mathbf{A}_{j,k}, \tag{1.83}$$

where the separate rotation matrices are

$$\mathbf{A}_{0,i} = \begin{bmatrix} \cos\psi & -\sin\psi & 0 \\ \sin\psi & \cos\psi & 0 \\ 0 & 0 & 1 \end{bmatrix},$$

$$\mathbf{A}_{i,j} = \begin{bmatrix} 1 & 0 & 0 \\ 0 & \cos\theta & -\sin\theta \\ 0 & \sin\theta & \cos\theta \end{bmatrix},$$

$$\mathbf{A}_{j,k} = \begin{bmatrix} \cos\phi & 0 & \sin\phi \\ 0 & 1 & 0 \\ -\sin\phi & 0 & \cos\phi \end{bmatrix}.$$

Let us now multiply the three matrices (1.83) and write only the second column and the third row

$$\mathbf{A}_{0,k} = \begin{bmatrix} \cdot & -\sin\psi\cos\theta & \cdot \\ \cdot & \cos\psi\cos\theta & \cdot \\ -\cos\theta\sin\phi & \sin\theta & \cos\theta\cos\phi \end{bmatrix}.$$

From here the YPR orientation angles will be derived. The angle θ has two solutions

$$\theta = \arcsin a_{32} \pm 2k\pi, \quad k = 0, 1, \ldots. \tag{1.84}$$

When choosing the first one it follows

$$\psi = \arctan_2 \frac{-a_{12}}{a_{22}} \pm 2k\pi, \quad k = 0, 1, \ldots, \tag{1.85}$$

$$\phi = \arctan_2 \frac{-a_{31}}{a_{33}} \pm 2k\pi, \quad k = 0, 1, \ldots. \tag{1.86}$$

When selecting the second solution

$$\theta = \pi - \arcsin a_{32} \pm 2k\pi, \quad k = 0, 1, \ldots, \tag{1.87}$$

we have

$$\psi = \arctan_2 \frac{a_{12}}{-a_{22}} \pm 2k\pi, \quad k = 0, 1, \ldots, \tag{1.88}$$

$$\phi = \arctan_2 \frac{a_{31}}{-a_{33}} \pm 2k\pi, \quad k = 0, 1, \ldots. \tag{1.89}$$

Both triples of the YPR angles are equivalent and belong to the same orientation of a body.

The above equations are valid only when at least one of the elements a_{12} and a_{22} and at least one of the elements a_{31} and a_{33} is nonzero. Otherwise the rotations ψ and ϕ are parallel. When $a_{32} = 1$ and $\theta = \pi/2 \pm 2k\pi$, $k = 0, 1, \ldots,$ we can only calculate the difference

$$\psi - \phi = \arctan_2 \frac{a_{12}}{a_{11}} \pm 2k\pi, \quad k = 0, 1, \ldots,$$

when $a_{32} = -1$ and $\theta = -\pi/2 \pm 2k\pi$, $k = 0, 1, \ldots,$ the following sum can be obtained

$$\psi + \phi = \arctan_2 \frac{a_{12}}{a_{11}} \pm 2k\pi, \quad k = 0, 1, \ldots.$$

1.5.5 Invariants of Rotation Matrices

When a matrix property is independent of the basis of the vector space, we call it an invariant of the matrix. An example of an invariant of a matrix is its determinant, which is a scalar. There are also invariant vectors. The invariants of rotation matrices can be used to describe the spatial orientation of a body.

The local coordinate frame $\mathbf{x}_j, \mathbf{y}_j, \mathbf{z}_j$ is attached to the body. Consider its orientation relative to the reference frame $\mathbf{x}_i, \mathbf{y}_i, \mathbf{z}_i$, defined by the rotation matrix $\mathbf{A}_{i,j}$. The invariants of this matrix are the rotation vector $\mathbf{e}_j^{(i)}$ and the rotation angle q_j. Through these invariants, other invariants of the rotation matrix $\mathbf{A}_{i,j}$ can be obtained. When defining the spatial orientation of a body, we often introduce four Euler parameters [1, 86]. The first three Euler parameters are the elements of the following vector

$$\mathbf{r} = \mathbf{e}_j^{(i)} \sin \frac{q_j}{2}, \tag{1.90}$$

while the fourth one is

$$r_0 = \cos \frac{q_j}{2}. \tag{1.91}$$

In order to determine the orientation of a body, three parameters are sufficient. This means that the four Euler parameters are not independent. It can be shown that their values lay on the surface of a sphere

$$\mathbf{r}^{\mathrm{T}}\mathbf{r} + r_0^2 = 1. \tag{1.92}$$

The Euler parameters $\mathbf{r}$ cannot be determined when $q_j = 0 \pm k\pi$, $k = 0, 1, \ldots$. In this case the rotation vector $\mathbf{e}_j$ is not defined.

The rotation matrix $\mathbf{A}_{i,j}$ can be written in terms of the Euler parameters as follows

$$\mathbf{A}_{i,j} = \mathbf{I}\left(2r_0^2 - 1\right) + 2\mathbf{r}\mathbf{r}^{\mathrm{T}} + 2r_0\mathbf{r} \otimes \mathbf{I}, \tag{1.93}$$

where $\mathbf{I}$ is the identity matrix, while the symbol $\otimes$ is the operation of the product of a vector and matrix as defined in (1.26). The relation (1.93) can be verified by replacing the Euler parameters $\mathbf{r}$ and r_0 by the expressions (1.90), (1.91). In this way the formula describing the rotation matrix, equation (1.48), is obtained. As $\mathbf{r}$ and r_0 are squared in (1.93), we say, that the Euler parameters are square invariants of the rotation matrix $\mathbf{A}_{i,j}$.

The inverse relations can be derived, when the values of at least the first three Euler parameters are known

$$q_j = \pm \arccos \sqrt{1 - \mathbf{r}^{\mathrm{T}}\mathbf{r}} \pm 2k\pi, \quad k = 0, 1, \ldots. \tag{1.94}$$

Both the positive and negative solutions are correct. However, the sign must also be transferred to the rotation vector

$$\mathbf{e}_j^{(i)} = \pm \frac{\mathbf{r}}{\sqrt{\mathbf{r}^{\mathrm{T}}\mathbf{r}}}. \tag{1.95}$$

The inequality $\mathbf{r}^{\mathrm{T}}\mathbf{r} > 0$ must hold, otherwise the rotation vector $\mathbf{e}_j$ is not determined.

In the literature we also find the Rodrigues parameters, which are the elements of the vector $\mathbf{r}'$

$$\mathbf{r}' = \frac{\mathbf{r}}{r_0} = \mathbf{e}_j^{(i)} \tan \frac{q_j}{2},$$

which are computationally more demanding than the Euler parameters.

Finally, let us examine another method for describing the orientation of a body. These are the linear vector invariants $\mathbf{s}$ and the linear scalar invariant s_0 of the rotation matrix $\mathbf{A}_{i,j}$ [1]. Both have already been determined by (1.60), (1.61). They can be written as

$$\mathbf{s} = \mathbf{e}_j^{(i)} \sin q_j, \tag{1.96}$$

$$s_0 = 1 + 2\cos q_j. \tag{1.97}$$

Instead of s_0 the following form is often used

$$s_1 = \cos q_j. \tag{1.98}$$

The values of linear vector and scalar invariants also lay on a sphere

$$\mathbf{s}^{\mathrm{T}}\mathbf{s} + s_1^2 = 1, \tag{1.99}$$

which means that three values are sufficient in order to determine the orientation of a body. The vector $\mathbf{s}$ cannot be determined, when $q_j = 0 \pm k\pi$, $k = 0, 1, \ldots$. It this case the rotation vector $\mathbf{e}_j$ is not defined.

The rotation matrix $\mathbf{A}_{i,j}$ can be expressed by the linear scalar and vector invariants in the following way

$$\mathbf{A}_{i,j} = \mathbf{I}s_1 + \frac{1 - s_1}{\mathbf{s}^{\mathrm{T}}\mathbf{s}}\mathbf{s}\mathbf{s}^{\mathrm{T}} + \mathbf{s} \otimes \mathbf{I}, \tag{1.100}$$

where $\mathbf{s}^{\mathrm{T}}\mathbf{s} > 0$. The first two summands in (1.100) are linearly dependent on the scalar s_1, while the third one is linearly dependent on the vector $\mathbf{s}$. The vector $\mathbf{s}$ is squared in the second summand, however it appears both in the numerator and denominator. In this way its quadratic influence is annihilated and we can claim that $\mathbf{s}$ and s_1 are linear invariants of the rotation matrix $\mathbf{A}_{i,j}$. The relation (1.100) can be verified, by inserting the expressions for $\mathbf{s}$ and s_1, yielding the rotation matrix formula (1.48).

The inverse relations can be derived when the values of the vector invariants are known

$$q_j = \pm \arccos\sqrt{1 - \mathbf{s}^{\mathrm{T}}\mathbf{s}} \pm 2k\pi, \quad k = 0, 1, \ldots \tag{1.101}$$

and the sign must be transferred to the rotation vector

$$\mathbf{e}_j^{(i)} = \pm \frac{\mathbf{s}}{\sqrt{\mathbf{s}^{\mathrm{T}}\mathbf{s}}}. \tag{1.102}$$

The inequality $\mathbf{s}^{\mathrm{T}}\mathbf{s} > 0$ must be valid, otherwise the rotation vector $\mathbf{e}_j$ is not defined.

Since multiplication by the matrix $\mathbf{A}_{i,j}$ does not change the rotation vector $\mathbf{e}_j^{(i)}$,

$$\mathbf{A}_{i,j}\mathbf{e}_j^{(i)} = \mathbf{e}_j^{(i)} \quad \Longrightarrow \quad (\mathbf{A}_{i,j} - \mathbf{I})\mathbf{e}_j^{(i)} = \mathbf{0},$$

we can see the rotation vector $\mathbf{e}_j^{(i)}$ as one of the eigenvectors of the rotation matrix $\mathbf{A}_{i,j}$. It corresponds to the real eigenvalue $\lambda_1 = 1$. The other two eigenvalues are a conjugate complex pair $\lambda_{2,3} = \cos q_j \pm i \sin q_j$.

1.6 Linear and Angular Velocity of a Body

The velocity distribution in a rigid body can be defined by the velocities of three non-collinear points on the body P_1, P_2, and P_3 as shown in Fig. 1.4. However, we prefer to describe it as a combination of translation and rotation and therefore we divide the body's velocity into linear and angular components.

1.6.1 Rotation Before Translation

The motion of a body depends on the sequence of translation and rotation. Let us first consider that the body is translationally and rotationally displaced, as shown in Fig. 1.19. The coordinate frame $\mathbf{x}_k$, $\mathbf{y}_k$, $\mathbf{z}_k$ is moving with the velocity $\dot{q}_k$ in the direction of the $\mathbf{e}_k$ axis and at a selected instant it is displaced by the distance q_k from the frame $\mathbf{x}_j$, $\mathbf{y}_j$, $\mathbf{z}_j$. The latter is rotating with angular velocity $\dot{q}_j$ around the $\mathbf{e}_j$ axis and is at a selected instant rotated by the angle q_j relative to the frame $\mathbf{x}_i$, $\mathbf{y}_i$, $\mathbf{z}_i$.

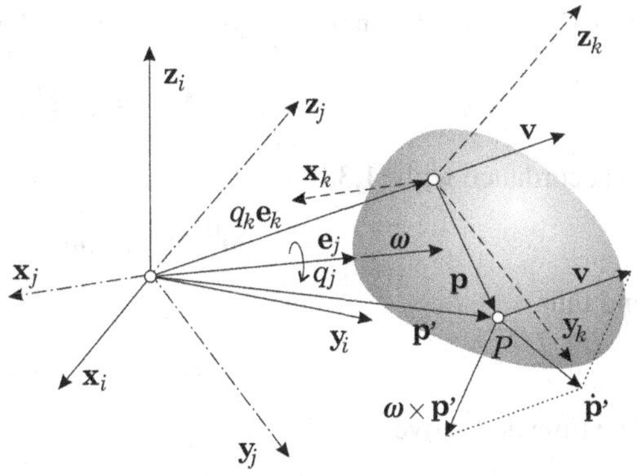

Fig. 1.19 Velocity of the point P, when rotation precedes translation

As shown in (1.66), the position of the point P belonging to the moving body, can be written as

$$\mathbf{p}'^{(i)} = \mathbf{A}_{i,j}\left(\mathbf{p}^{(j)} + q_k\mathbf{e}_k^{(j)}\right), \tag{1.103}$$

while its velocity is expressed by the time derivative

$$\dot{\mathbf{p}}'^{(i)} = \dot{\mathbf{A}}_{i,j}\left(\mathbf{p}^{(j)} + q_k\mathbf{e}_k^{(j)}\right) + \mathbf{A}_{i,j}\left(\dot{q}_k\mathbf{e}_k^{(j)}\right).$$

As $\mathbf{p}^{(j)}$ is not changing its position in time, we have $\dot{\mathbf{p}}^{(j)} = \mathbf{0}$. In order to express all vectors in the reference coordinate frame $\mathbf{x}_i$, $\mathbf{y}_i$, $\mathbf{z}_i$, we combine the last two equations

$$\dot{\mathbf{p}}'^{(i)} = \dot{\mathbf{A}}_{i,j}\mathbf{A}_{i,j}^{\mathrm{T}}\mathbf{p}'^{(i)} + \mathbf{A}_{i,j}\left(\dot{q}_k\mathbf{e}_k^{(j)}\right).$$

The equation obtained has two parts. On the far right side we have the velocity of the origin of the local coordinate frame, which is attached to the body. It is directed along the vector $\mathbf{e}_k$

$$\mathbf{v}^{(i)} = \mathbf{A}_{i,j}\left(\dot{q}_k\mathbf{e}_k^{(j)}\right).$$

Its direction is changing depending on the rotation angle q_j around the $\mathbf{e}_j$ axis or, in other words, it is dependent on the rotation matrix $\mathbf{A}_{i,j}$. The expression $\dot{\mathbf{A}}_{i,j}\mathbf{A}_{i,j}^{\mathrm{T}}\mathbf{p}'^{(i)}$ results from the rotational velocity of the body around the $\mathbf{e}_j$ axis. The matrix

$$\boldsymbol{\Omega}_{i,j} = \dot{\mathbf{A}}_{i,j}\mathbf{A}_{i,j}^{\mathrm{T}} \tag{1.104}$$

is called the matrix (or tensor) of the angular velocity of a body [1, 10]. The angular velocity of a body is defined as the vector invariant of the angular velocity matrix

$$\boldsymbol{\omega}^{(i)} = \mathrm{vect}(\boldsymbol{\Omega}_{i,j}). \tag{1.105}$$

Let us recall the relation $\mathbf{A}_{i,j}\mathbf{A}_{i,j}^{\mathrm{T}} = \mathbf{I}$ and find its time derivative. The result is

$$\dot{\mathbf{A}}_{i,j}\mathbf{A}_{i,j}^{\mathrm{T}} + \mathbf{A}_{i,j}\dot{\mathbf{A}}_{i,j}^{\mathrm{T}} = \mathbf{0}.$$

This shows that the angular velocity matrix is skew symmetrical

$$\boldsymbol{\Omega}_{i,j} = -\boldsymbol{\Omega}_{i,j}^{\mathrm{T}}. \tag{1.106}$$

In accordance with (1.38)

$$\boldsymbol{\omega}^{(i)} = \mathrm{vect}\big(\boldsymbol{\omega}^{(i)} \otimes \mathbf{I}\big),$$

and thus

$$\boldsymbol{\Omega}_{i,j} = \boldsymbol{\omega}^{(i)} \otimes \mathbf{I}. \tag{1.107}$$

The time derivative

$$\dot{\mathbf{A}}_{i,j} = \dot{q}_j \frac{\partial \mathbf{A}_{i,j}}{\partial q_j} = \dot{q}_j \mathbf{e}_j^{(i)} \otimes \mathbf{A}_{i,j},$$

which when multiplied by the matrix $\mathbf{A}_{i,j}^{\mathrm{T}}$, yields

$$\dot{\mathbf{A}}_{i,j}\mathbf{A}_{i,j}^{\mathrm{T}} = \boldsymbol{\Omega}_{i,j} = \dot{q}_j \mathbf{e}_j^{(i)} \otimes \mathbf{I},$$

and the angular velocity is

$$\boldsymbol{\omega}^{(i)} = \dot{q}_j \mathbf{e}_j^{(i)}.$$

The velocity of the point P in the space $\Re_i^3$ results from translation and rotation of a body. It can be presented in two ways, either by the help of the angular velocity vector

$$\dot{\mathbf{p}}'^{(i)} = \boldsymbol{\omega}^{(i)} \times \mathbf{p}'^{(i)} + \mathbf{v}^{(i)}, \tag{1.108}$$

which can easily be demonstrated in Cartesian space, or by the use of the angular velocity matrix

$$\dot{\mathbf{p}}'^{(i)} = \boldsymbol{\Omega}_{i,j}\mathbf{p}'^{(i)} + \mathbf{v}^{(i)}, \tag{1.109}$$

which is sometimes advantageous, as only standard vector and matrix operations are used in the formula. We should not overlook, however, that the vectors $\mathbf{p}'^{(i)}$ and $\mathbf{v}^{(i)}$ depend on the values of the rotation variable q_j.

1.6.2 Translation Before Rotation

Let us consider the motion of a body, when the translation precedes the rotation, as shown in Fig. 1.20. The coordinate frame $\mathbf{x}_k$, $\mathbf{y}_k$, $\mathbf{z}_k$ is rotated with the velocity $\dot{q}_k$ around the $\mathbf{e}_k$ axis. In a selected instant it is rotated by the angle q_k relative to the frame $\mathbf{x}_j$, $\mathbf{y}_j$, $\mathbf{z}_j$. The latter frame moves with the velocity $\dot{q}_j$ in the direction $\mathbf{e}_j$ and in a selected instant it is displaced by the distance q_j from the reference coordinate frame $\mathbf{x}_i$, $\mathbf{y}_i$, $\mathbf{z}_i$.

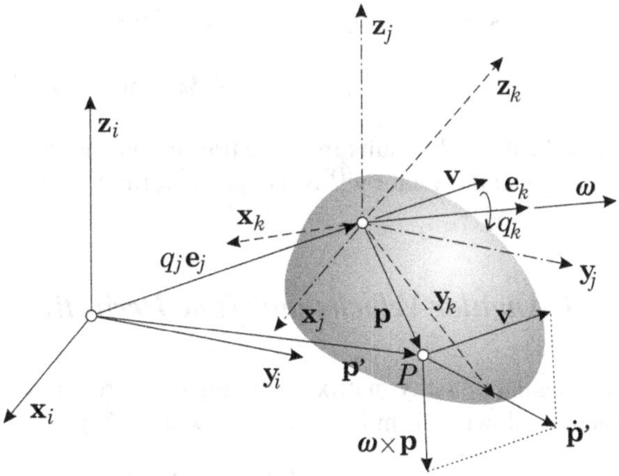

Fig. 1.20 Velocity of the point P, when translation precedes rotation

The position of a point P belonging to the moving body is determined by the following relation (1.67)

$$\mathbf{p}'^{(i)} = q_j\mathbf{e}_j^{(i)} + \mathbf{A}_{j,k}\mathbf{p}^{(k)}, \tag{1.110}$$

when considering $\dot{\mathbf{p}}^{(k)} = \mathbf{0}$, the time derivative is as follows

$$\dot{\mathbf{p}}'^{(i)} = \dot{q}_j\mathbf{e}_j^{(i)} + \dot{\mathbf{A}}_{j,k}\mathbf{p}^{(k)}.$$

By inserting $\mathbf{p}^{(k)} = \mathbf{A}_{j,k}^{\mathrm{T}}(\mathbf{p}'^{(i)} - q_j\mathbf{e}_j^{(i)})$ we obtain

$$\dot{\mathbf{p}}'^{(i)} = \dot{q}_j\mathbf{e}_j^{(i)} + \dot{\mathbf{A}}_{j,k}\mathbf{A}_{j,k}^{\mathrm{T}}\left(\mathbf{p}'^{(i)} - q_j\mathbf{e}_j^{(i)}\right).$$

In the above equation we recognize the linear velocity

$$\mathbf{v}^{(i)} = \dot{q}_j\mathbf{e}_j^{(i)}$$

and the angular velocity matrix

$$\boldsymbol{\Omega}_{j,k} = \dot{\mathbf{A}}_{j,k}\mathbf{A}_{j,k}^{\mathrm{T}},$$

which can be rewritten, as we already know, in the following form

$$\boldsymbol{\Omega}_{j,k} = \boldsymbol{\omega}^{(j)} \otimes \mathbf{I}.$$

The vector $\boldsymbol{\omega}^{(j)}$ represents the angular velocity of the body

$$\boldsymbol{\omega}^{(j)} = \dot{q}_k\mathbf{e}_k^{(j)}.$$

As the coordinate frames $\mathbf{x}_i$, $\mathbf{y}_i$, $\mathbf{z}_i$ and $\mathbf{x}_j$, $\mathbf{y}_j$, $\mathbf{z}_j$ are parallel through the entire motion, we have $\boldsymbol{\omega}^{(j)} = \boldsymbol{\omega}^{(i)}$. The velocity of the point P is

$$\dot{\mathbf{p}}'^{(i)} = \mathbf{v}^{(i)} + \boldsymbol{\omega}^{(i)} \times \left(\mathbf{p}'^{(i)} - q_j\mathbf{e}_j^{(i)}\right), \tag{1.111}$$

which can be rewritten by use of the angular velocity matrix

$$\dot{\mathbf{p}}'^{(i)} = \mathbf{v}^{(i)} + \boldsymbol{\Omega}_{j,k}\big(\mathbf{p}'^{(i)} - q_j\mathbf{e}_j^{(i)}\big). \tag{1.112}$$

The velocity of P results from the translation and the rotation of a body. It must be stressed that the vector $\mathbf{v}^{(i)}$ is independent from the rotation variable q_k.

1.6.3 Angular Velocity and Time Derivatives of the Euler Angles

The angular velocity matrix belonging to the rotational matrix of the Euler angles, has the following form in accordance with (1.77)

$$\boldsymbol{\Omega}_{0,k} = \frac{d(\mathbf{A}_{0,i}\mathbf{A}_{i,j}\mathbf{A}_{j,k})}{dt}(\mathbf{A}_{0,i}\mathbf{A}_{i,j}\mathbf{A}_{j,k})^{\mathrm{T}}.$$

After time differentiation and multiplication with the transposed matrix, it follows that

$$\boldsymbol{\Omega}_{0,k} = \dot{\mathbf{A}}_{0,i}\mathbf{A}_{0,i}^{\mathrm{T}} + \mathbf{A}_{0,i}\dot{\mathbf{A}}_{i,j}\mathbf{A}_{i,j}^{\mathrm{T}}\mathbf{A}_{0,i}^{\mathrm{T}} + \mathbf{A}_{0,i}\mathbf{A}_{i,j}\dot{\mathbf{A}}_{j,k}\mathbf{A}_{j,k}^{\mathrm{T}}\mathbf{A}_{i,j}^{\mathrm{T}}\mathbf{A}_{0,i}^{\mathrm{T}}.$$

Here we have the time derivatives of the individual rotation matrices, where the corresponding rotation axes are not changing with time. The individual time derivatives can therefore be written as follows

$$\dot{\mathbf{A}}_{0,i} = \frac{\partial\mathbf{A}_{0,i}}{\partial\psi}\dot{\psi}, \qquad \dot{\mathbf{A}}_{i,j} = \frac{\partial\mathbf{A}_{i,j}}{\partial\theta}\dot{\theta}, \qquad \dot{\mathbf{A}}_{j,k} = \frac{\partial\mathbf{A}_{j,k}}{\partial\phi}\dot{\phi}.$$

By considering the formula (1.57), we obtain

$$\dot{\mathbf{A}}_{0,i} = \dot{\psi}\mathbf{e}_i^{(0)} \otimes \mathbf{A}_{0,i}, \qquad \dot{\mathbf{A}}_{i,j} = \dot{\theta}\mathbf{e}_j^{(i)} \otimes \mathbf{A}_{i,j}, \qquad \dot{\mathbf{A}}_{j,k} = \dot{\phi}\mathbf{e}_k^{(j)} \otimes \mathbf{A}_{j,k}.$$

The above relation is substituted into the equation describing the angular velocity matrix

$$\begin{aligned}\boldsymbol{\Omega}_{0,k} &= \big(\dot{\psi}\mathbf{e}_i^{(0)} \otimes \mathbf{A}_{0,i}\big)\mathbf{A}_{0,i}^{\mathrm{T}} + \mathbf{A}_{0,i}\big(\dot{\theta}\mathbf{e}_j^{(i)} \otimes \mathbf{A}_{i,j}\big)\mathbf{A}_{i,j}^{\mathrm{T}}\mathbf{A}_{0,i}^{\mathrm{T}}\\ &\quad + \mathbf{A}_{0,i}\mathbf{A}_{i,j}\big(\dot{\phi}\mathbf{e}_k^{(j)} \otimes \mathbf{A}_{j,k}\big)\mathbf{A}_{j,k}^{\mathrm{T}}\mathbf{A}_{i,j}^{\mathrm{T}}\mathbf{A}_{0,i}^{\mathrm{T}}.\end{aligned}$$

When using the operation $\otimes$, we have the result

$$\boldsymbol{\Omega}_{0,k} = \dot{\psi}\mathbf{e}_i^{(0)} \otimes \mathbf{I} + \dot{\theta}\mathbf{e}_j^{(0)} \otimes \mathbf{I} + \dot{\phi}\mathbf{e}_k^{(0)} \otimes \mathbf{I},$$

where we recognize the individual angular velocities, expressed in the reference frame $\mathbf{x}_0, \mathbf{y}_0, \mathbf{z}_0$, as functions of the time derivatives of the Euler orientation angles $\dot{\psi}$, $\dot{\theta}$ and $\dot{\phi}$

$$\dot{\psi}\mathbf{e}_i^{(0)} = \boldsymbol{\omega}_i^{(0)}, \qquad \dot{\theta}\mathbf{e}_j^{(0)} = \boldsymbol{\omega}_j^{(0)}, \qquad \dot{\phi}\mathbf{e}_k^{(0)} = \boldsymbol{\omega}_k^{(0)}.$$

It follows

$$\boldsymbol{\Omega}_{0,k} = \boldsymbol{\omega}_i^{(0)} \otimes \mathbf{I} + \boldsymbol{\omega}_j^{(0)} \otimes \mathbf{I} + \boldsymbol{\omega}_k^{(0)} \otimes \mathbf{I}.$$

After factoring we obtain

$$\boldsymbol{\Omega}_{0,k} = \left(\boldsymbol{\omega}_i^{(0)} + \boldsymbol{\omega}_j^{(0)} + \boldsymbol{\omega}_k^{(0)}\right) \otimes \mathbf{I}.$$

The angular velocity of a body $\boldsymbol{\omega}^{(0)}$ is represented by the sum of the angular velocities belonging to the particular orientation angles

$$\boldsymbol{\omega}^{(0)} = \boldsymbol{\omega}_i^{(0)} + \boldsymbol{\omega}_j^{(0)} + \boldsymbol{\omega}_k^{(0)}. \tag{1.113}$$

Such a notation is possible, as the rotation axes bringing the body into a selected orientation, are determined in advance.

Let us consider more closely the above relation and rewrite it in the following way

$$\boldsymbol{\omega}^{(0)} = \dot{\psi}\mathbf{z}_0^{(0)} + \mathbf{A}_{0,i}\left(\dot{\theta}\mathbf{x}_i^{(i)}\right) + \mathbf{A}_{0,i}\mathbf{A}_{i,j}\left(\dot{\psi}\mathbf{z}_j^{(j)}\right),$$

where in accordance with the definition of the Euler angles, the following replacements were introduced $\mathbf{e}_i^{(0)} = \mathbf{z}_0^{(0)}$, $\mathbf{e}_j^{(i)} = \mathbf{x}_i^{(i)}$, $\mathbf{e}_k^{(j)} = \mathbf{z}_j^{(j)}$. As we are dealing with the base vectors, we have

$$\mathbf{z}_0^{(0)} = \begin{bmatrix} 0 \\ 0 \\ 1 \end{bmatrix}, \qquad \mathbf{x}_i^{(i)} = \begin{bmatrix} 1 \\ 0 \\ 0 \end{bmatrix}, \qquad \mathbf{z}_j^{(j)} = \begin{bmatrix} 0 \\ 0 \\ 1 \end{bmatrix}.$$

The multiplication of the rotational matrices and the unit vectors gives

$$\boldsymbol{\omega}^{(0)} = \dot{\psi}\begin{bmatrix} 0 \\ 0 \\ 1 \end{bmatrix} + \dot{\theta}\begin{bmatrix} \cos\psi \\ \sin\psi \\ 0 \end{bmatrix} + \dot{\psi}\begin{bmatrix} \sin\psi\sin\theta \\ -\cos\psi\sin\theta \\ \cos\theta \end{bmatrix},$$

which can be united into a single matrix

$$\boldsymbol{\omega}^{(0)} = \begin{bmatrix} 0 & \cos\psi & \sin\psi\sin\theta \\ 0 & \sin\psi & -\cos\psi\sin\theta \\ 1 & 0 & \cos\theta \end{bmatrix}\begin{bmatrix} \dot{\psi} \\ \dot{\theta} \\ \dot{\phi} \end{bmatrix}, \tag{1.114}$$

denoted as $\mathbf{W}$. Therefore

$$\boldsymbol{\omega}^{(0)} = \mathbf{W}\begin{bmatrix} \dot{\psi} \\ \dot{\theta} \\ \dot{\phi} \end{bmatrix}. \tag{1.115}$$

The relation between the angular velocity and the time derivatives of the Euler angles is linear. The direction of the angular velocity vector depends on the orientation of the body i.e. it changes with the values of the first two Euler angles ψ and

θ, presenting themselves in the matrix $\mathbf{W}$ as the arguments of the sine and cosine functions.

When the orientation of a body is expressed with the Euler angles and its rotation with the time derivatives of the Euler angles, the angular velocity vector can be calculated with the simple formula (1.115). The inverse calculation

$$\begin{bmatrix} \dot{\psi} \\ \dot{\theta} \\ \dot{\phi} \end{bmatrix} = \mathbf{W}^{-1}\boldsymbol{\omega}^{(0)} \tag{1.116}$$

is possible, when the matrix $\mathbf{W}$ is not singular. Singularities of $\mathbf{W}$ occur when

$$\det \mathbf{W} = -\sin\theta = 0, \tag{1.117}$$

which corresponds to the condition when the Euler angles themselves are not defined. As $\mathbf{W}$ is a relatively simple matrix, it is not difficult to solve the system of the algebraic equations (1.115), while considering the examples when $\sin\theta \neq 0$. Let us first find

$$\dot{\theta} = \omega_1^{(0)} \cos\psi + \omega_2^{(0)} \sin\psi,$$

followed by the angular velocities

$$\dot{\phi} = \frac{1}{\sin\theta}\left(\omega_1^{(0)} \sin\psi - \omega_2^{(0)} \cos\psi\right)$$

and

$$\dot{\psi} = \frac{\cos\theta}{\sin\theta}\left(-\omega_1^{(0)} \sin\psi + \omega_2^{(0)} \cos\psi\right) + \omega_3^{(0)}.$$

Here, $\omega_1^{(0)}$, $\omega_2^{(0)}$, and $\omega_3^{(0)}$ are the components of the angular velocity vector $\boldsymbol{\omega}^{(0)}$. For completeness, the inverse of $\mathbf{W}$ is given by

$$\mathbf{W}^{-1} = \frac{1}{\sin\theta}\begin{bmatrix} -\sin\psi\cos\theta & \cos\psi\cos\theta & \sin\theta \\ \cos\psi\sin\theta & \sin\psi\sin\theta & 0 \\ \sin\psi & -\cos\psi & 0 \end{bmatrix}, \tag{1.118}$$

which only holds when $\sin\theta \neq 0$.

1.6.4 Angular Velocity and Time Derivatives of the YPR Angles

Here again, we shall proceed from the observation, that the angular velocity of a body is equal to the sum of the individual time derivatives of the YPR angles. A different sequence of rotations must be considered now

$$\boldsymbol{\omega}^{(0)} = \dot{\psi}\mathbf{z}_0^{(0)} + \mathbf{A}_{0,i}\left(\dot{\theta}\mathbf{x}_i^{(i)}\right) + \mathbf{A}_{0,i}\mathbf{A}_{i,j}\left(\dot{\psi}\mathbf{z}_j^{(j)}\right),$$

as there is $\mathbf{e}_i^{(0)} = \mathbf{z}_0^{(0)}$, $\mathbf{e}_j^{(i)} = \mathbf{x}_i^{(i)}$, $\mathbf{e}_k^{(j)} = \mathbf{y}_j^{(j)}$. For these vectors we have

$$\mathbf{z}_0^{(0)} = \begin{bmatrix} 0 \\ 0 \\ 1 \end{bmatrix}, \qquad \mathbf{x}_i^{(i)} = \begin{bmatrix} 1 \\ 0 \\ 0 \end{bmatrix}, \qquad \mathbf{z}_j^{(j)} = \begin{bmatrix} 0 \\ 1 \\ 0 \end{bmatrix}.$$

After multiplying the rotation matrices and rotation vectors, we obtain

$$\boldsymbol{\omega}^{(0)} = \dot{\psi} \begin{bmatrix} 0 \\ 0 \\ 1 \end{bmatrix} + \dot{\theta} \begin{bmatrix} \cos\psi \\ \sin\psi \\ 0 \end{bmatrix} + \dot{\psi} \begin{bmatrix} -\sin\psi\cos\theta \\ \cos\psi\cos\theta \\ \sin\theta \end{bmatrix}.$$

The above equation is put in matrix form

$$\boldsymbol{\omega}^{(0)} = \begin{bmatrix} 0 & \cos\psi & -\sin\psi\cos\theta \\ 0 & \sin\psi & \cos\psi\cos\theta \\ 1 & 0 & \sin\theta \end{bmatrix} \begin{bmatrix} \dot{\psi} \\ \dot{\theta} \\ \dot{\phi} \end{bmatrix}, \tag{1.119}$$

in which the matrix will be again denoted as $\mathbf{W}$. It follows

$$\boldsymbol{\omega}^{(0)} = \mathbf{W} \begin{bmatrix} \dot{\psi} \\ \dot{\theta} \\ \dot{\phi} \end{bmatrix}. \tag{1.120}$$

There is a linear relation between the angular velocity and the time derivatives of the YPR angles. The direction of the angular velocity vector depends on the orientation of the body, i.e. it changes with the values of the angles ψ and θ, which appear in the matrix $\mathbf{W}$.

When the orientation of a body is expressed by the YPR angles and their time derivatives, the angular velocity vector can be calculated with the formula (1.120). The inverse relation is possible

$$\begin{bmatrix} \dot{\psi} \\ \dot{\theta} \\ \dot{\phi} \end{bmatrix} = \mathbf{W}^{-1}\boldsymbol{\omega}^{(0)}, \tag{1.121}$$

when the matrix $\mathbf{W}$ is not singular. $\mathbf{W}$ is singular, when

$$\det \mathbf{W} = \cos\theta = 0.$$

As shown earlier, this is the case when the orientation angles YPR are not defined.

Assuming $\cos\theta \neq 0$, the system of the algebraic equations (1.120) can be solved for the time derivatives of the YPR angles in terms of the components of the angular velocity vector $\omega_1^{(0)}$, $\omega_2^{(0)}$, and $\omega_3^{(0)}$,

$$\dot{\theta} = \omega_1^{(0)} \cos\psi + \omega_2^{(0)} \sin\psi,$$

$$\dot{\phi} = \frac{1}{\cos\theta}\left(-\omega_1^{(0)}\sin\psi + \omega_2^{(0)}\cos\psi\right)$$

and

$$\dot{\psi} = \frac{\sin\theta}{\cos\theta}\left(\omega_1^{(0)}\sin\psi - \omega_2^{(0)}\cos\psi\right) + \omega_3^{(0)},$$

which when written in a matrix form allows us to identify

$$\mathbf{W}^{-1} = \frac{1}{\cos\theta}\begin{bmatrix} \sin\psi\sin\theta & -\cos\psi\sin\theta & \cos\theta \\ \cos\psi\cos\theta & \sin\psi\cos\theta & 0 \\ -\sin\psi & \cos\psi & 0 \end{bmatrix}, \tag{1.122}$$

when $\cos\theta \neq 0$.

Expressing the angular velocity by the time derivatives of the Euler or YPR angles is possible, however it looses its purpose when approaching the singularities of the matrix $\mathbf{W}$. These singularities arise from the definitions of the orientation angles. In the vicinity of these singularities the time derivatives of the orientation angles approach infinity, even when the motion of a body is modest and smooth. These singularities are mathematical artifacts associated with these representations of orientation.

1.7 Linear and Angular Acceleration of a Body

The linear acceleration of a body can be described as the acceleration of a local coordinate frame attached to the body, describing the variation of the translational body velocity. The angular acceleration, which is the same for all points of the body, is the rotational acceleration describing the variation of the angular velocity of the body.

1.7.1 Rotation Before Translation

The coordinate frame $\mathbf{x}_k$, $\mathbf{y}_k$, $\mathbf{z}_k$, which is attached to the body in Fig. 1.21, moves with the velocity $\dot{q}_k$ and the acceleration $\ddot{q}_k$ in the direction of the vector $\mathbf{e}_k$ relative to the coordinate frame $\mathbf{x}_j$, $\mathbf{y}_j$, $\mathbf{z}_j$. The latter rotates with angular velocity $\dot{q}_j$ and angular acceleration $\ddot{q}_j$ around the $\mathbf{e}_j$ axis relative to the reference coordinate system $\mathbf{x}_i$, $\mathbf{y}_i$, $\mathbf{z}_i$. At a selected instant the coordinate frames $\mathbf{x}_k$, $\mathbf{y}_k$, $\mathbf{z}_k$ and $\mathbf{x}_j$, $\mathbf{y}_j$, $\mathbf{z}_j$ are displaced by the distance q_k, while the frames $\mathbf{x}_j$, $\mathbf{y}_j$, $\mathbf{z}_j$ and $\mathbf{x}_i$, $\mathbf{y}_i$, $\mathbf{z}_i$ are rotated by the angle q_j.

Let us observe the acceleration of the point P on the body. From the time derivative of the velocity, we obtain

$$\ddot{\mathbf{p}}'^{(i)} = \dot{\mathbf{\Omega}}_{i,j}\mathbf{p}'^{(i)} + \mathbf{\Omega}_{i,j}\dot{\mathbf{p}}'^{(i)} + \dot{\mathbf{v}}^{(i)}.$$

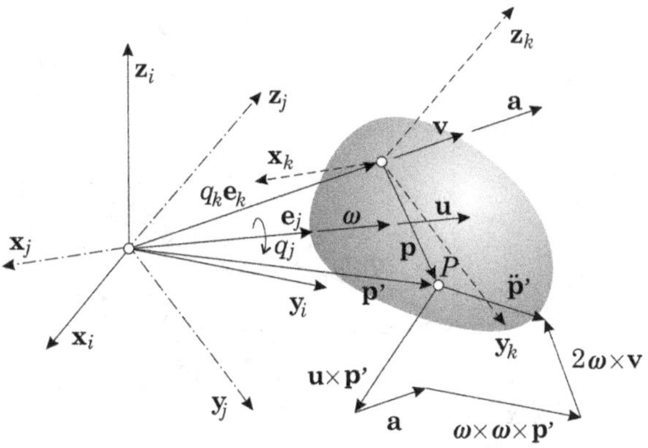

Fig. 1.21 Acceleration of the point P, when rotation precedes translation

When inserting (1.109) we have

$$\ddot{\mathbf{p}}'^{(i)} = \left(\dot{\mathbf{\Omega}}_{i,j} + \mathbf{\Omega}_{i,j}^2\right)\mathbf{p}'^{(i)} + \mathbf{\Omega}_{i,j}\mathbf{v}^{(i)} + \dot{\mathbf{v}}^{(i)}.$$

The matrix

$$\mathbf{\Psi}_{i,j} = \dot{\mathbf{\Omega}}_{i,j} + \mathbf{\Omega}_{i,j}^2 \tag{1.123}$$

is called the angular acceleration matrix of a body [1, 10]. It is composed from two matrices, the skew symmetrical matrix $\dot{\mathbf{\Omega}}_{i,j}$ and the symmetrical matrix $\mathbf{\Omega}_{i,j}^2$. The angular acceleration of a body $\mathbf{u}$ is defined as the vector invariant of this matrix

$$\mathbf{u}^{(i)} = \dot{\boldsymbol{\omega}}^{(i)} = \text{vect}(\mathbf{\Psi}_{i,j}). \tag{1.124}$$

As the vector invariants of the symmetrical matrix equals zero, the following relation also holds

$$\mathbf{u}^{(i)} = \text{vect}(\dot{\mathbf{\Omega}}_{i,j}). \tag{1.125}$$

Taking into account

$$\mathbf{u}^{(i)} = \text{vect}\left(\mathbf{u}^{(i)} \otimes \mathbf{I}\right)$$

the previous equation becomes

$$\dot{\mathbf{\Omega}}_{i,j} = \mathbf{u}^{(i)} \otimes \mathbf{I}. \tag{1.126}$$

Let us step back and see what is hidden in the vector $\dot{\mathbf{v}}^{(i)}$, whose direction depends on the angle q_j

$$\dot{\mathbf{v}}^{(i)} = \dot{\mathbf{A}}_{i,j}\mathbf{v}^{(j)} + \mathbf{A}_{i,j}\dot{\mathbf{v}}^{(j)}.$$

Since $\mathbf{v}^{(j)} = \dot{q}_k\mathbf{e}_k^{(j)}$, $\mathbf{v}^{(i)} = \mathbf{A}_{i,j}\mathbf{v}^{(j)}$, and $\dot{\mathbf{A}}_{i,j} = \boldsymbol{\omega}^{(i)} \otimes \mathbf{A}_{i,j}$, we can write

$$\dot{\mathbf{v}}^{(i)} = \boldsymbol{\omega}^{(i)} \times \mathbf{v}^{(i)} + \mathbf{a}^{(i)} = \mathbf{\Omega}_{i,j}\mathbf{v}^{(i)} + \mathbf{a}^{(i)},$$

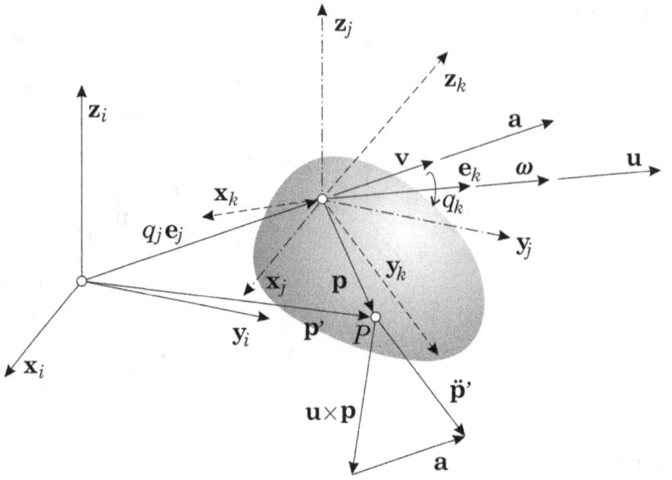

Fig. 1.22 Acceleration of the point P, when translation precedes rotation

where

$$\mathbf{a}^{(i)} = \mathbf{A}_{i,j}\mathbf{a}^{(j)} = \mathbf{A}_{i,j}\left(\ddot{q}_k \mathbf{e}_k^{(j)}\right)$$

is a contribution of the translational variable. All the above relations are inserted into the expression describing the acceleration of the point P

$$\ddot{\mathbf{p}}'^{(i)} = \mathbf{u}^{(i)} \times \mathbf{p}'^{(i)} + \boldsymbol{\omega}^{(i)} \times \left(\boldsymbol{\omega}^{(i)} \times \mathbf{p}'^{(i)}\right) + 2\boldsymbol{\omega}^{(i)} \times \mathbf{v}^{(i)} + \mathbf{a}^{(i)}. \tag{1.127}$$

In the above equation we can recognize the tangential acceleration $\mathbf{u}^{(i)} \times \mathbf{p}'^{(i)}$, produced by the angular acceleration $\mathbf{u}$, radial acceleration $\boldsymbol{\omega}^{(i)} \times (\boldsymbol{\omega}^{(i)} \times \mathbf{p}'^{(i)})$, induced by the angular velocity $\boldsymbol{\omega}$, and the Coriolis acceleration $2\boldsymbol{\omega}^{(i)} \times \mathbf{v}^{(i)}$, which is the consequence of the variation of the direction of translation during the rotation of a body.

The last equation can be written in a more compact form

$$\ddot{\mathbf{p}}'^{(i)} = \boldsymbol{\Psi}_{i,j}\mathbf{p}'^{(i)} + 2\boldsymbol{\Omega}_{i,j}\mathbf{v}^{(i)} + \mathbf{a}^{(i)}. \tag{1.128}$$

1.7.2 Translation Before Rotation

The coordinate frame $\mathbf{x}_k, \mathbf{y}_k, \mathbf{z}_k$ attached to the moving body in Fig. 1.22, is rotating with angular velocity $\dot{q}_k$ and angular acceleration $\ddot{q}_k$ around the vector $\mathbf{e}_k$ relative to the frame $\mathbf{x}_j, \mathbf{y}_j, \mathbf{z}_j$. The latter is displaced with velocity $\dot{q}_j$ and acceleration $\ddot{q}_j$ in the direction $\mathbf{e}_j$ relative to the reference coordinate frame $\mathbf{x}_i, \mathbf{y}_i, \mathbf{z}_i$. In a selected instant the coordinate frames $\mathbf{x}_k, \mathbf{y}_k, \mathbf{z}_k$ and $\mathbf{x}_j, \mathbf{y}_j, \mathbf{z}_j$ are rotated by the angle q_k, while the frames $\mathbf{x}_j, \mathbf{y}_j, \mathbf{z}_j$ and $\mathbf{x}_i, \mathbf{y}_i, \mathbf{z}_i$ are displaced by the distance q_j.

The time derivative of (1.112) is

$$\ddot{\mathbf{p}}'^{(i)} = \dot{\mathbf{v}}^{(i)} + \dot{\boldsymbol{\Omega}}_{j,k}\big(\mathbf{p}'^{(i)} - q_j\mathbf{e}_j^{(i)}\big) + \boldsymbol{\Omega}_{j,k}\big(\dot{\mathbf{p}}'^{(i)} - \dot{q}_j\mathbf{e}_j^{(i)}\big).$$

The expression for $\dot{\mathbf{p}}'^{(i)}$ (1.110) will be inserted in the above equation. Considering that $\mathbf{v}^{(i)} = \dot{q}_j\mathbf{e}_j^{(i)}$ then

$$\mathbf{a}^{(i)} = \dot{\mathbf{v}}^{(i)} = \ddot{q}_j\mathbf{e}_j^{(i)},$$

where $\mathbf{a}^{(i)}$ is the acceleration of the translational variable. It follows that

$$\ddot{\mathbf{p}}'^{(i)} = \mathbf{a}^{(i)} + \big(\dot{\boldsymbol{\Omega}}_{j,k} + \boldsymbol{\Omega}_{j,k}^2\big)\big(\mathbf{p}'^{(i)} - q_j\mathbf{e}_j^{(i)}\big).$$

Here we notice the angular acceleration matrix

$$\boldsymbol{\Psi}_{j,k} = \dot{\boldsymbol{\Omega}}_{j,k} + \boldsymbol{\Omega}_{j,k}^2,$$

from the preceding section. The angular acceleration is a vector invariant

$$\mathbf{u}^{(j)} = \dot{\boldsymbol{\omega}}^{(j)} = \mathrm{vect}(\boldsymbol{\Psi}_{j,k}),$$

which in our example has the form

$$\mathbf{u}^{(j)} = \dot{\boldsymbol{\omega}}^{(j)} = \ddot{q}_k\mathbf{e}_k^{(j)}.$$

As the coordinate frames $\mathbf{x}_i$, $\mathbf{y}_i$, $\mathbf{z}_i$ and $\mathbf{x}_j$, $\mathbf{y}_j$, $\mathbf{z}_j$ are always parallel, we can write $\mathbf{u}^{(i)} = \mathbf{u}^{(j)}$. The final formula has the following form

$$\ddot{\mathbf{p}}'^{(i)} = \mathbf{a}^{(i)} + \mathbf{u}^{(i)} \times \big(\mathbf{p}'^{(i)} - q_j\mathbf{e}_j^{(i)}\big). \tag{1.129}$$

Instead of the angular acceleration vector the corresponding matrix can be used

$$\ddot{\mathbf{p}}'^{(i)} = \mathbf{a}^{(i)} + \boldsymbol{\Psi}_{j,k}\big(\mathbf{p}'^{(i)} - q_j\mathbf{e}_j^{(i)}\big). \tag{1.130}$$

1.7.3 Angular Acceleration and Time Derivatives of the Euler Angles

In Sect. 1.6.3 we learned the relation between the angular velocity and the time derivative of the Euler angles. According to (1.115), we are dealing with a linear transformation. It is defined by the matrix $\mathbf{W}$, which depends on the angles ψ and θ. The angular acceleration can be easily obtained as time derivative of (1.115)

$$\mathbf{u} = \dot{\boldsymbol{\omega}} = \dot{\mathbf{W}}\begin{bmatrix}\dot{\psi}\\ \dot{\theta}\\ \dot{\phi}\end{bmatrix} + \mathbf{W}\begin{bmatrix}\ddot{\psi}\\ \ddot{\theta}\\ \ddot{\phi}\end{bmatrix},$$

where

$$\dot{\mathbf{W}} = \frac{\partial \mathbf{W}}{\partial \psi}\dot{\psi} + \frac{\partial \mathbf{W}}{\partial \theta}\dot{\theta}.$$

The partial derivatives are

$$\frac{\partial \mathbf{W}}{\partial \theta} = \begin{bmatrix} 0 & -\sin\psi & \cos\psi\sin\theta \\ 0 & \cos\psi & \sin\psi\sin\theta \\ 0 & 0 & 0 \end{bmatrix},$$

$$\frac{\partial \mathbf{W}}{\partial \psi} = \begin{bmatrix} 0 & 0 & \sin\psi\cos\theta \\ 0 & 0 & -\cos\psi\cos\theta \\ 0 & 0 & -\sin\theta \end{bmatrix}.$$

The original equation can be rewritten in a more compact form

$$\mathbf{u} = \mathbf{W}_0 \begin{bmatrix} \dot{\psi}\dot{\theta} \\ \dot{\psi}\dot{\phi} \\ \dot{\theta}\dot{\phi} \end{bmatrix} + \mathbf{W} \begin{bmatrix} \ddot{\psi} \\ \ddot{\theta} \\ \ddot{\phi} \end{bmatrix}, \tag{1.131}$$

where

$$\mathbf{W}_0 = \begin{bmatrix} -\sin\psi & \cos\psi\sin\theta & \sin\psi\cos\theta \\ \cos\psi & \sin\psi\sin\theta & -\cos\psi\cos\theta \\ 0 & 0 & -\sin\theta \end{bmatrix}. \tag{1.132}$$

From here the inverse relation can be obtained

$$\begin{bmatrix} \ddot{\psi} \\ \ddot{\theta} \\ \ddot{\phi} \end{bmatrix} = \mathbf{W}^{-1} \left(\mathbf{u} - \mathbf{W}_0 \begin{bmatrix} \dot{\psi}\dot{\theta} \\ \dot{\psi}\dot{\phi} \\ \dot{\theta}\dot{\phi} \end{bmatrix} \right), \tag{1.133}$$

which is valid only when the matrix **W** is not singular, i.e. when $\sin\theta \neq 0$.

1.7.4 Angular Acceleration and Time Derivatives of the YPR Angles

The relation between the angular velocity and the time derivatives of the YPR orientation angles was explained in Sect. 1.6.4. The transformation is given by the **W** matrix, which depends on the angles ψ and θ. The angular acceleration is the time derivative of (1.120)

$$\mathbf{u} = \dot{\omega} = \dot{\mathbf{W}} \begin{bmatrix} \dot{\psi} \\ \dot{\theta} \\ \dot{\phi} \end{bmatrix} + \mathbf{W} \begin{bmatrix} \ddot{\psi} \\ \ddot{\theta} \\ \ddot{\phi} \end{bmatrix},$$

where

$$\dot{\mathbf{W}} = \frac{\partial \mathbf{W}}{\partial \psi}\dot{\psi} + \frac{\partial \mathbf{W}}{\partial \theta}\dot{\theta}$$

and

$$\frac{\partial \mathbf{W}}{\partial \theta} = \begin{bmatrix} 0 & -\sin\psi & -\cos\psi\cos\theta \\ 0 & \cos\psi & -\sin\psi\cos\theta \\ 0 & 0 & 0 \end{bmatrix},$$

$$\frac{\partial \mathbf{W}}{\partial \psi} = \begin{bmatrix} 0 & 0 & \sin\psi\sin\theta \\ 0 & 0 & -\cos\psi\sin\theta \\ 0 & 0 & \cos\theta \end{bmatrix}.$$

The original equation can be rewritten in a more compact form

$$\mathbf{u} = \mathbf{W}_0 \begin{bmatrix} \dot{\psi}\dot{\theta} \\ \dot{\psi}\dot{\phi} \\ \dot{\theta}\dot{\phi} \end{bmatrix} + \mathbf{W} \begin{bmatrix} \ddot{\psi} \\ \ddot{\theta} \\ \ddot{\phi} \end{bmatrix}, \tag{1.134}$$

where

$$\mathbf{W}_0 = \begin{bmatrix} -\sin & -\sin\psi\cos\theta & \sin\psi\sin\theta \\ \cos\psi & -\sin\psi\cos\theta & -\cos\psi\sin\theta \\ 0 & 0 & \cos\theta \end{bmatrix}. \tag{1.135}$$

Finally, let us write the inverse relation

$$\begin{bmatrix} \ddot{\psi} \\ \ddot{\theta} \\ \ddot{\phi} \end{bmatrix} = \mathbf{W}^{-1} \left(\mathbf{u} - \mathbf{W}_0 \begin{bmatrix} \dot{\psi}\dot{\theta} \\ \dot{\psi}\dot{\phi} \\ \dot{\theta}\dot{\phi} \end{bmatrix} \right), \tag{1.136}$$

which is valid only when the matrix $\mathbf{W}$ is not singular, i.e. when $\cos\theta \neq 0$.

1.8 Homogeneous Transformations

A homogeneous transformation is a 4×4 matrix operator [17]. It is widely used in robotics, biomechanics, and computer graphics [15, 67]. The homogeneous transformation matrix leads to a compact formulation of the kinematic equations. The homogeneous matrix combines the translational displacement of a body, represented by the elements d_1, d_2 and d_3, and the rotational displacement defined by the elements $a_{11}, a_{12}, \ldots, a_{33}$ into a single operator

$$\mathbf{H} = \begin{bmatrix} a_{11} & a_{12} & a_{13} & d_1 \\ a_{21} & a_{22} & a_{23} & d_2 \\ a_{31} & a_{32} & a_{33} & d_3 \\ f_1 & f_2 & f_3 & f_0 \end{bmatrix}. \tag{1.137}$$

The elements f_1, f_2 and f_3 of the matrix can be used for perspective transformations in the area of computer graphics, while f_0 is used as a scaling factor. As these elements are of no importance to us, their values will be $f_1 = f_2 = f_3 = 0$ and $f_0 = 1$ throughout the text. The matrix operates with 4-dimensional vectors

$$\begin{bmatrix} a_1 \\ a_2 \\ a_3 \\ f_0 \end{bmatrix}, \tag{1.138}$$

where the first three elements belong to the vector $\mathbf{a} = \Re^3$.

1.8.1 Rotation Before Translation

Consider an example of three coordinate frames. The coordinate frame $\mathbf{x}_k$, $\mathbf{y}_k$, $\mathbf{z}_k$, which is attached to the body, is displaced by the distance q_k in the direction $\mathbf{e}_k$ relative to the frame $\mathbf{x}_j$, $\mathbf{y}_j$, $\mathbf{z}_j$, which is rotated by the angle q_j around $\mathbf{e}_j$ relative to the reference frame $\mathbf{x}_i$, $\mathbf{y}_i$, $\mathbf{z}_i$. In the homogeneous transformation matrix $\mathbf{H}_{i,k}$ we are uniting the rotational matrix $\mathbf{A}_{i,j}$ and the translational vector $q_k\mathbf{e}_k^{(j)}$

$$\mathbf{H}_{i,k} = \begin{bmatrix} & \mathbf{A}_{i,j} & & \mathbf{A}_{i,j}(q_k\mathbf{e}_k^{(j)}) \\ 0 & 0 & 0 & 1 \end{bmatrix}. \tag{1.139}$$

By adding the fourth element to the vector $\mathbf{p}^{(k)}$ and multiplying it by the homogeneous transformation matrix $\mathbf{H}_{i,k}$, we obtain

$$\mathbf{H}_{i,k} \begin{bmatrix} \mathbf{p}^{(k)} \\ 1 \end{bmatrix} = \begin{bmatrix} \mathbf{A}_{i,j}(\mathbf{p}^{(k)} + q_k\mathbf{e}_k^{(j)}) \\ 1 \end{bmatrix} = \begin{bmatrix} \mathbf{p}'^{(i)} \\ 1 \end{bmatrix}. \tag{1.140}$$

When comparing the above equation with (1.103), we can observe, that two kinematic operations are simultaneously accomplished by multiplication with the homogeneous transformation matrix, rotation and translation. They are presented by the rotation matrix $\mathbf{A}_{i,j}$ and the translation vector $q_k\mathbf{e}_k^{(j)}$.

The two parts are hidden in the homogeneous transformation matrix

$$\mathbf{H}_{i,k} = \mathbf{R}_{i,j}\mathbf{T}_{j,k}, \tag{1.141}$$

where the first part describes the rotation

$$\mathbf{R}_{i,j} = \begin{bmatrix} & \mathbf{A}_{i,j} & & \mathbf{0} \\ 0 & 0 & 0 & 1 \end{bmatrix}, \tag{1.142}$$

while the second part belongs to the translation

$$\mathbf{T}_{j,k} = \begin{bmatrix} & \mathbf{I} & & q_k\mathbf{e}_k^{(j)} \\ 0 & 0 & 0 & 1 \end{bmatrix}. \tag{1.143}$$

Let us derive the inverse homogeneous transformation $\mathbf{H}_{i,k}^{-1}$. There must be

$$\mathbf{H}_{i,k}^{-1}\mathbf{H}_{i,k} = \mathbf{I}.$$

By multiplying the submatrices and equating the left and the right side of the equation, we obtain

$$\mathbf{H}_{i,k}^{-1} = \begin{bmatrix} & \mathbf{A}_{i,j}^{\mathrm{T}} & & -q_k\mathbf{e}_k^{(j)} \\ 0 & 0 & 0 & 1 \end{bmatrix}. \tag{1.144}$$

Observe each factor separately

$$\mathbf{H}_{i,k}^{-1} = \mathbf{T}_{j,k}^{-1}\mathbf{R}_{i,j}^{-1},$$

where the rotational part is

$$\mathbf{R}_{i,j}^{-1} = \begin{bmatrix} & \mathbf{A}_{i,j}^{\mathrm{T}} & & 0 \\ 0 & 0 & 0 & 1 \end{bmatrix}, \tag{1.145}$$

and the translational

$$\mathbf{T}_{j,k}^{-1} = \begin{bmatrix} & \mathbf{I} & & -q_k\mathbf{e}_k^{(j)} \\ 0 & 0 & 0 & 1 \end{bmatrix}. \tag{1.146}$$

The homogeneous transformation matrix apparently is not orthogonal. However, the rotational part of the matrix is orthogonal

$$\mathbf{R}_{i,j}^{\mathrm{T}}\mathbf{R}_{i,j} = \mathbf{I},$$

therefore

$$\mathbf{R}_{i,j}^{-1} = \mathbf{R}_{i,j}^{\mathrm{T}}. \tag{1.147}$$

The transformations of velocities and accelerations are obtained by the time derivatives. Let us first differentiate (1.140)

$$\begin{bmatrix} \dot{\mathbf{p}}'^{(i)} \\ 0 \end{bmatrix} = \dot{\mathbf{H}}_{i,k}\begin{bmatrix} \mathbf{p}^{(k)} \\ 1 \end{bmatrix},$$

where

$$\dot{\mathbf{H}}_{i,k} = \begin{bmatrix} & \dot{\mathbf{A}}_{i,j} & & \dot{\mathbf{A}}_{i,j}(q_k\mathbf{e}_k^{(j)}) + \mathbf{A}_{i,j}(\dot{q}_k\mathbf{e}_k^{(j)}) \\ 0 & 0 & 0 & 0 \end{bmatrix}.$$

By inserting the inverse equation (1.140)

$$\begin{bmatrix} \mathbf{p}^{(k)} \\ 1 \end{bmatrix} = \mathbf{H}_{i,k}^{-1}\begin{bmatrix} \mathbf{p}'^{(i)} \\ 1 \end{bmatrix},$$

we obtain

$$\begin{bmatrix} \dot{\mathbf{p}}'^{(i)} \\ 0 \end{bmatrix} = \mathbf{V}_{i,k} \begin{bmatrix} \mathbf{p}'^{(i)} \\ 1 \end{bmatrix}. \tag{1.148}$$

The following abbreviation is used

$$\mathbf{V}_{i,k} = \dot{\mathbf{H}}_{i,k}\mathbf{H}_{i,k}^{-1} = \begin{bmatrix} & \dot{\mathbf{A}}_{i,j}\mathbf{A}_{i,j}^{\mathrm{T}} & & \mathbf{A}_{i,j}(\dot{q}_k\mathbf{e}_k^{(j)}) \\ 0 & 0 & 0 & 0 \end{bmatrix}$$

or

$$\mathbf{V}_{i,k} = \begin{bmatrix} & \boldsymbol{\Omega}_{i,j} & & \mathbf{v}^{(i)} \\ 0 & 0 & 0 & 0 \end{bmatrix}. \tag{1.149}$$

The matrix $\mathbf{V}_{i,k}$ encompasses the angular velocity matrix $\boldsymbol{\Omega}_{i,j}$ and the velocity along the translational axis $\mathbf{v}^{(i)}$ [84]. From here we have

$$\begin{bmatrix} \dot{\mathbf{p}}'^{(i)} \\ 0 \end{bmatrix} = \begin{bmatrix} \boldsymbol{\Omega}_{i,j}\mathbf{p}'^{(i)} + \mathbf{v}^{(i)} \\ 0 \end{bmatrix}.$$

Equation (1.148) is identical to (1.109), both of which represent the velocity of point P when rotation precedes translation.

By differentiating (1.148) and inserting $\dot{\mathbf{p}}'^{(i)} = \mathbf{V}_{i,k}\mathbf{p}'^{(i)}$, we obtain

$$\ddot{\mathbf{p}}'^{(i)} = \dot{\mathbf{V}}_{i,k}\mathbf{p}'^{(i)} + \mathbf{V}_{i,k}\dot{\mathbf{p}}'^{(i)} = (\dot{\mathbf{V}}_{i,k} + \mathbf{V}_{i,k}^2)\mathbf{p}'^{(i)}.$$

Defining $\mathbf{P}_{i,k}$ as

$$\mathbf{P}_{i,k} = \dot{\mathbf{V}}_{i,k} + \mathbf{V}_{i,k}^2 \tag{1.150}$$

we can write

$$\ddot{\mathbf{p}}'^{(i)} = \mathbf{P}_{i,k}\mathbf{p}'^{(i)}. \tag{1.151}$$

The above equation is equivalent to expression (1.128) describing the acceleration of the point P on the body, when rotation is before the translation. This can easily be proven, as there is

$$\dot{\mathbf{V}}_{i,k} = \begin{bmatrix} & \dot{\boldsymbol{\Omega}}_{i,j} & & \dot{\mathbf{v}}^{(i)} \\ 0 & 0 & 0 & 0 \end{bmatrix}$$

and when adding $\mathbf{V}_{i,k}^2$, it follows that

$$\mathbf{P}_{i,k} = \begin{bmatrix} & \dot{\boldsymbol{\Omega}}_{i,j} + \boldsymbol{\Omega}_{i,j}^2 & & 2\boldsymbol{\Omega}_{i,j}\mathbf{v}^{(i)} + \mathbf{a}^{(i)} \\ 0 & 0 & 0 & 0 \end{bmatrix}$$

and

$$\mathbf{P}_{i,k} = \begin{bmatrix} & \boldsymbol{\Psi}_{i,j} & & 2\boldsymbol{\Omega}_{i,j}\mathbf{v}^{(i)} + \mathbf{a}^{(i)} \\ 0 & 0 & 0 & 0 \end{bmatrix}. \tag{1.152}$$

The matrix $\mathbf{P}_{i,k}$ includes the angular acceleration matrix $\boldsymbol{\Psi}_{i,j}$ and the acceleration along the translational coordinate $\mathbf{a}^{(i)}$ together with the Coriolis acceleration $2\boldsymbol{\Omega}_{i,j}\mathbf{v}^{(i)}$. When applying this matrix for calculation of the acceleration of the point P, the result

$$\begin{bmatrix} \ddot{\mathbf{p}}'^{(i)} \\ 0 \end{bmatrix} = \mathbf{P}_{i,k} \begin{bmatrix} \mathbf{p}'^{(i)} \\ 1 \end{bmatrix} = \begin{bmatrix} \boldsymbol{\Psi}_{i,j}\mathbf{p}'^{(i)} + 2\boldsymbol{\Omega}_{i,j}\mathbf{v}^{(i)} + \mathbf{a}^{(i)} \\ 0 \end{bmatrix}$$

which is the same as derived in (1.128).

1.8.2 Translation Before Rotation

Let us consider how the sequence of the translation before the rotation can be included in the homogeneous transformation matrix. Now the coordinate frame $\mathbf{x}_k$, $\mathbf{y}_k$, $\mathbf{z}_k$, which is attached to the body, is rotated by the angle q_k around the $\mathbf{e}_k$ axis relative to the frame $\mathbf{x}_j$, $\mathbf{y}_j$, $\mathbf{z}_j$, which is displaced by the distance q_j in the direction $\mathbf{e}_j$ towards the reference fame $\mathbf{x}_i$, $\mathbf{y}_i$, $\mathbf{z}_i$. The translation vector $q_j\mathbf{e}_j^{(i)}$ and the rotation matrix $\mathbf{A}_{j,k}$ are united into the homogeneous transformation matrix $\mathbf{H}_{i,k}$

$$\mathbf{H}_{i,k} = \begin{bmatrix} & \mathbf{A}_{j,k} & & q_j\mathbf{e}_j^{(i)} \\ 0 & 0 & 0 & 1 \end{bmatrix}. \tag{1.153}$$

As the rotation precedes the translation, we have

$$\mathbf{H}_{i,k} = \mathbf{T}_{i,j}\mathbf{R}_{j,k}, \tag{1.154}$$

with the translational part

$$\mathbf{T}_{i,j} = \begin{bmatrix} & \mathbf{I} & & q_j\mathbf{e}_j^{(i)} \\ 0 & 0 & 0 & 1 \end{bmatrix}, \tag{1.155}$$

and the rotational part

$$\mathbf{R}_{j,k} = \begin{bmatrix} & \mathbf{A}_{j,k} & & \mathbf{0} \\ 0 & 0 & 0 & 1 \end{bmatrix}. \tag{1.156}$$

When multiplying the vector $\mathbf{p}^{(k)}$ with the homogeneous transformation matrix $\mathbf{H}_{i,k}$, we obtain

$$\mathbf{H}_{i,k} \begin{bmatrix} \mathbf{p}^{(k)} \\ 1 \end{bmatrix} = \begin{bmatrix} \mathbf{A}_{j,k}\mathbf{p}^{(k)} + q_j\mathbf{e}_j^{(i)} \\ 1 \end{bmatrix} = \begin{bmatrix} \mathbf{p}'^{(i)} \\ 1 \end{bmatrix}, \tag{1.157}$$

which is in accordance with (1.110).

We have the following inverse matrix

$$\mathbf{H}_{i,k}^{-1} = \begin{bmatrix} & \mathbf{A}_{j,k}^{\mathrm{T}} & & -\mathbf{A}_{j,k}^{\mathrm{T}}(q_j\mathbf{e}_j^{(i)}) \\ 0 & 0 & 0 & 1 \end{bmatrix}. \tag{1.158}$$

As

$$\mathbf{H}_{i,k}^{-1} = \mathbf{R}_{j,k}^{-1}\mathbf{T}_{i,j}^{-1},$$

both matrices have the following form

$$\mathbf{R}_{j,k}^{-1} = \begin{bmatrix} & \mathbf{A}_{j,k}^{\mathrm{T}} & & \mathbf{0} \\ 0 & 0 & 0 & 1 \end{bmatrix}, \tag{1.159}$$

$$\mathbf{T}_{i,j}^{-1} = \begin{bmatrix} & \mathbf{I} & & -q_j\mathbf{e}_j^{(i)} \\ 0 & 0 & 0 & 1 \end{bmatrix}. \tag{1.160}$$

Differentiating (1.157) gives

$$\begin{bmatrix} \dot{\mathbf{p}}'^{(i)} \\ 0 \end{bmatrix} = \dot{\mathbf{H}}_{i,k}\begin{bmatrix} \mathbf{p}^{(k)} \\ 1 \end{bmatrix}.$$

As

$$\dot{\mathbf{H}}_{i,k} = \begin{bmatrix} & \dot{\mathbf{A}}_{j,k} & & \dot{q}_j\mathbf{e}_j^{(i)} \\ 0 & 0 & 0 & 0 \end{bmatrix}$$

and

$$\begin{bmatrix} \mathbf{p}^{(k)} \\ 1 \end{bmatrix} = \mathbf{H}_{i,k}^{-1}\begin{bmatrix} \mathbf{p}'^{(i)} \\ 1 \end{bmatrix},$$

we can write

$$\begin{bmatrix} \dot{\mathbf{p}}'^{(i)} \\ 0 \end{bmatrix} = \mathbf{V}_{i,k}\begin{bmatrix} \mathbf{p}'^{(i)} \\ 1 \end{bmatrix}, \tag{1.161}$$

where we have $\mathbf{V}_{i,k} = \dot{\mathbf{H}}_{i,k}\mathbf{H}_{i,k}^{-1}$ as in the preceding section. After multiplication we obtain

$$\dot{\mathbf{H}}_{i,k}\mathbf{H}_{i,k}^{-1} = \begin{bmatrix} & \dot{\mathbf{A}}_{j,k}\mathbf{A}_{j,k}^{\mathrm{T}} & & \dot{q}_j\mathbf{e}_j^{(i)} - \dot{\mathbf{A}}_{j,k}\mathbf{A}_{j,k}^{\mathrm{T}}(q_j\mathbf{e}_j^{(i)}) \\ 0 & 0 & 0 & 0 \end{bmatrix}$$

or

$$\mathbf{V}_{i,k} = \begin{bmatrix} & \boldsymbol{\Omega}_{j,k} & & \mathbf{v}^{(i)} - \boldsymbol{\Omega}_{j,k}(q_j\mathbf{e}_j^{(i)}) \\ 0 & 0 & 0 & 0 \end{bmatrix}. \tag{1.162}$$

The matrix $\mathbf{V}_{i,k}$ encompasses the angular velocity matrix $\boldsymbol{\Omega}_{j,k}$ and the velocity vector along the translational axis $\mathbf{v}^{(i)} = \dot{q}_j \mathbf{e}_j^{(i)}$ [84]. Therefore

$$\begin{bmatrix} \dot{\mathbf{p}}'^{(i)} \\ 0 \end{bmatrix} = \begin{bmatrix} \mathbf{v}^{(i)} + \boldsymbol{\Omega}_{j,k}(\mathbf{p}'^{(i)} - q_j \mathbf{e}_j^{(i)}) \\ 0 \end{bmatrix}. \tag{1.163}$$

Equation (1.161) is equivalent to (1.112) which determines the velocity of point P on the body, when translation precedes rotation.

Finally let us differentiate (1.161). Again we shall use the abbreviation $\mathbf{P}_{i,k} = \dot{\mathbf{V}}_{i,k} + \mathbf{V}_{i,k}^2$ and write $\ddot{\mathbf{p}}'^{(i)} = \mathbf{P}_{i,k}\mathbf{p}'^{(i)}$. The equation is equivalent to the expression (1.130) representing the acceleration of point P on the body, when the rotation is before the translation. This can be proven in the following way. First we calculate

$$\dot{\mathbf{V}}_{i,k} = \begin{bmatrix} \dot{\boldsymbol{\Omega}}_{j,k} & \dot{\mathbf{v}}^{(i)} - \dot{\boldsymbol{\Omega}}_{j,k}(q_j \mathbf{e}_j^{(i)}) - \boldsymbol{\Omega}_{j,k}\mathbf{v}^{(i)} \\ 0 \quad 0 \quad 0 & 0 \end{bmatrix},$$

and afterwards add $\mathbf{V}_{i,k}^2$. Now we have

$$\mathbf{P}_{i,k} = \begin{bmatrix} \dot{\boldsymbol{\Omega}}_{j,k} + \boldsymbol{\Omega}_{j,k}^2 & \mathbf{a}^{(i)} - (\dot{\boldsymbol{\Omega}}_{j,k} + \boldsymbol{\Omega}_{j,k}^2)(q_j \mathbf{e}_j^{(i)}) \\ 0 \quad 0 \quad 0 & 0 \end{bmatrix}$$

and finally

$$\mathbf{P}_{i,k} = \begin{bmatrix} \boldsymbol{\Psi}_{j,k} & \mathbf{a}^{(i)} - \boldsymbol{\Psi}_{j,k}(q_j \mathbf{e}_j^{(i)}) \\ 0 \quad 0 \quad 0 & 0 \end{bmatrix}. \tag{1.164}$$

This time the matrix $\mathbf{P}_{i,k}$ incorporates the angular acceleration matrix $\boldsymbol{\Psi}_{j,k}$ and the acceleration along the translational axis $\mathbf{a}^{(i)}$. When inserting the matrix into the expression describing the acceleration of the point P, we obtain a result which is the same as in (1.130)

$$\begin{bmatrix} \ddot{\mathbf{p}}'^{(i)} \\ 0 \end{bmatrix} = \mathbf{P}_{i,k} \begin{bmatrix} \mathbf{p}'^{(i)} \\ 1 \end{bmatrix} = \begin{bmatrix} \mathbf{a}^{(i)} + \boldsymbol{\Psi}_{j,k}(\mathbf{p}'^{(i)} - q_j \mathbf{e}_j^{(i)}) \\ 0 \end{bmatrix}.$$

1.8.3 Characteristics of a Planar Motion

In Sect. 1.2.1 it was shown that only three coordinates are necessary to determine the pose of a body in a plane. They can be defined in different ways. Usually we use two elements of the positional vector and one orientation angle.

Assume that the body is displaced in a plane perpendicular to the $\mathbf{z}$ axis. This is the plane of motion. The point $\mathbf{p}$, defining the position of the body, changes its elements p_x and p_y with respect to the reference coordinate frame. The change in orientation can only be seen in the angle θ_z around the **z** axis. The mathematical tools describing general spatial motion developed in the preceding sections, can also be used to describe planar motion. In this section we shall introduce several mathematical simplifications which arise when applying the equations describing spatial motion to those describing planar motion.

In 2-dimensional Cartesian space (without the z coordinate) the rotation matrix is a 2×2 matrix with the following elements

$$\mathbf{A}_{i,j} = \begin{bmatrix} \cos q_j & -\sin q_j \\ \sin q_j & \cos q_j \end{bmatrix}, \tag{1.165}$$

where q_j is the rotation angle around the **z** axis (1.53). This matrix is orthogonal with all the pertaining properties. When the elements of the rotation matrix are given, the angle of rotation can be calculated

$$\cos q_j = \arctan \frac{a_{21}}{a_{11}}, \tag{1.166}$$

while the rotation vector

$$\mathbf{e}_j^{(i)} = \begin{bmatrix} 0 \\ 0 \end{bmatrix} \tag{1.167}$$

vanishes since it has no **x** or **y** component.

The translation is described by the sum (1.65), where the vectors are 2-dimensional. Nothing can be added to the combination of the translation and rotation as given by (1.66) and (1.67). A simplification in calculation occurs with two successive rotations, which in planar motion will always be parallel (1.69)

$$\mathbf{A}_{i,k} = \mathbf{A}_{i,j}\mathbf{A}_{j,k} = \begin{bmatrix} \cos(q_j + q_k) & -\sin(q_j + q_k) \\ \sin(q_j + q_k) & \cos(q_j + q_k) \end{bmatrix}. \tag{1.168}$$

Not much can be said about orientation angles in the plane, as there exists only one orientation angle which is about the axis perpendicular to the plane of motion. This is true for both Euler and YPR angles, where $\psi = \theta_z$ and $\theta = \phi = 0$. The two Euler parameters in the plane are

$$r = \sin \frac{q_j}{2}, \qquad r_0 = \cos \frac{q_j}{2}, \tag{1.169}$$

where

$$r^2 + r_0^2 = 1.$$

Equation (1.93) has the following form for planar motion

$$\mathbf{A}_{i,j} = (2r_0^2 - 1)\begin{bmatrix} 1 & 0 \\ 0 & 1 \end{bmatrix} + 2r_0 r \begin{bmatrix} 0 & -1 \\ 1 & 0 \end{bmatrix},$$

with unit and skew symmetrical matrices in space 2×2. The rotation variable is

$$q_j = \pm \arccos \sqrt{1 - r^2} \pm 2k\pi, \quad k = 0, 1, \ldots.$$

Similar simplified relations can be derived for planar example of Rodrigues parameters and in an analogue way also for linear vector and scalar invariants of the rotation matrix, which are defined in the following way

$$s = \sin q_j, \qquad s_1 = \cos q_j \tag{1.170}$$

where

$$s^2 + s_1^2 = 1.$$

The rotation matrix can be obtained in accordance with (1.100)

$$\mathbf{A}_{i,j} = s_1 \begin{bmatrix} 1 & 0 \\ 0 & 1 \end{bmatrix} + s \begin{bmatrix} 0 & -1 \\ 1 & 0 \end{bmatrix},$$

where the rotation angle is

$$q_j = \pm \arccos \sqrt{1 - s^2} \pm 2k\pi, \quad k = 0, 1, \ldots.$$

Now consider velocities and accelerations in planar motion. It is not difficult to derive the angular velocity matrix (1.104) for the planar example

$$\boldsymbol{\Omega}_{i,j} = \dot{q}_j \begin{bmatrix} 0 & -1 \\ 1 & 0 \end{bmatrix}. \tag{1.171}$$

In a similar way the angular acceleration matrix can be obtained from (1.123)

$$\boldsymbol{\Psi}_{i,j} = \ddot{q}_j \begin{bmatrix} 0 & -1 \\ 1 & 0 \end{bmatrix} - \dot{q}_j^2 \begin{bmatrix} 1 & 0 \\ 0 & 1 \end{bmatrix}. \tag{1.172}$$

Both expressions can be applied to simplify the equations for velocities and accelerations derived for spatial motion in the preceding section to the case of planar motion. Here we must take into account that multiplying a vector **a** by skew symmetrical matrix results in a perpendicular vector **b**, such as

$$\mathbf{b} = \begin{bmatrix} 0 & -1 \\ 1 & 0 \end{bmatrix} \mathbf{a},$$

then

$$\mathbf{b} = \begin{bmatrix} -a_y \\ a_x \end{bmatrix}.$$

When considering planar motion, the homogeneous transformation matrix can be reduced to the following

$$\mathbf{H}_{i,j} = \begin{bmatrix} \cos q_j & -\sin q_j & d_{i,j\mathrm{x}} \\ \sin q_j & \cos q_j & d_{i,j\mathrm{y}} \\ 0 & 0 & 1 \end{bmatrix}. \tag{1.173}$$

Such a matrix must operate on 3-dimensional vectors such as

$$\mathbf{a} = \begin{bmatrix} a_\mathrm{x} \\ a_\mathrm{y} \\ 1 \end{bmatrix}.$$

Chapter 2
Mechanisms

Abstract This chapter begins with a description of the different types of mechanisms that are generally used, especially in industrial robots. The parameters and variables of the mechanisms are defined and the degrees of freedom are calculated. Two methods to model a mechanism are presented. We show that in the Denavit-Hartenberg method, the attachment of local coordinate frames to the links is precisely specified, and relative to these frames a minimum number of translational and rotational parameters that describe the relative pose of two neighboring links are defined. In the so-called method of Vector Parameters, link and joint vectors are used to determine the geometry of the mechanism. As the reference position of a mechanism is a free choice, this method enables us to select the most appropriate reference position with respect to the requirements of the robot task.

In this chapter we shall consider a system of rigid bodies interconnected by joints. Joints allow particular types of relative motions between the connected bodies. For example, a rotational joint acts as a hinge and allows only a relative rotation between the connected bodies about the axis of the joint. The relative movements allowed by a joint are referred to as the joint variables or the internal coordinates. The rotational joint has only one joint variable and that is the relative rotation between the connected bodies.

A system of rigid bodies interconnected by joints is called a kinematic chain. Individual rigid bodies within the kinematic chain are called links. A kinematic chain can be serial, parallel, or serial and parallel combined together, i.e. the kinematic chain can be open, closed, or branched.

A mechanism results when a kinematic chain has one of its links fixed to ground, that link then being made immobile. The layout of the links and joints within a kinematic chain determines the motion properties of the mechanism that results from that kinematic chain.

In this chapter we first determine how the motion of a mechanism can be determined by considering the number of links in its kinematic chain and the constraints imposed by the joints connecting the links. We then describe several characteristic types of joints and pay special attention to the kinematic pair, which is the simplest kinematic chain, consisting of only two links connected by one joint. We then introduce the parameters of the kinematic pair and use them in modeling of mechanisms.

J. Lenarčič et al., *Robot Mechanisms*,
Intelligent Systems, Control and Automation: Science and Engineering 60,
DOI 10.1007/978-94-007-4522-3_2, © Springer Science+Business Media Dordrecht 2013

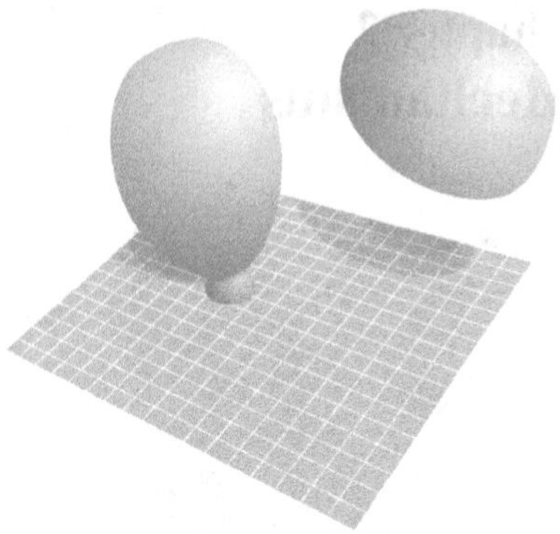

Fig. 2.1 Free body and body attached to the base by the help of a joint

We also become acquainted with the layout of links and joints which are frequently used in robotics.

2.1 Joints and Degrees of Freedom

A basic property of a mechanism is its ability to move. A mechanism results from a kinematic chain when, one of the links in the chain becomes fixed to ground. Thus a mechanism has a fixed base link. This fixed link connects to one or more links by one or more joints, and these links are then connected by joints to the remaining links. The moving links in a mechanism are accomplishing various tasks, such as grasping, picking and placing of objects, drilling, or grinding.

From a mathematical point of view, the degrees of freedom within a mechanism equals the minimum number of the mechanism's independent joint variables which must be specified in order to uniquely determine the spatial pose of all bodies belonging to the mechanism. This is also referred to as the number of degrees of freedom in the mechanism [21].

2.1.1 Types of Joints

A rigid body which is free to move in a 3-dimensional space has six degrees of freedom. The pose of a body is determined by $\lambda = 6$ parameters, where there are three positional and three orientational coordinates. A body free to move in a plane has $\lambda = 3$ degrees of freedom, two describing its position and one belonging to the orientation of a body.

When a body is attached to the base by virtue of a joint, as shown in Fig. 2.1, its number of degrees of freedom will be less than or equal to the number of degrees of freedom of a free body. Suppose there are f independent joint variables associated with a joint. We would say that the joint allows f degrees of freedom.

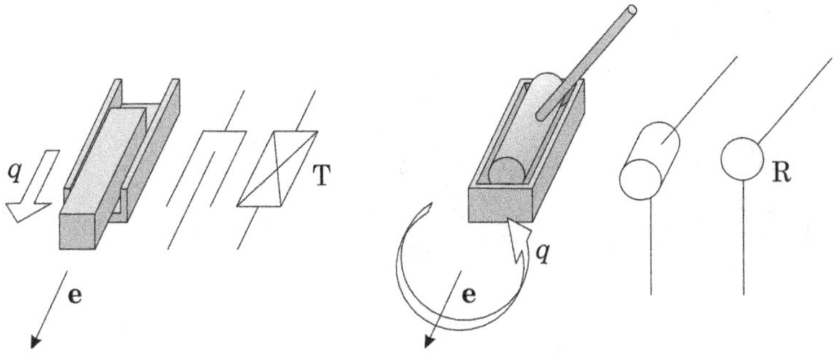

Fig. 2.2 Translational and rotational joint

The number of degrees of freedom of a body rigidly connected to the base is zero. When connected to the base by a joint, the body has at most as many degrees of freedom as allowed by the joint, which is always less then λ. Typically a joint is considered as a connection that allows certain motion of the body. Inversely, a joint can be considered as a connection which limits the motion of a body by virtue of constraints imposed by the joint. The difference between the possible degrees of freedom (λ) and the number of degrees of freedom allowed by a joint (f) is called the number of constraints

$$c = \lambda - f. \tag{2.1}$$

For example, in a rotational joint we have seen that $f = 1$ and hence in spatial motion ($\lambda = 6$), $c = 6 - 1 = 5$. That is to say that a rotational joint eliminates 5 degrees of freedom of relative motion between the connected bodies.

The number of degrees of freedom allowed by a joint, and the nature of those degrees of freedom, are determined by the shape of the contact areas between the two bodies that are associated with the particular type of joint. The two simplest joints allow $f = 1$ degrees of freedom and impose $c = 5$ constraints in spatial motion, or $c = 2$ constraints in planar motion. These are the translational and rotational joints, shown in Fig. 2.2, often denoted in the literature by the letters T and R. Both of these fundamental joints can be described by a unit vector $\mathbf{e}$, which defines the axis of either the linear displacement or rotation. The joint variable is the coordinate q describing either the distance of translation or the angle of rotation. In Fig. 2.2 the graphical representations of translational and rotational joints are shown. These representations will be adopted in the remainder of the text.

All other joints can be modeled as combinations of these two fundamental joints. In some cases there is an interdependence between the rotation(s) and translation(s). Let us examine some of the basic joints.

In addition to the translational and rotational joints there are the two degree of freedom cylindrical joint and universal joint, which have $f = 2$, and the three degree of freedom spherical joint, which has $f = 3$. The variety of joints in mechanical engineering is much larger than these, but these are the joints fundamental to robotics.

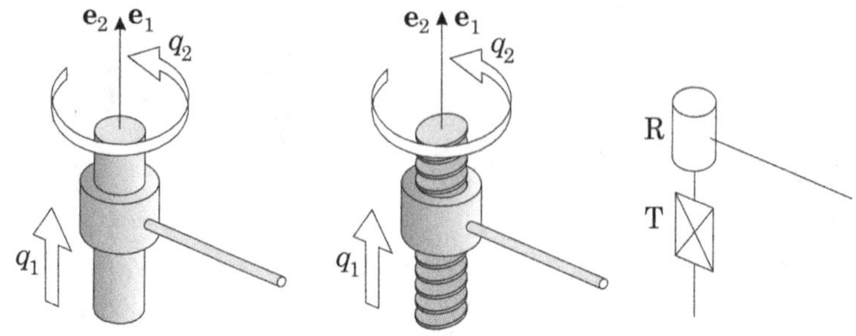

Fig. 2.3 Cylindrical and screw joints

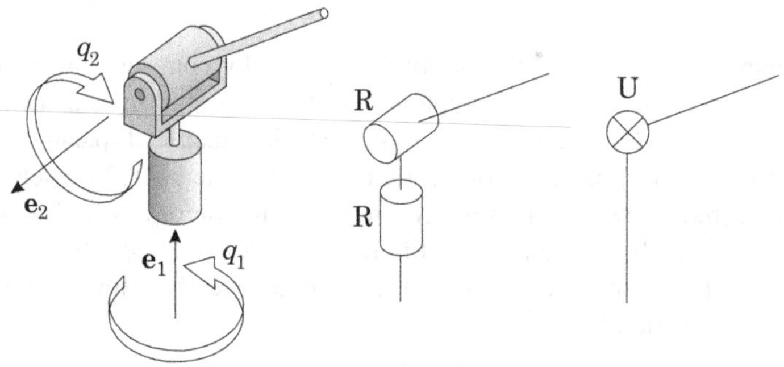

Fig. 2.4 Universal joint

Figure 2.3 presents two joints types of joints found in spatial mechanisms which can be represented by combination of a translational joint and a rotational joint, shown on the right hand side of the figure. The first joint on the left is called a cylindrical joint. Translation takes place in the direction of $\mathbf{e}_1$, rotation is about the vector $\mathbf{e}_2$, and $\mathbf{e}_1$ and $\mathbf{e}_2$ are coincident. The joint variables are the distance q_1 and the angle of rotation q_2, and they are independent. Hence $f = 2$ and $c = 4$. The second joint is called a screw joint. This joint is also represented by a translation in the direction of the vector $\mathbf{e}_1$ and rotation about the axis $\mathbf{e}_2$ and again $\mathbf{e}_1$ and $\mathbf{e}_2$ are coincident. However, in the screw joint there is an interdependence between rotation q_2 and translation displacement q_1. This interdependence can be described by the relation $\Delta q_1 = \xi \Delta q_2$, where Δq_1 and Δq_2 represent changes in the joint variables q_1 and q_2 and ξ is a known constant called the pitch of the screw. Thus although the screw joint has two joint variables, it has only one independent joint variable, either q_1 or q_2, and therefore it has $f = 1$ and $c = 5$.

Figure 2.4 shows a universal joint, or so called U-joint. This type of joint is also found in spatial mechanisms and it is equivalent to two rotational joints whose axes are intersecting. Kinematically it is represented by the rotational axes $\mathbf{e}_1$ and $\mathbf{e}_2$ and the angles q_1 and q_2 which are independent. Hence the U-joint has $f = 2$ and $c = 4$. A possible realization is shown in Fig. 2.4.

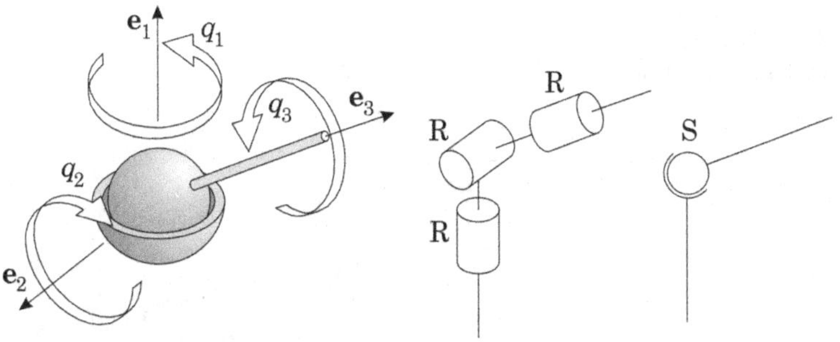

Fig. 2.5 Spherical joint

Among the three degree of freedom joints, the spherical joint is encountered most often. It is usually presented as the ball joint shown in Fig. 2.5. It is denoted by the letter S. For this joint we have $f = 3$ and $c = 3$. The joint allows independent rotations about three axes and is modeled by three rotations whose axes are intersecting at the same point. In Fig. 2.5 these are the axes $\mathbf{e}_1$, $\mathbf{e}_2$, and $\mathbf{e}_3$ with corresponding rotations q_1, q_2, and q_3. We must be aware that the spherical joint with three sequential rotations is not completely equivalent to the ball joint. The joint with three sequential rotations looses one degree of freedom when the third axis is collinear with the first axis. Such a situation does not occur with the ball joint. However, due to a constraint of force closure, the socket which encapsulates the ball must wrap over more than a hemisphere of the ball, hence the spherical joint has a workspace which is limited to significantly less than a hemisphere, and this is effectively as limited a workspace as the joint with three sequential rotations.

As in robotics, the mechanics of joints is important when studying human movements [93]. Among the so called synovial joints, where two bones are in contact, we have ellipsoidal and saddle joints. These resemble rotational joints, but are not completely equivalent. There also exist joints which do not have firm contact areas. Here, the two bones are connected through elastic bands (ligaments). A unique joint is that between the shoulder blade and the trunk, the scapulothoracic joint. In this non-synovial joint, the shoulder blade actually slides and rotates between layers of muscles on the back.

2.1.2 Types of Mechanisms

Links connected by joints create kinematic chains. These can be open, closed, or branched [86] kinematic chains. Figure 2.6 shows some mechanisms coming from different kinematic chains. The mechanism on the left results from a six link open kinematic chain where one of the links is the fixed base. The five remaining links of the chain move in space. The mechanism in the back results an eleven link closed loop kinematic chain which has one of it links fixed to ground. The mechanism in front and on the right results from a branched open kinematic chain.

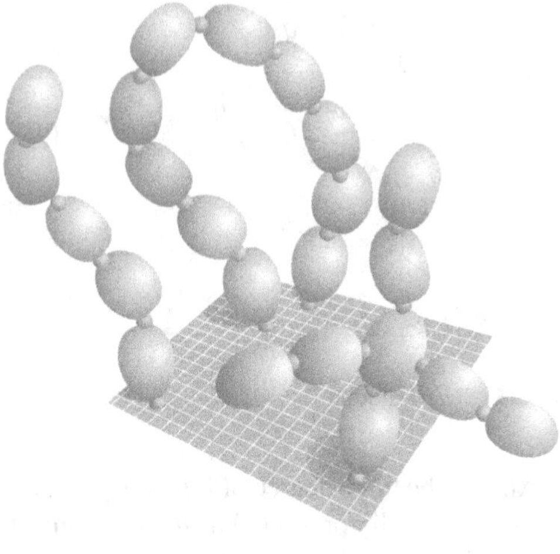

Fig. 2.6 Mechanisms from open, closed and hybrid open/closed kinematic chains

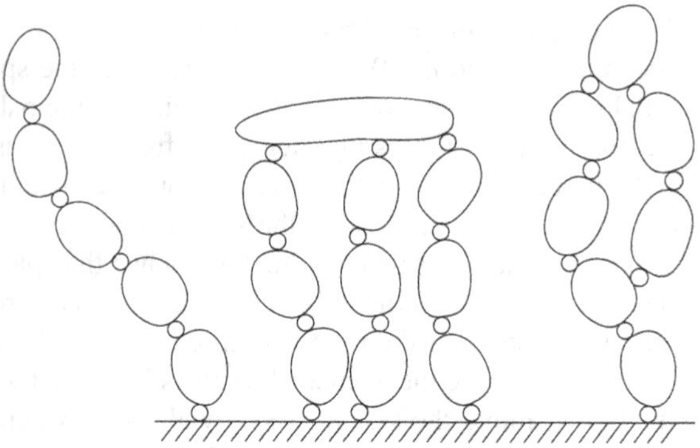

Fig. 2.7 Serial, parallel, and hybrid kinematic chains

Figure 2.7 shows additional examples of mechanisms that result from open, closed and hybrid open/closed kinematic chains. Mechanisms that result from open kinematic chains, such as that on the left, are referred to as serial mechanisms. Serial mechanisms typically have larger reachable workspace, but are less rigid. Mechanisms that result from closed kinematic chains, or groups of closed kinematic chains, such as that in the center, are referred to as parallel mechanisms. Parallel mechanisms typically are more rigid and exhibit larger load capacity and improved accuracy, but have smaller workspace. In parallel mechanisms, many times the multiple closed loops carry a single common link, referred to as the moving platform. In these cases, typically the fixed link is also a common element and is referred to as the fixed platform, or base. The links which connect between the moving and fixed platforms are referred to as legs. The example of a parallel mechanism in Fig. 2.7 has an identifiable moving platform, fixed platform (base) and three legs. Mechanisms that result from hybrid open/closed kinematic chains are referred as hybrid serial/parallel mechanisms. Hybrid serial/parallel mechanisms exhibit a combina-

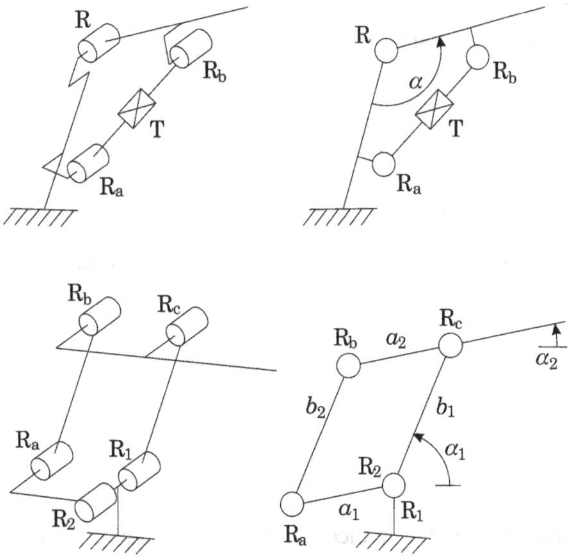

Fig. 2.8 Remotely actuated joints in a mechanism

tion of the traits of serial and parallel mechanisms. Both in robotics and with living organisms we can find examples of hybrid parallel/serial mechanisms.

Industrial robots are usually serial mechanisms. It was only recently that parallel robots (i.e. robots which are parallel mechanisms) were introduced into the industrial arena. Within serial mechanisms all joints are actuated by the motors and the motors closer to the base carry the motors farther from the base. This causes the motors closer to the base to be larger which has negative effects of increasing weight and cost. Many times in parallel manipulators, the driving motors can all be located on the base. The motors are smaller, less expensive and the machine is lighter.

Two simple examples of mechanisms often used in practice are shown in Fig. 2.8. The first mechanism represents a situation where actuation of the rotation joint R is accomplished through the translation T. We can think of R as being remotely actuated by T. This is accomplished by connecting two additional rotational joints R_a and R_b at both sides of translational joint. All together the mechanism now has four joints, however, only one is actuated. As will be shown, this mechanism has only one degree of freedom. Remote actuation of the rotation in a rotational joint is often used in mechanical engineering. In robots with hydraulic and pneumatic actuation, the T joint is an actuated hydraulic or pneumatic cylinder. In electrically driven robots, the T joint is a lead screw driven by an electric motor. The joints of living organisms are similarly remotely actuated by muscles, and in some cases several joints are actuated by the same muscle. This causes coupling in human joint motions. For example, when you curl your finger, your distal two knuckle joints are actuated simultaneously. This is because the muscle actuating this distal pair of knuckle joints spans across the two joints.

The second mechanism, shown in Fig. 2.8 is a parallelogram or pantograph mechanism. The pantograph is a device originally intended to copy drawings in

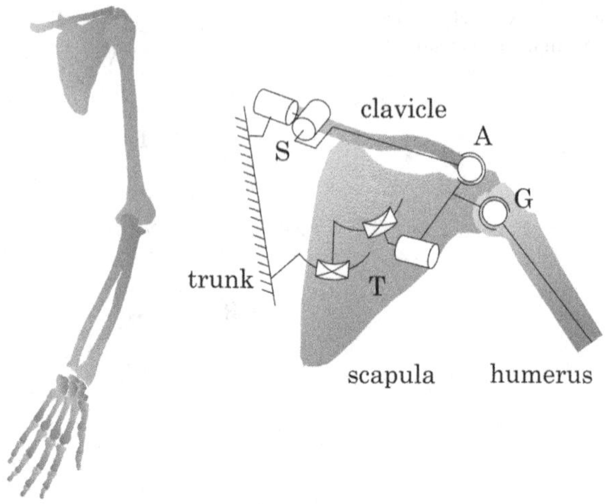

Fig. 2.9 Skeleton of human arm and the shoulder girdle

either enlarged or reduced scale. Office lamps commonly make use of a pantograph mechanism. The mechanism has five rotational joints, but only two degrees of freedom. It has interesting properties when $a_1 = a_2$ and $b_1 = b_2$. In that case, links a_1 and a_2 are parallel throughout the motion of the mechanism. The same is also true for links b_1 and b_2. Assume that the joints R_1 and R_2 are actuated. When the rotation R_1 is held constant, the angle α_1 is constant. Then by rotating the joint R_2, only the angle α_2 is changing. When the rotation R_2 is held constant, the angle α_2 is constant. Then by rotating R_1, only the angle α_1 varies. Despite rotation R_1, the orientation of link a_2 remains constant and only angle α_1 varies. This alternate type of remote actuation is used in many industrial robots, particularly those with high payloads. The mechanism also has the advantage that the actuators for R_1 and R_2 do not translate.

The skeleton of a human arm consists of a large number of bones, see Fig. 2.9, creating serial and parallel chains. The human arm is attached to the trunk through the shoulder girdle, this being a parallel mechanism. The clavicle is connected to the breastbone (a.k.a. sternum) via the sternoclavicular joint (S) from one side, and from the other side to the scapula via the acromioclavicular joint (A). The scapula glides along the back of the rib cage via the scapulothoracic joint (T) and at the lateral end it is connected through the glenohumeral joint (G) to the upper arm. A simplified model of a mechanism emulating the shoulder girdle is shown on the right hand side of Fig. 2.9. The joints of the model all together have eleven degrees of freedom. Later in the text we shall show however that this shoulder mechanism has only five independent degrees of freedom.

The upper arm (humeral bone) can be considered as part of a serial mechanism. In the elbow joint the humeral bone connects to radial bone and ulna, creating a parallel mechanism. Every finger together with the corresponding metacarpal bone represents a serial mechanism. In this way the palm is an example of the branched kinematic chain. When the palm and the fingers grasp an object, the kinematic chains of the fingers are closed and the hand is transformed into a parallel mechanism.

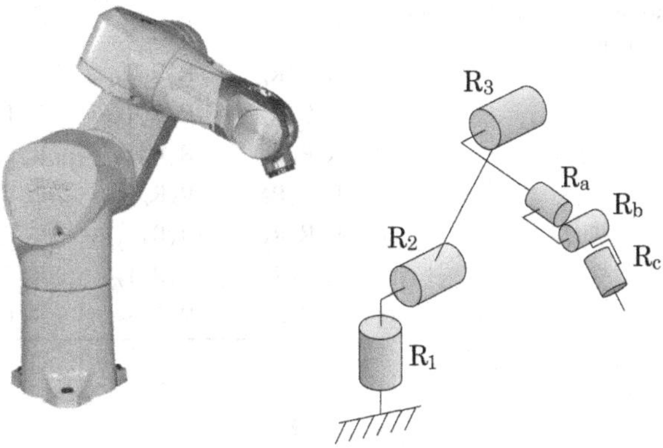

Fig. 2.10 Mechanism of industrial manipulator (courtesy of Stäubli)

The kinematic structure of a human arm is adapted to three principal functions. The primary function of the shoulder and elbow joints is to bring the hand into a desired position in space. The function of the wrist is to rotate the hand into a desired orientation, while grasping is the function of the palm and fingers.

The great majority of today's robots are industrial manipulators. An example is shown in Fig. 2.10. These are mechanical arms, which, through the shoulder joint, are attached to a fixed base. A robot gripper or another tool is attached to the distal end of the manipulator. The mechanisms of most industrial robots can be divided into two mechanisms, one for positioning and another for orientating. The positional part of the mechanism in Fig. 2.10 includes the rotations R_1, R_2, and R_3 and ranges from the shoulder to the lower arm. Its main task is to position the robot end-point into the desired position. The orientational part of the mechanism, which is in Fig. 2.10 denoted by rotations R_a, R_b, and R_c, represents the robot wrist, whose main task is to bring the gripper or end-point tool into the desired orientation. Each of these two parts requires three degrees of freedom.

Let us examine first the positional part of the mechanism. In this positional part of the mechanism, both translational and rotational can be used, since both types of joints can contribute to the positioning a point belonging to the mechanism. The axes of these joints can be directed arbitrarily in space. However, we shall limit ourselves to those structures where the successive translational or rotational axes are either parallel or perpendicular and at times intersecting, which is the case in industrial manipulators. Such positional mechanisms can always be placed in a pose such that all its joint axes are parallel to one of the axes of the fixed coordinate frame **x**, **y**, **z**. Therefore we can select among translational joints T_x, T_y, and T_z in the **x**, **y**, **z** directions, and rotational joints R_x, R_y, and R_z about the **x**, **y**, **z** directions, for the three joints of the mechanism. The number of all possible combinations of three of these six joints is $6^3 = 216$. All of these combinations do not result in a spatial mechanism. For example, structures, such as $T_xT_xT_x$ and $T_xT_yR_z$, do not permit motion in at least one of the directions of the coordinate frame **x**, **y**, **z**. If we wish to have a spatial mechanism, its joint variables must enable motion in all three directions. Thus each joint variable must provides for a component of motion

Table 2.1 Spatial positional mechanisms

$R_xR_xR_y$	$R_xR_xR_x$	$R_xT_xR_y$	$R_xT_yT_x$	$T_xR_xT_z$
$R_xR_xR_z$	$R_xR_yT_x$	$R_xT_xR_z$	$R_xT_zT_x$	$T_xR_yT_y$
$R_xR_yR_z$	$R_xR_yT_y$	$R_xT_yR_y$	$T_xR_xR_x$	$T_xT_yR_x$
$R_xR_yR_y$	$R_xR_yT_z$	$R_xT_yR_z$	$T_xR_xR_y$	$T_xT_yR_y$
$R_xR_yR_z$	$R_xR_zT_x$	$R_xT_zR_y$	$T_xR_xR_x$	$T_xT_yT_z$
$R_xR_zR_x$	$R_xR_zT_y$	$R_xT_zR_z$	$T_xR_yR_x$	
$R_xR_zR_y$	$R_xR_zT_z$	$R_xT_xT_y$	$T_xR_yR_z$	
$R_xR_zR_z$	$R_xT_xR_x$	$R_xT_xT_z$	$T_xR_xT_y$	

in one direction which is independent of the motions caused by the other two joint variables.

The letter in parentheses denotes the direction of the action of each degree of freedom. In this way we have $T_x \rightarrow (x)$, $T_y \rightarrow (y)$, $T_z \rightarrow (z)$, $R_x \rightarrow (y, z)$, $R_y \rightarrow (x, z)$ and $R_z \rightarrow (x, y)$. The mechanism $R_zR_xT_y \rightarrow (x, y)(y, z)(y)$ is a spatial mechanism, as each degree of freedom enables displacement of the mechanism in a unique direction. With such analysis of all 216 mechanisms, 129 spatial mechanism are selected. Some of them appear twice in the analysis and some differ only in the orientation with respect to the coordinate frame. Such examples are $R_zR_xR_x$ and $R_zR_yR_y$. When excluding all repeated variations, 37 different positional spatial mechanism remain [46], as shown in Table 2.1.

Only some of these mechanism are used as industrial robots. The International Federation of Robotics classifies in its statistical reports five types of positioning mechanisms that are found in industrial robots. They are shown in Fig. 2.11. Together with five serial mechanisms, we are presenting also a parallel mechanism, whose structure is not included in the classification. The parallel mechanism in Fig. 2.11 has three degrees of freedom. In general, parallel mechanisms differ considerably from serial mechanisms, in that the positioning and orienting functions in parallel mechanisms are typically completely coupled whereas in serial mechanisms they are typically only partially decoupled.

The orientational part of the robot mechanism needs to include at least three degrees of freedom in order to be able to bring the end-point tool into the desired orientation. By combining the rotations R_x, R_y, and R_z, 27 different wrist structures can be created. However, only the structures with successively perpendicular axes are considered. If the first rotation has the **x** direction, the next must be in the **y** or **z** direction. If the second rotation goes along **y** axis, the third must be about **z** or **x** and when the direction of the second joint axis is **z**, the third must be aligned with **y** or **x** axes. We have 12 such structures, which are all kinematically equivalent. They differ only with respect to the orientation of their attachment to the terminal link of the positioning mechanism.

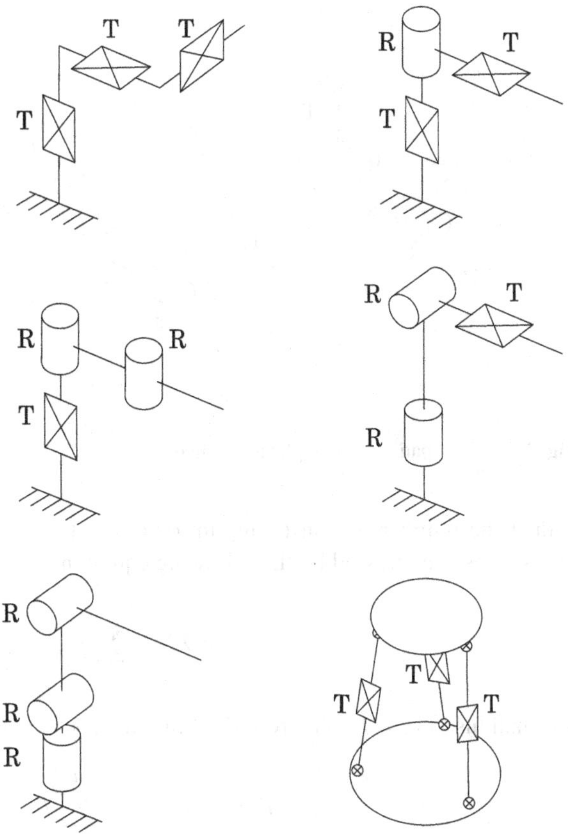

Fig. 2.11 Positional mechanisms of industrial manipulators according to the classification of International Federation of Robotics include TTT Cartesian manipulator, TRT cylindrical manipulator, TRR Scara manipulator, RRT spherical manipulator, RRR articulated manipulator and parallel manipulator

2.1.3 Degrees of Freedom in Mechanisms

The links of a mechanism are interconnected with various types of joints, either single or multiple points. Some examples of mechanisms are shown in Fig. 2.12. The number of independent degrees of freedom allowed by a joint i will be denoted as f_i, while the number of constraints imposed by the joints by c_i. All together the mechanism has $i = 1, 2, \ldots, n$ joints and $i = 0, 1, 2, \ldots, N$ links. The number 0 belongs to the reference link, i.e. the base. This is the fixed link in the kinematic chain from which the mechanism was derived. The base link does not move, while the remaining links $i = 1, 2, \ldots, N$ are in motion.

The number of degrees of freedom F belonging to a mechanism is defined as the number of independent joint variables which need to be specified in order to uniquely define the spatial configuration of the mechanism, i.e. in order to uniquely define the pose of every link in the mechanism. The number of degrees of freedom in a mechanism is obtained by first summing up the degrees of freedom made available by all mobile links of a mechanism, which would be λN. This would be the number of degrees of freedom in the mechanism if there were no joints. From this we must

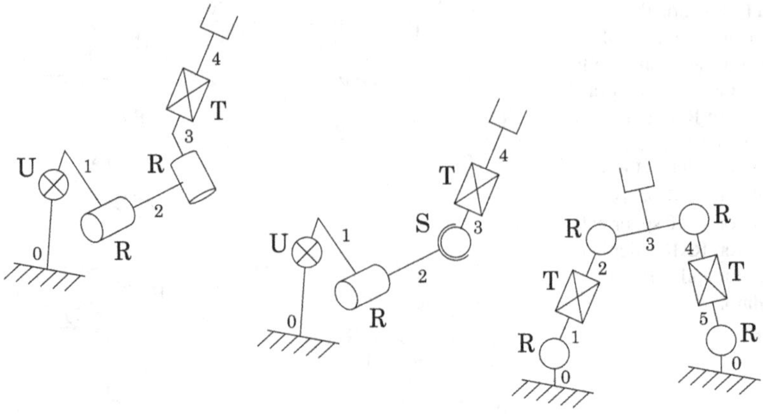

Fig. 2.12 Two spatial and one planar mechanism

deduct the number of constraints introduced into the mechanism by each of the joints. This is expressed by the following equation

$$F = \lambda N - \sum_{i=1}^{n} c_i .$$

We shall insert $c_i = \lambda - f_i$ from (2.1) into the above equation. It follows

$$F = \lambda(N - n) + \sum_{i=1}^{n} f_i . \tag{2.2}$$

The expression above is well known as Grübler's formula [84].

Let us now calculate the number of degrees of freedom of the mechanisms shown in Fig. 2.12. On the far left we have a serial mechanism in space ($\lambda = 6$) with four joints ($n = 4$), a two degree of freedom universal joint ($f_1 = 2$), two rotational joints ($f_2 = f_3 = 1$), and a translational joint ($f_4 = 1$). The mechanism has five links (including the base), four of them are mobile ($N = 4$). According to Grübler's formula the number of degrees of freedom in the mechanism is

$$F = 6(4 - 4) + 5 = 5.$$

Therefore, the gripper of the robot end-point has five degrees of freedom. We can also observe, that in serial mechanisms the number of mobile links N always equals the number of the joints n

$$n = N. \tag{2.3}$$

Therefore, the expression in parentheses in Grübler's formula, $(N - n)$, is always zero for serial mechanisms. The number of degrees of freedom of a serial mechanism is equal to the sum of the independent joint variables associated with each joint

$$F = \sum_{i=1}^{n} f_i \tag{2.4}$$

for both planar and spatial mechanisms.

Let us now examine the degrees of freedom of the mechanism shown in the middle of Fig. 2.12. This mechanism is also serial and spatial ($\lambda = 6$) and has four joints ($n = 4$), a two degree of freedom universal joint ($f_1 = 2$), a rotational joint ($f_2 = 1$), a spherical joint ($f_3 = 3$) and a translational joint ($f_4 = 1$). In accordance with (2.4) the mechanism has

$$F = 2 + 1 + 3 + 1 = 7$$

degrees of freedom. The robot gripper cannot have more than $\lambda = 6$ degrees of freedom, the mechanism is therefore kinematically redundant. A redundant mechanism can hold the gripper in a desired position and orientation, while the rest of the mechanism is still movable. This is referred to as a self-motion of the mechanism and it is a characteristic of a redundant mechanism.

In the right side of Fig. 2.12 is shown a planar parallel mechanism, hence $\lambda = 3$. The mechanism has six joints ($n = 6$), all of them are either rotational or translational, and six links, where five are mobile ($N = 5$). In accordance with Grübler's formula, the number of degrees of freedom is

$$F = 3(5 - 6) + 6 = 3.$$

This is also the number of the degrees of freedom of the gripper. In parallel mechanisms we are dealing with the following inequality

$$n > N, \tag{2.5}$$

and the simplification (2.4) does not hold.

By using Grübler's formula we shall calculate the number of degrees of freedom for the mechanisms from the previous section. The mechanisms from Fig. 2.8 are both planar, therefore $\lambda = 3$. With the upper mechanism we have $n = 4$ and $f_1 = f_2 = f_3 = f_4 = 1$. The mechanism has four links one of them is selected as a reference link, therefore $N = 3$. It follows

$$F = 3(3 - 4) + 4 = 1,$$

which further proves the statement that this mechanism has only a single degree of freedom. In the lower mechanism there is $n = 5$ and $f_1 = f_2 = f_3 = f_4 = f_5 = 1$. This mechanism has five links, from which one is a reference one, therefore $N = 4$. It follows

$$F = 3(4 - 5) + 5 = 2.$$

The pantographic mechanism has two degrees of freedom and can be used to position a point on its terminal link.

The mechanism of the shoulder girdle from Fig. 2.9 is a spatial mechanism, hence $\lambda = 6$. The mechanism has five one degree o freedom joints $f_1 = f_2 = f_3 = f_4 = f_5 = 1$ and two three degree of freedom joints $f_6 = f_7 = 3$, therefore $n = 7$. After subtracting the reference link, we have all together $N = 6$ links. There is

$$F = 6(6-7) + 11 = 5.$$

Thus the upper arm has five degrees of freedom. The shoulder girdle, the mechanism without the spherical glenohumeral joint G, has only two degrees of freedom. This can be calculated by subtracting one spherical joint and one link in the preceding equation, giving

$$F = 6(5-6) + 8 = 2.$$

The calculation of the number of degrees of freedom of the industrial manipulator from Fig. 2.10 is even simpler as the expression (2.4) can be used. This spatial serial mechanism has only rotations and $f_1 = f_2 = f_3 = f_4 = f_5 = f_6 = 1$, therefore

$$F = 6.$$

The gripper at the end of the robot has six degrees of freedom and can be placed in an arbitrary position and orientation inside its workspace.

2.2 Parameters and Variables of a Kinematic Pair

A kinematic pair is the basic element of a kinematic chain. It consists of two links connected by a joint with a translational or rotational degree of freedom. In the literature there exist two approaches to the mathematical description of a kinematic pair. The difference between them is in the attachment of the coordinate frames to both links. The so called Denavit and Hartenberg method [17] is based on an adapted homogeneous transformation matrix. This method makes use of four scalars, which will be called the Denavit and Hartenberg parameters of a kinematic pair. Four is the minimum number of parameters required for describing the link geometry and the relative joint displacement of the links in a kinematic pair. This leads to a minimum number of arithmetic operations in computations, which is an advantage of the method. Denavit and Hartenberg establish precise rules on how to position and orient the two coordinate frames and this first step in modeling of a mechanism can be cumbersome. If the rules are not followed precisely, the resulting kinematic equations are incorrect.

In the second approach, the method of Vector Parameters, vectors are used to describe a kinematic pair [46]. This method is based on the general rotational matrix and is therefore computationally more complex. The benefit of the method is in a simpler determination of the parameters of a kinematic pair. A similar vectorial method was introduced by [79] in modeling robot dynamics.

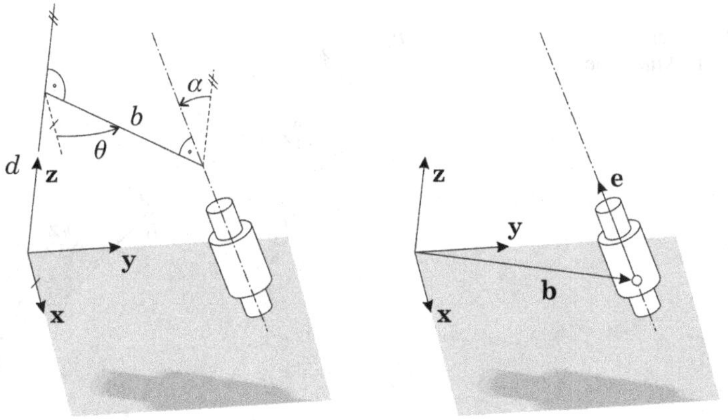

Fig. 2.13 Description of the position and direction of a cylindrical joint in a Cartesian coordinate frame

2.2.1 Cylindrical Joint in a Cartesian Space

The motion characteristics of the joints described in the beginning of this chapter can be illustrated with the aid of various combinations of rotational or translational joints, whose coordinates are appropriately mathematically related. As the basis for many of these joints, we use the cylindrical joint, wherein the rotation and translation between connected links takes place about and along a single axis, see Fig. 2.3. In order to describe this joint mathematically, we must know the position and direction of this axis with respect to the selected coordinate frame, and the values of the translational and rotational joint variables.

In the left side of Fig. 2.13 the position and direction of the joint axis is determined with respect to the Cartesian frame using the minimum possible number of parameters. To do this we introduce the common normal between the joint axis and the **z** axis of the coordinate frame. We know the common normal between two arbitrary non-parallel lines is unique. The location of the common normal can be determined by assessing the length, d, measured along **z** axis from the origin to the foot of the common normal on the **z** axis. The angle θ, represents the rotation about the **z** axis and is measured from the **x** axis to the direction of the common normal. The length of the common normal is b. The angle α represents the rotation of the joint axis about the common normal. These four scalar parameters d, θ, b, and α are the minimum number of parameters required to describe the location of the axis of a cylindrical joint in a Cartesian coordinate system. Denavit and Hartenberg's scalar parameters of a kinematic pair is based on these four parameters. Denavit and Hartenberg notation will be explained later in the text.

The right side of Fig. 2.13 the location of a joint axis is defined by the position of an arbitrary point on the axis. This is represented by the vector $\mathbf{b} = (b_x, b_y, b_z)^T$. The direction of the axis is defined by the unit vector **e**, which points in a direction along the axis. Here, the position and direction of the line is determined by the two

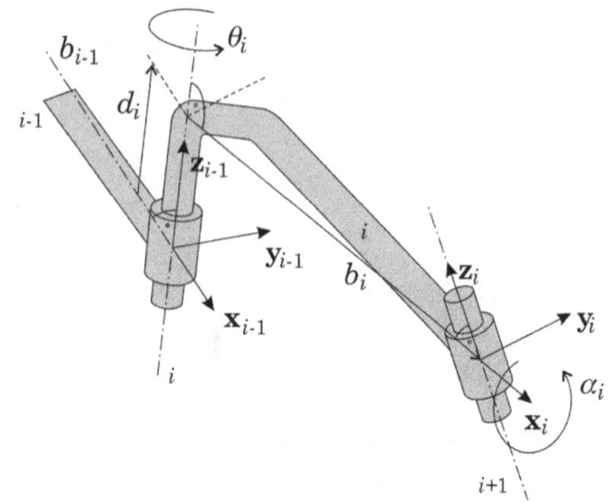

Fig. 2.14 Denavit-Hartenberg parameters of a kinematic pair

vectors, **b** and **e**, where $\mathbf{e}^T\mathbf{e} = 1$ and the component of **b** along **e** is arbitrary. The method of Vector Parameters is based on these two vectors and a general rotation matrix. Note that vectors **b** and **e** include information about the direction of the joint axis, which is not directly described by the four scalar parameters d, θ, b, and α.

2.2.2 Scalar Parameters of a Kinematic Pair

Let us examine first the characteristic properties of the Denavit-Hartenberg method in modeling a kinematic pair. The links and joints will be enumerated in a way which is widely accepted in robotics [67]. The links and joints can be enumerated in a second way [15] without changing the important properties of the approach. Here we follow [67].

Figure 2.14 shows a kinematic pair consisting of links $i-1$ and i connected by cylindrical joint i. At the end of the link i we have the joint $i+1$. The coordinate frame, which is attached to link i, is oriented in such a way, that its $\mathbf{z}_i$ axis is aligned with the joint axis $i+1$, while the $\mathbf{x}_i$ axis goes along the common normal between the joint axes i and $i+1$. The origin of this frame is positioned at the intersection of this common normal and the $i+1$ joint axis. The third axis is represented by the $\mathbf{y}_i = \mathbf{z}_i \times \mathbf{x}_i$.

The Denavit-Hartenberg parameters, describing the geometry and the relative displacement between the bodies of a kinematic pair, are the following:

d_i: translational coordinate—the distance between the origin of the coordinate frame $\mathbf{x}_{i-1}$, $\mathbf{y}_{i-1}$, $\mathbf{z}_{i-1}$ or the end-point of the normal b_{i-1} and the starting point of the normal b_i. The distance d_i is positive in direction of the axis $\mathbf{z}_{i-1}$;

θ_i: rotational coordinate—the angle of rotation between the links i and $i-1$, which is measured from b_{i-1} to b_i, i.e. from $\mathbf{x}_{i-1}$ to $\mathbf{x}_i$, about the $\mathbf{z}_{i-1}$ vector;

b_i: length of the link i—the length of the common normal of joint axes i and $i+1$, representing the length of the i-th link, measured positively in the direction from the joint axis i to the joint axis $i+1$;

Fig. 2.15 Denavit-Hartenberg parameters shown as sequence of two translations and two rotations

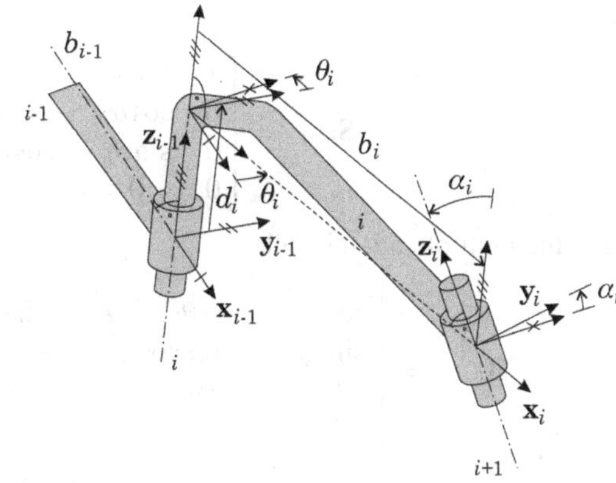

α_i: angle of inclination between two consecutive axes—the angle measured about the $\mathbf{x}_i$ axis between the vectors $\mathbf{z}_{i-1}$ and $\mathbf{z}_i$.

The scalars d_i, θ_i, b_i, and α_i represent the minimal number of the parameters required for a complete description of an arbitrary kinematic pair with the cylindric joint. When a joint is rotational, the joint variable of the kinematic pair is θ_i, while the rest of the parameters are constant. When a joint is translational, the joint variable of the kinematic pair is d_i, while the other three parameters are constant.

While using the Denavit-Hartenberg parameters, the kinematic pair is considered as a sequence of two translations and two rotations. As shown in Fig. 2.15, the coordinate frame $\mathbf{x}_{i-1}, \mathbf{y}_{i-1}, \mathbf{z}_{i-1}$ is superimposed over the frame $\mathbf{x}_i, \mathbf{y}_i, \mathbf{z}_i$ through translation by the distance d_i along the i-th joint axis and rotation by the angle θ_i about the joint axis i This is followed by a translation b_i, which is perpendicular to the joint axes i and $i+1$, and rotation by the angle α_i about the common normal between the joint axes i and $i+1$.

The pose between the coordinate frames $\mathbf{x}_{i-1}, \mathbf{y}_{i-1}, \mathbf{z}_{i-1}$ and $\mathbf{x}_i, \mathbf{y}_i, \mathbf{z}_i$ is given by the homogeneous transformation matrix $\mathbf{H}_{i-1,i}$, which always has the same form. It is a product of a homogeneous transformation matrix $\mathbf{Q}_{i-1,i}$, describing the joint translation d_i and rotation θ_i, and the homogeneous transformation matrix $\mathbf{S}_{i-1,i}$, describing the length b_i and the link inclination α_i. We have

$$\mathbf{H}_{i-1,i} = \mathbf{Q}_{i-1,i}\mathbf{S}_{i-1,i}, \tag{2.6}$$

with the matrices

$$\mathbf{Q}_{i-1,i} = \begin{bmatrix} \cos\theta_i & -\sin\theta_i & 0 & 0 \\ \sin\theta_i & \cos\theta_i & 0 & 0 \\ 0 & 0 & 1 & d_i \\ 0 & 0 & 0 & 1 \end{bmatrix} \tag{2.7}$$

and

$$\mathbf{S}_{i-1,i} = \begin{bmatrix} 1 & 0 & 0 & b_i \\ 0 & \cos\alpha_i & -\sin\alpha_i & 0 \\ 0 & \sin\alpha_i & \cos\alpha_i & 0 \\ 0 & 0 & 0 & 1 \end{bmatrix}. \tag{2.8}$$

After the multiplication it follows

$$\mathbf{H}_{i-1,i} = \begin{bmatrix} \cos\theta_i & -\sin\theta_i\cos\alpha_i & \sin\theta_i\sin\alpha_i & b_i\cos\theta_i \\ \sin\theta_i & \cos\theta_i\cos\alpha_i & -\cos\theta_i\sin\alpha_i & b_i\sin\theta_i \\ 0 & \sin\alpha_i & \cos\alpha_i & d_i \\ 0 & 0 & 0 & 1 \end{bmatrix}. \tag{2.9}$$

This matrix can be adapted to any kinematic pair which consists of one rotation and one translation. In robot systems the axes of consecutive joints i and $i+1$ are often either parallel or orthogonal. In these cases the transformation matrix has a simpler form. When the axes are parallel, the inclination angle between the joint axes is $\alpha_i = 0$ and we have

$$\mathbf{H}_{i-1,i} = \begin{bmatrix} \cos\theta_i & -\sin\theta_i & 0 & b_i\cos\theta_i \\ \sin\theta_i & \cos\theta_i & 0 & b_i\sin\theta_i \\ 0 & 0 & 1 & d_i \\ 0 & 0 & 0 & 1 \end{bmatrix}. \tag{2.10}$$

When the axes are orthogonal, we have $\alpha_i = \pm\pi/2$. The simplified homogeneous transformation matrix has the following form

$$\mathbf{H}_{i-1,i} = \begin{bmatrix} \cos\theta_i & 0 & \pm\sin\theta_i & b_i\cos\theta_i \\ \sin\theta_i & 0 & \mp\cos\theta_i & b_i\sin\theta_i \\ 0 & \pm 1 & 0 & d_i \\ 0 & 0 & 0 & 1 \end{bmatrix}. \tag{2.11}$$

With the Denavit-Hartenberg method it is important to draw attention to the case when axes i and $i+1$ are orthogonal and intersecting. In this case parameter b_i is zero and the common normal $\mathbf{x}_i$ is the normal to the plane defined by the intersecting axes. There are two oppositely directed possibilities for the direction of $\mathbf{x}_i$. Either is acceptable. As well, in the case when neighboring axes i and $i+1$ are parallel, the direction $\mathbf{x}_i$ is defined, but its position is not. In this case parameter b_i is arbitrary and is usually selected so as to reduce the total number of parameters describing the mechanism.

2.2.3 Vector Parameters of a Kinematic Pair

Consider the method of Vector Parameters, which is primarily used in the remainder of this book, and the procedure for determining the parameters of a kinematic pair.

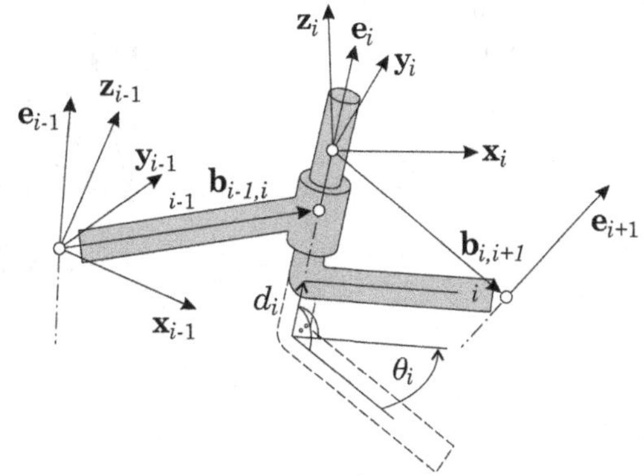

Fig. 2.16 Vector parameters of a kinematic pair

Figure 2.16 shows links $i-1$ and i connected by a cylindrical joint i. The position and direction of the joint axis is determined by the link vector $\mathbf{b}_{i-1,1}$ and the unit joint vector $\mathbf{e}_i$. Link i can be translated by a distance d_i in the direction $\mathbf{e}_i$, and rotated through an angle θ_i about $\mathbf{e}_i$, relative to link $i-1$. The coordinate frame $\mathbf{x}_i$, $\mathbf{y}_i$, $\mathbf{z}_i$ is attached to link i and frame $\mathbf{x}_{i-1}$, $\mathbf{y}_{i-1}$, $\mathbf{z}_{i-1}$ is attached to link $i-1$.

In the reference position of the kinematic pair, both joint variables of the kinematic pair are assigned zero values, i.e. $\theta_i = 0$ and $d_i = 0$, and the frame is taken to be parallel to the preceding frame $\mathbf{x}_{i-1}$, $\mathbf{y}_{i-1}$, $\mathbf{z}_{i-1}$.

With this method it is not necessary that one of the axes $\mathbf{x}_i$, $\mathbf{y}_i$, $\mathbf{z}_i$ is parallel to the joint vector $\mathbf{e}_i$, as is the case with the Denavit-Hartenberg method. In the method of Vector Parameters the geometry of a kinematic pair and the relative displacement between the link are defined by the following parameters:

- $\mathbf{e}_i$: joint vector—the unit vector defining the rotational or translational axis of the i-th joint;
- $\mathbf{b}_{i-1,i}$: link vector—the vector describing the length and direction of the link $i-1$ (the length and the direction of the link i is given by the vector $\mathbf{b}_{i,i+1}$);
- θ_i: rotational coordinate—the angle measured about the $\mathbf{e}_i$ axis in the plane perpendicular to the vector $\mathbf{e}_i$ (the rotation coordinate is zero when the kinematic pair is in its reference position);
- d_i: translation coordinate—the distance measured in the direction $\mathbf{e}_i$ (the translation coordinate is zero when the kinematic pair is in its reference position).

The cylindrical joint in Fig. 2.17 can be reduced to either a rotational joint, or a translational joint. When the kinematic pair is rotational (upper example in Fig. 2.17), the joint variable is the rotational coordinate θ_i, while $d_i = 0$. When the mechanism is in its reference position, then $\theta_i = 0$ and coordinate frames $\mathbf{x}_i$, $\mathbf{y}_i$, $\mathbf{z}_i$ and $\mathbf{x}_{i-1}$, $\mathbf{y}_{i-1}$, $\mathbf{z}_{i-1}$ are parallel. When the kinematic pair is translational (lower example in Fig. 2.17), the joint variable is translational coordinate d_i, while $\theta_i = 0$. When the kinematic pair is in the reference position, we have $d_i = 0$. The coordinate frames $\mathbf{x}_i$, $\mathbf{y}_i$, $\mathbf{z}_i$ and $\mathbf{x}_{i-1}$, $\mathbf{y}_{i-1}$, $\mathbf{z}_{i-1}$ are now parallel irrespective of the value of translational coordinate d_i.

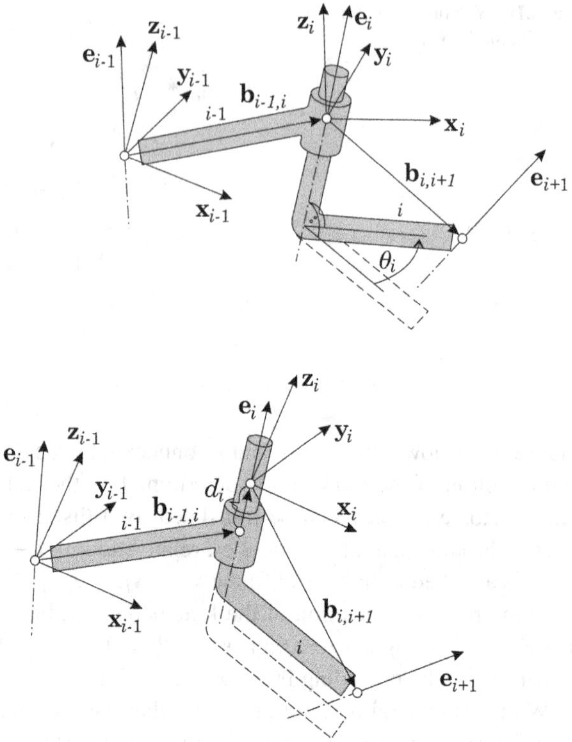

Fig. 2.17 Vector parameters of a kinematic pair

By changing the values of the rotational coordinate θ_i, the coordinate frame $\mathbf{x}_i$, $\mathbf{y}_i$, $\mathbf{z}_i$ is rotating together with the i-th link with respect to the preceding link $i-1$ i.e. with respect to the preceding frame $\mathbf{x}_{i-1}$, $\mathbf{y}_{i-1}$, $\mathbf{z}_{i-1}$. When changing the values of the translational coordinate d_i, the displacement is translational and only the distance between the origins of both frames is changing.

Let us mathematically define the transformation between coordinate frame $\mathbf{x}_{i-1}$, $\mathbf{y}_{i-1}$, $\mathbf{z}_{i-1}$ and coordinates frame $\mathbf{x}_i$, $\mathbf{y}_i$, $\mathbf{z}_i$. Frame $\mathbf{x}_i$, $\mathbf{y}_i$, $\mathbf{z}_i$ is translated relative to frame $\mathbf{x}_{i-1}$, $\mathbf{y}_{i-1}$, $\mathbf{z}_{i-1}$ along the vector $\mathbf{b}_{i-1,i}$ and along the vector $d_i\mathbf{e}_i$. Then it is rotated through an angle θ_i about the unit vector $\mathbf{e}_i$

$$\mathbf{H}_{i-1,i} = \begin{bmatrix} & \mathbf{A}_{i-1,i} & & d_i\mathbf{e}_i^{(i-1)} + \mathbf{b}_{i-1,i}^{(i-1)} \\ 0 & 0 & 0 & 1 \end{bmatrix}. \tag{2.12}$$

The rotation matrix $\mathbf{A}_{i-1,i}$, defining the transformation between the vector spaces $\Re^3_{i-1}$ and $\Re^3_i$, is obtained by the use of general formula (1.48), where the rotational vector is the joint vector $\mathbf{e}_i^{(i-1)}$ and the rotation coordinate is θ_i. In practical applications the kinematic pair can be placed in a reference position where joint vector $\mathbf{e}_i$ is parallel to one of the axes of the frame $\mathbf{x}_i$, $\mathbf{y}_i$, $\mathbf{z}_i$. In this case the rotation matrix $\mathbf{A}_{i-1,i}$ can be calculated with one of the simplified expressions (1.51), (1.52) or (1.53).

In the reference position, coordinate frames $\mathbf{x}_{i-1}$, $\mathbf{y}_{i-1}$, $\mathbf{z}_{i-1}$ and $\mathbf{x}_i$, $\mathbf{y}_i$, $\mathbf{z}_i$ are parallel ($\theta_i = 0$ in $d_i = 0$), and we have

$$\mathbf{H}_{i-1,i} = \begin{bmatrix} & \mathbf{I} & & \mathbf{b}_{i-1,i}^{(i-1)} \\ 0 & 0 & 0 & 1 \end{bmatrix}. \tag{2.13}$$

When a joint is only rotational ($d_i = 0$), then

$$\mathbf{H}_{i-1,i} = \begin{bmatrix} & \mathbf{A}_{i-1,i} & & \mathbf{b}_{i-1,i}^{(i-1)} \\ 0 & 0 & 0 & 1 \end{bmatrix}, \tag{2.14}$$

and when a joint is only translation ($\theta_i = 0$), we have

$$\mathbf{H}_{i-1,i} = \begin{bmatrix} & \mathbf{I} & & d_i\mathbf{e}_i^{(i-1)} + \mathbf{b}_{i-1,i}^{(i-1)} \\ 0 & 0 & 0 & 1 \end{bmatrix}. \tag{2.15}$$

In general, all the components of the vector parameters $\mathbf{e}_i$ and $\mathbf{b}_{i-1,i}$ are non-zero. In special cases, depending on the selection of a joint center and the reference position of a kinematic pair, some of the components of these vector parameters can be zero. The scalar Denavit-Hartenberg parameters of a kinematic pair represent a special case of vector parameters. The advantages of using the method of Vector Parameters is in the simplicity of the coordinate frame assignments and in the freedom of choosing any relative position of the two links as the position where the joint variables are zero.

When comparing the Vector Parameters of a kinematic pair with scalar Denavit-Hartenberg parameters, the following important difference can be noticed. With Vector Parameters the length of the translation d_i is measured from the selected reference position in the direction of vector $\mathbf{e}_i$. With Denavit and Hartenberg parameters the length of the translation d_i is the distance between the intersection of vector b_{i-1} with joint axis i and the intersection of vector b_i with joint axis i. With Vector Parameters the angle of rotation θ_i is assessed from the selected reference position about vector $\mathbf{e}_i$. With Denavit and Hartenberg parameters the angle of rotation θ_i is defined as the angle between vectors $\mathbf{x}_{i-1}$ and $\mathbf{x}_i$.

2.3 Parameters and Variables of a Mechanism

In this section we extend the description of a kinematic pair to the description of an entire mechanism which consists of several links and joints. The extension is relatively straightforward. First we consider this description in terms of the Denavit-Hartenberg parameters and then by the Vector Parameters. The two methods are compared via the example robot mechanism shown in Fig. 2.18. Examples of mechanisms with various kinematic arrangements can be found in [6, 67, 77].

The serial spatial mechanism shown in Fig. 2.18 has $n = 4$ degrees of freedom. At one end it is attached to the base, while a gripper is mounted to the other end.

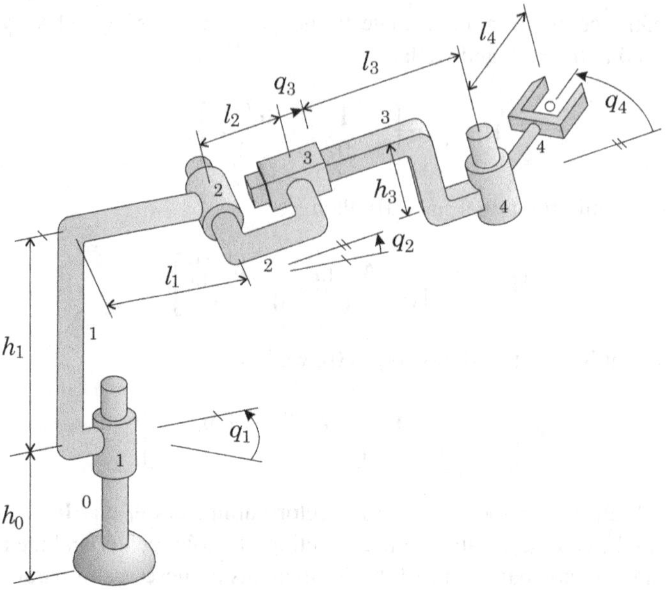

Fig. 2.18 Example of mechanism with four degrees of freedom

The mechanism has three rotations with joint variables q_1, q_2, and q_4 and one translation with joint variable q_3. The links are denoted as $0, 1, \ldots, 4$, and the joints are numbered $1, 2, 3, 4$. The mechanism has $N = n = 4$ of mobile links. Their lengths are given by distances h_0, h_1, l_1, l_2, l_3, h_3, and l_4. The link 0 is the fixed base.

2.3.1 Denavit and Hartenberg Parameters of a Mechanism

In the Denavit and Hartenberg method, a local coordinate frame is attached to each mobile link of a mechanism. The motion of this frame is observed relative to a fixed reference frame $\mathbf{x}_0, \mathbf{y}_0, \mathbf{z}_0$ which is attached to the fixed base. The i-th joint connects the links $i-1$ and i, $i = 1, 2, \ldots, n$. Recall, with serial mechanisms the number of joints n is equal to the number of mobile links N. The Denavit-Hartenberg parameters of a mechanism are determined in the five following steps [77, 84]:

Step 1 In order to reduce the number of parameters, the fixed frame $\mathbf{x}_0$, $\mathbf{y}_0$, $\mathbf{z}_0$ is typically attached to body 0 in such a way that the $\mathbf{z}_0$ axis is coincident with the axis of joint 1. Axis $\mathbf{x}_0$ is arbitrarily directed in the plane perpendicular to the axis of joint 1. Typically this is a direction considered to be the forward reaching direction of the robot. The origin of the frame is positioned anywhere along the axis of joint 1. The choice of origin is typically made to try and eliminate additional parameters. Being a right hand coordinate system, the third axis of the frame, $\mathbf{y}_0$, is known from $\mathbf{x}_0$ and $\mathbf{z}_0$, $\mathbf{y}_0 = \mathbf{z}_0 \times \mathbf{x}_0$;

Step 2 Attach to each subsequent joint axis $i = 2, 3, \ldots, n$ the axis $\mathbf{z}_{i-1}$, which belongs to coordinate frame $\mathbf{x}_{i-1}, \mathbf{y}_{i-1}, \mathbf{z}_{i-1}$. The origin of this frame is located on the axis of i-th joint at the foot of the common normal from the axis of joint $i-1$ to the axis of joint i. When the axes of the joints $i-1$ and i are parallel and the i-th joint is rotational, the origin of the frame is positioned so that $d_i = 0$, reducing the number of non-zero parameters. When the joint is translational, the origin of the frame can be positioned anywhere along the i-th joint axis. When joint axes $i-1$ and i intersect, the origin of the frame is positioned at the point of intersection;

Step 3 For each moving link $i-1$ we define axis $\mathbf{x}_{i-1}$, $i = 2, 3, \ldots, n$, in the direction of the common normal between joint axes $i-1$ and i, in the direction from joint axis $i-1$ towards joint axis i. When the axes of joints $i-1$ and i intersect, axis $\mathbf{x}_{i-1}$ is perpendicular to the plane containing the intersecting axes, and can be directed in either direction perpendicular to the plane. Be aware that which normal direction is chosen influences the value of angle θ_i, which is measured between the vectors $\mathbf{x}_{i-1}$ and $\mathbf{x}_i$ about axis $\mathbf{z}_{i-1}$. Being a right hand coordinate system, the third axis of the frame is determined as $\mathbf{y}_{i-1} = \mathbf{z}_{i-1} \times \mathbf{x}_{i-1}$;

Step 4 The robot end effector coordinate frame $\mathbf{x}_n, \mathbf{y}_n, \mathbf{z}_n$ has its origin placed at a reference point on the gripper. The axis $\mathbf{z}_n$ lies anywhere in the plane perpendicular to $\mathbf{x}_n$, where in Step 3, $\mathbf{x}_n$ was aligned along the common normal between $\mathbf{z}_{n-1}$ and $\mathbf{z}_n$. When the last joint n is rotational, axis $\mathbf{z}_n$ is taken to be parallel to the n-th joint axis. The $\mathbf{y}_n$ axis is found as, $\mathbf{y}_n = \mathbf{z}_n \times \mathbf{x}_n$;

Step 5 With coordinate frames for all links $i = 1, 2, \ldots, n$ assigned in the manner described above, we can determine the values of the Denavit-Hartenberg parameters, which are usually presented in a tabular form.

Following the above rules, we place coordinate frames onto the links of the mechanism shown in Fig. 2.18, resulting in Fig. 2.19. To make determination of the kinematic parameters easier, we position the mechanism in such a way that the consecutive joint axes are either perpendicular or parallel. We begin with the fixed body 0, and frame $\mathbf{x}_0, \mathbf{y}_0, \mathbf{z}_0$. Axis $\mathbf{z}_0$ is placed along the axis of joint 1. The origin of the frame can be selected anywhere along this axis, most conveniently at $h_0 + h_1$. In this case h_0 and h_1 will not appear in the kinematic equations. It may be more convenient however, to place the frame's origin on the robot base. The axis $\mathbf{x}_0$ is taken perpendicular to $\mathbf{z}_0$, in what we consider to be the forward facing direction of the robot.

Axis $\mathbf{z}_1$ is placed along the axis of joint 2. Its origin is at the intersection of $\mathbf{z}_1$ with the common normal between the axes of joints 1 and 2. The length of the common normal is $b_1 = l_1$. The offset distance between the frames in the $\mathbf{z}_0$ direction is $d_1 = h_0 + h_1$ and is constant. Axis $\mathbf{x}_1$ lies in the direction of the common normal and the angle measured between the axes $\mathbf{z}_0$ and $\mathbf{z}_1$ about the axis $\mathbf{x}_1$ is the inclination angle $\alpha_1 = \pi/2$. Consistent with Fig. 2.18, the variable of the first joint is $\theta_1 = q_1 + \pi/2$. It is measured from axis $\mathbf{x}_0$ to axis $\mathbf{x}_1$ about $\mathbf{z}_0$.

Axis $\mathbf{z}_2$ is directed along joint axis 3. Its origin is at the intersection of the axes of joints 2 and 3. The length of the common normal is $b_2 = 0$ and the offset distance between the frames in the $\mathbf{z}_1$ direction is zero, so $d_2 = 0$. As the joint axes intersect,

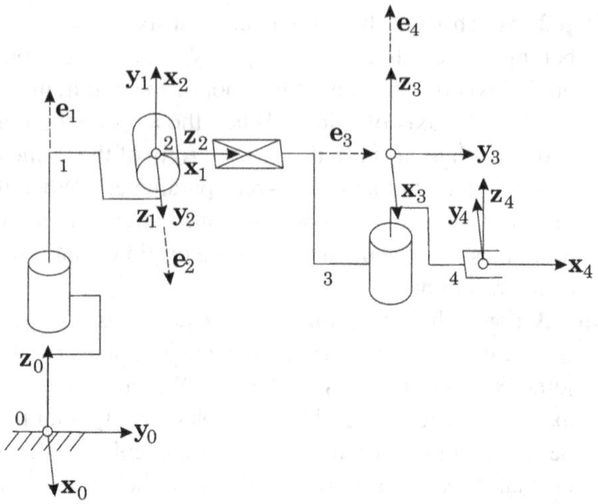

Fig. 2.19 Placement of the coordinate frames for the mechanism with four degrees of freedom

Table 2.2 Denavit-Hartenberg parameters of the mechanism from Fig. 2.19

i	b_i	α_i	d_i	θ_i
1	l_1	$\pi/2$	$h_0 + h_1$	$q_1 + \pi/2$
2	0	$\pi/2$	0	$q_2 + \pi/2$
3	0	$\pi/2$	$q_3 + l_2 + l_3$	$\pi/2$
4	l_4	0	$-h_3$	$q_4 + \pi/2$

we have $\mathbf{x}_2 = \mathbf{z}_1 \times \mathbf{z}_2$ (this vector could have been taken in the opposite direction). The angle of inclination α_2 is measured from $\mathbf{z}_1$ to $\mathbf{z}_2$, about axis $\mathbf{x}_2$, and $\alpha_2 = \pi/2$. The variable of the second joint is $\theta_2 = q_2 + \pi/2$. It is measured from $\mathbf{x}_1$ to $\mathbf{x}_2$ about $\mathbf{z}_1$.

Axis $\mathbf{z}_3$ lies on joint axis 4. Its origin is at the intersection of the axes of joints 3 and 4. The length of the common normal is $b_3 = 0$. The offset distance between the frames in the $\mathbf{z}_2$ direction is the variable of the third joint, and is given by $d_3 = q_3 + l_2 + l_3$. As the joint axes intersect, we have $\mathbf{x}_3 = \mathbf{z}_2 \times \mathbf{z}_3$ (this vector could have been selected in the opposite direction). The inclination angle α_3 is measured from $\mathbf{z}_2$ and $\mathbf{z}_3$ about $\mathbf{x}_3$, and $\alpha_3 = \pi/2$. Observe that the angle between $\mathbf{x}_2$ and $\mathbf{x}_3$ about the $\mathbf{z}_2$ axis does not change and is a constant $\theta_3 = \pi/2$.

Attachment of axis $\mathbf{z}_4$ on the gripper is arbitrary, since there is no joint 5. Take $\mathbf{z}_4$ parallel to axis 4. The origin of this frame is placed on a reference point on the gripper, which might represent a tool tip, or the midpoint of the gripper. Axis $\mathbf{x}_4$ is in the direction of the common normal from $\mathbf{z}_3$ to $\mathbf{z}_4$. The length of the normal is $b_4 = l_4$ and the offset distance between the frames in the direction of $\mathbf{z}_3$ is $d_4 = -h_3$. The inclination angle between $\mathbf{z}_3$ and $\mathbf{z}_4$, measured about $\mathbf{x}_4$ is $\alpha_4 = 0$. The variable of the fourth joint is $\theta_4 = q_4 + \pi/2$. It is measured from $\mathbf{x}_3$ to $\mathbf{x}_4$ about $\mathbf{z}_3$. The Denavit-Hartenberg parameters, belonging to the mechanism from Fig. 2.19, are given in Table 2.2.

Even in this case of a simple robot mechanism, we observe that placement of the coordinate frames according to Denavit-Hartenberg method is rather complex. An advantage of the Denavit and Hartenberg method is that the transformation matrices between the frames always have the same general form. We only have to enter the values of the parameters from Table 2.2 into (2.9). In our case we have

$$\mathbf{H}_{0,1} = \begin{bmatrix} -\sin q_1 & 0 & \cos q_1 & -l_1 \sin q_1 \\ \cos q_1 & 0 & \sin q_1 & l_1 \cos q_1 \\ 0 & 1 & 0 & h_0 + h_1 \\ 0 & 0 & 0 & 1 \end{bmatrix},$$

$$\mathbf{H}_{1,2} = \begin{bmatrix} -\sin q_2 & 0 & \cos q_2 & 0 \\ \cos q_2 & 0 & \sin q_2 & 0 \\ 0 & 1 & 0 & 0 \\ 0 & 0 & 0 & 1 \end{bmatrix},$$

$$\mathbf{H}_{2,3} = \begin{bmatrix} 0 & 0 & 1 & 0 \\ 1 & 0 & 0 & 0 \\ 0 & 1 & 0 & q_3 + l_2 + l_3 \\ 0 & 0 & 0 & 1 \end{bmatrix},$$

$$\mathbf{H}_{3,4} = \begin{bmatrix} -\sin q_4 & -\cos q_4 & 0 & -l_4 \sin q_4 \\ \cos q_4 & -\sin q_4 & 0 & l_4 \cos q_4 \\ 0 & 0 & 1 & -h_3 \\ 0 & 0 & 0 & 1 \end{bmatrix}.$$

We have substituted $\sin(q_i + \pi/2) = \cos q_i$ and $\cos(q_i + \pi/2) = -\sin q_i$.

2.3.2 Vector Parameters of a Mechanism

Consider applying the Method of Vector Parameters to the same example. The method of Vector Parameters allows arbitrary placement of the coordinate frames. The method is applied in the following five steps [46], with the coordinate frames attached to the bodies as follows:

Step 1 The mechanism is placed into the desired reference position (initial pose). One can consider the reference position as the zero position, where the values of all joint variables are taken as zero, $\theta_i = 0$, $d_i = 0$, $i = 1, 2, \ldots, n$. The fixed coordinate frame $\mathbf{x}_0$, $\mathbf{y}_0$, $\mathbf{z}_0$ which is attached to the base, is arbitrarily located in the space. Usually it is attached to a reference point on the 0 link, which may be the point where the mechanism is attached to the ground;

Step 2 The centers of the joints $i = 1, 2, \ldots, n$ are selected. The center of the i-th joint can be taken anywhere along the axis of joint i. A local coordinate frame $\mathbf{x}_i$, $\mathbf{y}_i$, $\mathbf{z}_i$ is placed on the i-th joint center in such a way that its axes are parallel to the axes of the fixed coordinate frame $\mathbf{x}_0$, $\mathbf{y}_0$, $\mathbf{z}_0$. The local frame $\mathbf{x}_i$, $\mathbf{y}_i$, $\mathbf{z}_i$ is attached to link i and displaces with it;

Step 3 A joint vector $\mathbf{e}_i$ is placed onto each axis of the mechanism. The translational variable d_i is measured in the direction of the joint vector, while the rotational variable θ_i is measured about it;

Step 4 Link vectors $\mathbf{b}_{i-1,i}$ are directed from the origin of coordinate frame $\mathbf{x}_{i-1}$, $\mathbf{y}_{i-1}$, $\mathbf{z}_{i-1}$ to the origin of frame $\mathbf{x}_i$, $\mathbf{y}_i$, $\mathbf{z}_i$ $i = 1, 2, \ldots, n$, locating the origin of frame i relative to frame $i-1$. Link vector $\mathbf{b}_{n,n+1}$ lies between the origin of the coordinate frame $\mathbf{x}_n$, $\mathbf{y}_n$, $\mathbf{z}_n$ and the robot end-point;

Step 5 Joint vectors $\mathbf{e}_i$ and the link vectors $\mathbf{b}_{i-1,i}$ are expressed in the coordinate frame $\mathbf{x}_{i-1}$, $\mathbf{y}_{i-1}$, $\mathbf{z}_{i-1}$ (the vector $\mathbf{b}_{n,n+1}$ in the frame $\mathbf{x}_n$, $\mathbf{y}_n$, $\mathbf{z}_n$), as the vectors $\mathbf{e}_i^{(i-1)}$, $\mathbf{b}_{i-1,i}^{(i-1)}$, and $\mathbf{b}_{n,n+1}^{(n)}$ do not depend upon the variables θ_i and d_i, $i = 1, 2, \ldots, n$. Taking into account that in the reference (initial) pose of the mechanism all local frames $\mathbf{x}_i$, $\mathbf{y}_i$, $\mathbf{z}_i$, $i = 1, 2, \ldots, n$, are parallel to the reference frame $\mathbf{x}_0$, $\mathbf{y}_0$, $\mathbf{z}_0$, the joint and link vectors can be determined in the following way

$$\mathbf{e}_i^{(i-1)} = \mathbf{e}_i^{(0)}, \tag{2.16}$$

$$\mathbf{b}_{i-1,i}^{(i-1)} = \mathbf{b}_{i-1,i}^{(0)}, \qquad \mathbf{b}_{n,n+1}^{(n)} = \mathbf{b}_{n,n+1}^{(0)}. \tag{2.17}$$

Sometimes another frame, $\mathbf{x}_{n+1}$, $\mathbf{y}_{n+1}$, $\mathbf{z}_{n+1}$, is introduced at the reference point on the robot gripper. There is no movement between the frames $\mathbf{x}_n$, $\mathbf{y}_n$, $\mathbf{z}_n$ and $\mathbf{x}_{n+1}$, $\mathbf{y}_{n+1}$, $\mathbf{z}_{n+1}$ since both are attached to the same link. Thus the transformation between them is constant. From the point of view of the kinematics the frame $\mathbf{x}_{n+1}$, $\mathbf{y}_{n+1}$, $\mathbf{z}_{n+1}$ is not necessary, but it may assist in defining the grasp of a tool in the gripper of the mechanism.

In order to simplify the development and form of the kinematic equations, it is desirable that the maximum number of the components of vectors $\mathbf{e}_i^{(0)}$, $\mathbf{b}_{i-1,i}^{(0)}$, and $\mathbf{b}_{n,n+1}^{(0)}$ are zero in the reference position. When a mechanism has consecutively perpendicular or parallel joint axes, which is common in practical robot structures, the mechanism can be placed into a reference position in such a way that particular joint axes are parallel to the axes of the fixed reference frame. As well, with appropriate choice of the joint centers, the link vectors can be chosen so that they are directed parallel to one of the axes of the fixed reference frame.

The method of Vector Parameters will be demonstrated for the same example of a four degree of freedom mechanism shown in Fig. 2.18. The result is presented in Fig. 2.20. The reference position of the mechanism, which corresponds to when the joint variables are zero, $q_1 = q_2 = q_3 = q_4 = 0$, is shown in Fig. 2.20. The vector parameters and the joint variables corresponding to the reference position of the mechanism and the selected positions of the joint centers as shown in Fig. 2.20, are presented in Table 2.3.

The rotational variables θ_1, θ_2, and θ_4 are measured in planes which are perpendicular to axes $\mathbf{e}_1$, $\mathbf{e}_2$, and $\mathbf{e}_4$ respectively, while translational variable d_3 is measured in the direction of axis $\mathbf{e}_3$. All joint variables are zero when the mechanism is in its reference position (initial pose). In Fig. 2.21 the mechanism is shown in a pose, where all four variables are nonzero and positive. The variable θ_1 is the angle between the initial (reference position) direction and the current direction of the

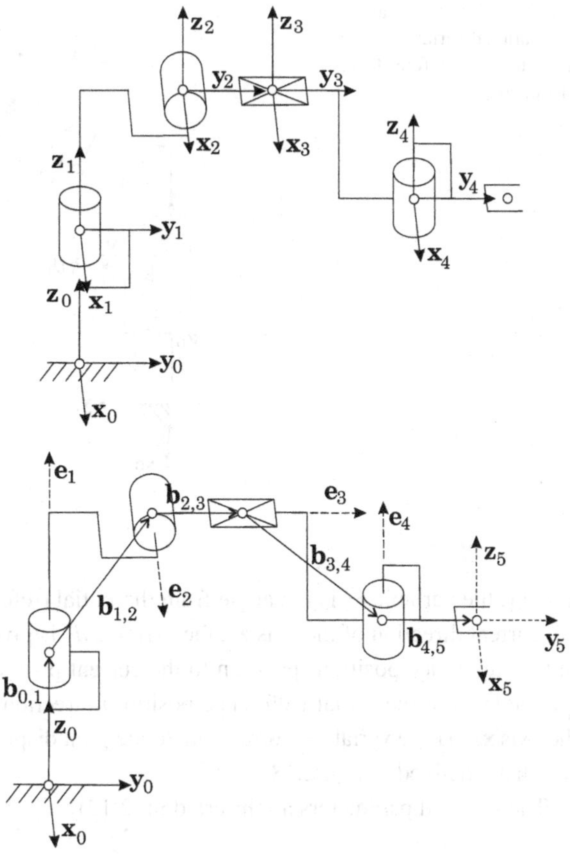

Fig. 2.20 Placement of the coordinate frames for the mechanism with four degrees of freedom

Table 2.3 Vector parameters and variables of the mechanism from Fig. 2.20

i	1	2	3	4
θ_i	q_1	q_2	0	q_4
d_i	0	0	q_3	0

i	1	2	3	4
$\mathbf{e}_i^{(i-1)}$	0	1	0	0
	0	0	1	0
	1	0	0	1

i	1	2	3	4	5
$\mathbf{b}_{i-1,i}^{(i-1)}$	0	0	0	0	0
	0	l_1	l_2	l_3	l_4
	h_0	h_1	0	$-h_3$	0

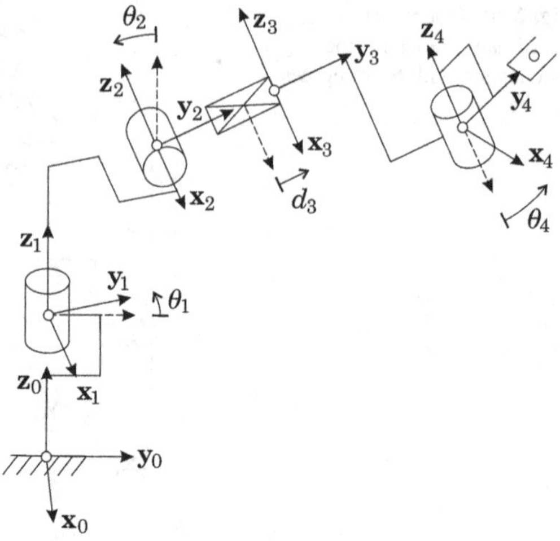

Fig. 2.21 Rotational and translational variables for the mechanism with four degrees of freedom

axis $\mathbf{y}_1$, the variable θ_2 is the angle from the initial (reference position) direction to the current direction of the axis $\mathbf{z}_2$, the variable d_3 is given by the distance from the initial (reference position) position to the current position of the axis $\mathbf{x}_3$, and θ_4 is the angle from the initial (reference position) direction to the current direction of the axis $\mathbf{x}_4$. These variables correspond to the joint displacements q_1, q_2, q_3, and q_4, which are defined in Fig. 2.18.

The selected parameters are inserted in (2.12)

$$\mathbf{H}_{0,1} = \begin{bmatrix} \cos q_1 & -\sin q_1 & 0 & 0 \\ \sin q_1 & \cos q_1 & 0 & 0 \\ 0 & 0 & 1 & h_0 \\ 0 & 0 & 0 & 1 \end{bmatrix},$$

$$\mathbf{H}_{1,2} = \begin{bmatrix} 1 & 0 & 0 & 0 \\ 0 & \cos q_2 & -\sin q_2 & l_1 \\ 0 & \sin q_2 & \cos q_2 & h_1 \\ 0 & 0 & 0 & 1 \end{bmatrix},$$

$$\mathbf{H}_{2,3} = \begin{bmatrix} 1 & 0 & 0 & 0 \\ 0 & 1 & 0 & q_3 + l_2 \\ 0 & 0 & 1 & 0 \\ 0 & 0 & 0 & 1 \end{bmatrix},$$

$$\mathbf{H}_{3,4} = \begin{bmatrix} \cos q_4 & -\sin q_4 & 0 & 0 \\ \sin q_4 & \cos q_4 & 0 & l_3 \\ 0 & 0 & 1 & -h_3 \\ 0 & 0 & 0 & 1 \end{bmatrix}.$$

Here, the simplified expressions for the rotation matrices were used (1.51), (1.52), (1.53), describing the rotations about the axes of the coordinate frame. This is made possible by our choice of reference frame and reference position.

Placing a coordinate frame $\mathbf{x}_5, \mathbf{y}_5, \mathbf{z}_5$ onto the reference point of the robot gripper, we obtain the following additional homogeneous transformation matrix

$$\mathbf{H}_{4,5} = \begin{bmatrix} 1 & 0 & 0 & 0 \\ 0 & 1 & 0 & l_4 \\ 0 & 0 & 1 & 0 \\ 0 & 0 & 0 & 1 \end{bmatrix},$$

which is independent of the joint variables, as the frames $\mathbf{x}_4, \mathbf{y}_4, \mathbf{z}_4$ and $\mathbf{x}_5, \mathbf{y}_5, \mathbf{z}_5$ are both attached to body 4 and are parallel and only displaced for by a constant distance l_4. In a purely kinematic sense the additional frame $\mathbf{x}_5, \mathbf{y}_5, \mathbf{z}_5$ need not be attached to the mechanism, as the position and orientation of the gripper can be described by the $\mathbf{x}_4, \mathbf{y}_4, \mathbf{z}_4$ frame alone. This additional frame however may be used to define the grasp of a tool, body 5, by the gripper, body 4, since in many applications control of the tool's position and orientation is the ultimate goal.

This chapter has introduced the Denavit and Hartenberg method and the method of Vector Parameters. Both methods were applied to a kinematic pair and then to an example of a four degree of freedom mechanism. In the Denavit-Hartenberg method, the attachment of local coordinate frames to the links is precisely specified and relative to these frames a minimum number of translational and rotational parameters which describe the relative pose of two neighboring links are defined. In this method the transformation matrices between the coordinate frames have a generic form, requiring fewer algebraic operations in developing the kinematic equations of motion The disadvantage of the approach is in the forced placements of the coordinate frames on the links, placements which may seem irregular or unnatural. The origins of these frames are often at points which are distant from the corresponding joint centers or links. In many cases the configuration of a robot mechanism corresponding to when the joint variables are zero is unnatural and because of physical constraints may be inaccessible.

The method of Vector Parameters uses link and joint vectors to describe the geometry of a link and the variables at the joints. It is important in this method to select an appropriate reference position of the mechanism where all the coordinate frames are parallel to the reference coordinate frame and the translational and rotational joint variables are zero. As the reference position of a mechanism is a free choice, we can select the most appropriate reference position from the point of view of clarity of the approach, requirements of the robot task, or the number of mathematical operations included in the transformation matrices. In general the matrix describing the rotation about an arbitrary axis must be used in the transformation matrices. However, when selecting the reference position in such a way that particular joint axes are parallel to one of the axes of the reference frame, the rotational matrices are simplified and the number of required arithmetic operations is not higher than with the Denavit-Hartenberg method.

Chapter 3
Serial Mechanisms

Abstract Equations for the kinematic analysis of serial mechanisms are introduced. In order to reduce the necessary numerical computations, they are expressed in the form of iterative procedures. The direct and the inverse kinematics problems are formulated and solved. It is shown that in serial mechanisms, the direct kinematics problem is relatively simple to solve, while the inverse kinematics problem includes difficulties associated with the existence of a real solution, with multiple solutions, and with the convergence of iterative numerical procedures.

The kinematic equations of a mechanism describe the position and orientation (i.e. the pose) of the mechanism's links, along with the angular velocities and accelerations of the links, as well as the linear velocities and accelerations of points of interest on the links. In non redundant mechanisms, the pose of the terminal link (the end-effector), determines the pose of each of the previous links of the mechanism. Kinematics of serial mechanisms can be described with either internal or external coordinates. The internal coordinates are the rotational or translational joint variables, while the external coordinates are usually associated with the pose of the end-effector. The external and internal coordinates are related through a system of nonlinear algebraic equations [90]. The first time derivatives of the internal and external coordinates are linearly related by the Jacobian matrix. The Jacobian matrix appears together with the Hessian matrix in the kinematic equations describing the second time derivatives.

When the internal coordinates appear in the kinematic equations as the independent variables, we are dealing with the direct kinematic equations. When the external variables are the independent variables, the equations describe the inverse kinematics of a mechanism [3, 84]. Usually, calculating the direct kinematics is simple with serial mechanisms, while the solution of inverse kinematics is often so complex that numerical methods must be applied [22]. Additional difficulty is presented by the multiple solutions of the inverse kinematic equations, which degenerate at certain values of the internal and external coordinates.

J. Lenarčič et al., *Robot Mechanisms*,
Intelligent Systems, Control and Automation: Science and Engineering 60,
DOI 10.1007/978-94-007-4522-3_3, © Springer Science+Business Media Dordrecht 2013

3.1 Kinematic Equations

In this section we develop the kinematic equations of a mechanism using the method of Vector Parameters, while placing the coordinate frames as explained in Sect. 2.3.2 and described in the literature [46]. The links of the mechanism will be numbered sequentially as $i = 0, 1, 2, \ldots, n$. A coordinate system which is the $\mathbf{x}_i, \mathbf{y}_i, \mathbf{z}_i$ frame, is attached to each link i. Link 0 corresponds to the fixed base and link n corresponds to the robot end-effector. The joints of the mechanism will be sequentially numbered as $i = 1, 2, \ldots, n$, in such a way that joint i serves as the connection between links $i-1$ and i. We assume that all parameters of the mechanism are known, i.e. the link vectors $\mathbf{b}_{0,1}^{(0)}, \mathbf{b}_{1,2}^{(1)}, \ldots, \mathbf{b}_{n,n+1}^{(n)}$, the joint vectors $\mathbf{e}_1^{(0)}, \mathbf{e}_2^{(1)}, \ldots, \mathbf{e}_n^{(n-1)}$, and all mechanism variables $d_1, d_2, \ldots, d_n$ and $\theta_1, \theta_2, \ldots, \theta_n$, are known.

3.1.1 Orientation and Position of a Mechanism

Before developing the kinematic equations of a mechanism, we write the rotation matrix $\mathbf{A}_{i-1,i}$ describing the relative orientation of two neighboring links $i-1$ and i or coordinate frames $\mathbf{x}_{i-1}, \mathbf{y}_{i-1}, \mathbf{z}_{i-1}$ and $\mathbf{x}_i, \mathbf{y}_i, \mathbf{z}_i$ (Fig. 3.1). We shall use the general formula for calculating the rotation matrix (1.48). Into the formula we substitute the rotation vector $\mathbf{e}_i$, expressed in the vector space $\mathfrak{R}_{i-1}^3$, and the rotational variable θ_i. The rotation matrix $\mathbf{A}_{i-1,i}$ (for all $i = 1, 2, \ldots, n$) has the following form

$$\mathbf{A}_{i-1,i} = \mathbf{\Delta}_{i-1,i} \sin\theta_i + (\mathbf{I} - \mathbf{\Lambda}_{i-1,i}) \cos\theta_i + \mathbf{\Lambda}_{i-1,i}, \tag{3.1}$$

with the matrices

$$\mathbf{\Delta}_{i-1,i} = \mathbf{e}_i^{(i-1)} \otimes \mathbf{I},$$
$$\mathbf{\Lambda}_{i-1,i} = \mathbf{e}_i^{(i-1)} \mathbf{e}_i^{\mathrm{T}(i-1)},$$

defined in (1.49) and (1.50).

According to (1.68), the orientation between links i and j, where $j > i$, is given as product of rotation matrices describing all intermediate rotations

$$\mathbf{A}_{i,j} = \mathbf{A}_{i,i+1}\mathbf{A}_{i+1,i+2} \cdots \mathbf{A}_{j-1,j}.$$

The orientation of link i with respect to the base link is determined by the product of rotation matrices

$$\mathbf{A}_{0,i} = \mathbf{A}_{0,1}\mathbf{A}_{1,2} \cdots \mathbf{A}_{i-1,i}.$$

The orientation of the robot end-effector is described by the rotation matrix $\mathbf{A}_{0,n}$. Knowing $\mathbf{A}_{0,n}$, the orientation of the end-effector can be expressed in various ways,

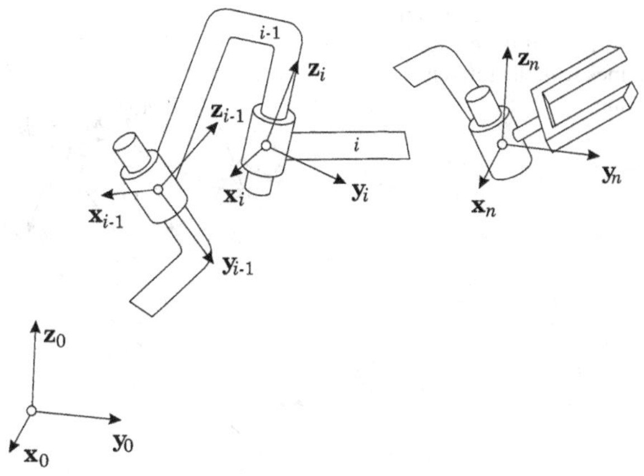

Fig. 3.1 The pose of mechanism is determined by the coordinate frames attached to individual links

either with Euler or YPR angles or Euler parameters, for example. The previous equation can be written in the recursive form

$$\mathbf{A}_{0,i} = \mathbf{A}_{0,i-1}\mathbf{A}_{i-1,i}, \quad i = 2, 3, \ldots, n. \tag{3.2}$$

The calculation proceeds in the following steps

$$\begin{aligned}
\mathbf{A}_{0,2} &= \mathbf{A}_{0,1}\mathbf{A}_{1,2}, \\
\mathbf{A}_{0,3} &= \mathbf{A}_{0,2}\mathbf{A}_{2,3}, \\
\mathbf{A}_{0,4} &= \mathbf{A}_{0,3}\mathbf{A}_{3,4}, \\
&\vdots
\end{aligned}$$

where the orientations of all the mechanism's links are calculated sequentially.

Also of interest is the orientation of link $i-1$ relative to link n, the end-effector. This can be obtained by successive multiplication of the intermediate rotation matrices

$$\mathbf{A}_{i-1,n} = \mathbf{A}_{i-1,i}\mathbf{A}_{i,i+1}\cdots\mathbf{A}_{n-1,n}.$$

In this case the recursive calculation goes in the opposite direction

$$\begin{aligned}
\mathbf{A}_{n-2,n} &= \mathbf{A}_{n-2,n-1}\mathbf{A}_{n-1,n}, \\
\mathbf{A}_{n-3,n} &= \mathbf{A}_{n-3,n-2}\mathbf{A}_{n-2,n}, \\
\mathbf{A}_{n-4,n} &= \mathbf{A}_{n-4,n-3}\mathbf{A}_{n-3,n}, \\
&\vdots
\end{aligned}$$

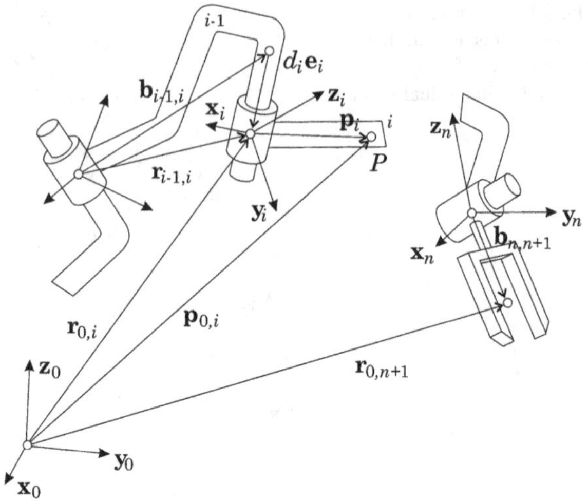

Fig. 3.2 Vectors connecting the origins of the coordinate frames with the reference frame

or

$$\mathbf{A}_{i-1,n} = \mathbf{A}_{i-1,i}\mathbf{A}_{i,n}, \quad i = n-1, n-2, \ldots, 1. \tag{3.3}$$

We can also develop formulas for calculating the positions of the mechanism's links. First we connect the centers of two neighboring joints $i-1$ and i, which lay on opposite ends of link $i-1$, with the vector $\mathbf{r}_{i-1,i}$, as shown in Fig. 3.2

$$\mathbf{r}_{i-1,i}^{(i-1)} = \mathbf{b}_{i-1,i}^{(i-1)} + d_i\mathbf{e}_i^{(i-1)}. \tag{3.4}$$

This equation is valid for all $i = 1, 2, \ldots, n$, where for the case of the end-effector (i.e. $i = n$), $\mathbf{r}_{n,n+1}^{(n)} = \mathbf{b}_{n,n+1}^{(n)}$.

The position of coordinate frame $\mathbf{x}_j$, $\mathbf{y}_j$, $\mathbf{z}_j$ with respect to the frame $\mathbf{x}_i$, $\mathbf{y}_i$, $\mathbf{z}_i$, given by the vector $\mathbf{r}_{i,j}$, is obtained by adding the intermediate link vectors and the translation vectors in the intermediate joints. Here, all the vectors must be expressed in the same vector space, e.g. the vector space $\Re_i^3$, as described by the following expression

$$\mathbf{r}_{i,j}^{(i)} = \mathbf{r}_{i,i+1}^{(i)} + \mathbf{r}_{i+1,i+2}^{(i)} + \cdots + \mathbf{r}_{j-1,j}^{(i)}.$$

The position of the joint center i with respect to the base is given by the vector $\mathbf{r}_{0,i}$ which can be calculated in the following way

$$\mathbf{r}_{0,i}^{(0)} = \mathbf{r}_{0,1}^{(0)} + \mathbf{A}_{0,1}\mathbf{r}_{1,2}^{(1)} + \mathbf{A}_{0,2}\mathbf{r}_{2,3}^{(2)} + \cdots + \mathbf{A}_{0,i-1}\mathbf{r}_{i-1,i}^{(i-1)}.$$

Here, all the vectors were expressed in the reference coordinate frame, i.e. vector space $\Re_0^3$, by use of the rotation matrices $\mathbf{A}_{0,1}, \mathbf{A}_{0,2}, \ldots, \mathbf{A}_{0,i-1}$. The equation is solved in steps

$$\mathbf{r}_{0,2}^{(0)} = \mathbf{r}_{0,1}^{(0)} + \mathbf{A}_{0,1}\mathbf{r}_{1,2}^{(1)},$$
$$\mathbf{r}_{0,3}^{(0)} = \mathbf{r}_{0,2}^{(0)} + \mathbf{A}_{0,2}\mathbf{r}_{2,3}^{(2)},$$
$$\mathbf{r}_{0,4}^{(0)} = \mathbf{r}_{0,3}^{(0)} + \mathbf{A}_{0,3}\mathbf{r}_{3,4}^{(3)},$$
$$\vdots$$

which can be written in the following form

$$\mathbf{r}_{0,i}^{(0)} = \mathbf{r}_{0,i-1}^{(0)} + \mathbf{A}_{0,i-1}\mathbf{r}_{i-1,i}^{(i-1)}, \quad i = 2, 3, \ldots, n. \tag{3.5}$$

In this way we gradually calculate the positions of the joint centers of the entire mechanism.

Let us consider the point P on link i, as shown in Fig. 3.2. We are interested in its position with respect to the base of the mechanism, which is given by the vector $\mathbf{p}_i$ with respect to the coordinate frame $\mathbf{x}_i$, $\mathbf{y}_i$, $\mathbf{z}_i$. The vector $\mathbf{p}_i$ is given in $\Re_i^3$ by $\mathbf{p}_i^{(i)}$ and is known. The calculation is simple, as we only have to add vector $\mathbf{p}_i$ to the vector $\mathbf{r}_{0,i}$ belonging to the center of the i-th joint. We must take care that both vectors are expressed in the same vector space, i.e. $\Re_0^3$

$$\mathbf{p}_{0,i}^{(0)} = \mathbf{r}_{0,i}^{(0)} + \mathbf{p}_i^{(0)} = \mathbf{r}_{0,i}^{(0)} + \mathbf{A}_{0,i}\mathbf{p}_i^{(i)}. \tag{3.6}$$

In the same way, the position of the robot's end-point, i.e. the reference point of the robot gripper, can be calculated. It will be denoted by vector $\mathbf{r}_{0,n+1}$ shown in Fig. 3.2

$$\mathbf{r}_{0,n+1}^{(0)} = \mathbf{r}_{0,n}^{(0)} + \mathbf{A}_{0,n}\mathbf{r}_{n,n+1}^{(n)}.$$

We must be aware that calculating the robot's end-point position by this equation is most appropriate when the rotation matrices $\mathbf{A}_{0,1}, \mathbf{A}_{0,2}, \ldots, \mathbf{A}_{0,n}$ are already calculated. This happens when the orientations of all the links of the mechanism are known before calculating the position of the robot's end-point. Otherwise the equation can be simplified by exposing the rotation matrices

$$\mathbf{r}_{0,n+1}^{(0)} = \mathbf{r}_{0,1}^{(0)} + \mathbf{A}_{0,1}\big(\mathbf{r}_{1,2}^{(1)} + \mathbf{A}_{1,2}\big(\mathbf{r}_{2,3}^{(2)} + \cdots$$
$$+ \mathbf{A}_{n-2,n-1}\big(\mathbf{r}_{n-1,n}^{(n-1)} + \mathbf{A}_{n-1,n}\mathbf{r}_{n,n+1}^{(n)}\big)\cdots\big)\big).$$

Here, only the rotation matrices between the neighboring frames $\mathbf{A}_{0,1}, \mathbf{A}_{1,2}, \ldots,$ $\mathbf{A}_{n-1,n}$ appear. In this way it is not necessary to calculate the orientations of the links before calculating the position of the robot's end-point. The calculation goes

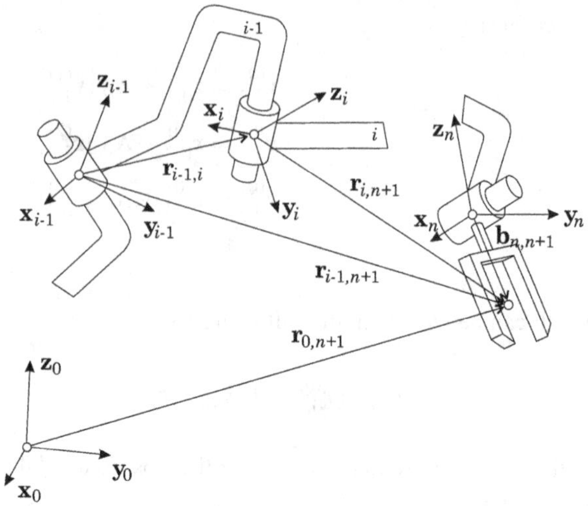

Fig. 3.3 Vectors connecting origins of coordinate frame with the robot's end-point

in the following steps from the robot's end-point to its base

$$\begin{aligned}
\mathbf{r}^{(n-1)}_{n-1,n+1} &= \mathbf{r}^{(n-1)}_{n-1,n} + \mathbf{A}_{n-1,n}\mathbf{r}^{(n)}_{n,n+1},\\
\mathbf{r}^{(n-2)}_{n-2,n+1} &= \mathbf{r}^{(n-2)}_{n-2,n-1} + \mathbf{A}_{n-2,n-1}\mathbf{r}^{(n-1)}_{n-1,n+1},\\
\mathbf{r}^{(n-3)}_{n-3,n+1} &= \mathbf{r}^{(n-3)}_{n-3,n-2} + \mathbf{A}_{n-3,n-2}\mathbf{r}^{(n-2)}_{n-2,n+1},\\
&\vdots
\end{aligned}$$

which can be written in recursive form

$$\mathbf{r}^{(i-1)}_{i-1,n+1} = \mathbf{r}^{(i-1)}_{i-1,i} + \mathbf{A}_{i-1,i}\mathbf{r}^{(i)}_{i,n+1}, \quad i = n, n-1, \ldots, 1. \tag{3.7}$$

The vector $\mathbf{r}_{i-1,n+1}$ connects the center of joint $i-1$ with the reference end-point of the robot mechanism as shown in Fig. 3.3.

When comparing the vectors from Fig. 3.3 with those in Fig. 3.2, it is not difficult to see that $\mathbf{r}_{0,n+1} = \mathbf{r}_{0,i} + \mathbf{r}_{i,n+1}$, which can be formalized as follows

$$\mathbf{r}^{(0)}_{0,n+1} = \mathbf{r}^{(0)}_{0,i} + \mathbf{A}_{0,i}\mathbf{r}^{(i)}_{i,n+1}, \quad i = 1, 2, \ldots, n. \tag{3.8}$$

As we already know, the homogeneous transformation matrix is an operator including both translation and rotation. Let us consider the following product of homogeneous matrices in more details

$$\mathbf{H}_{0,i} = \mathbf{H}_{0,1}\mathbf{H}_{1,2}\cdots\mathbf{H}_{i-1,i}.$$

These matrices are given by (2.12), so that we have

$$\mathbf{H}_{0,i} = \begin{bmatrix} & \mathbf{A}_{0,1} & & \mathbf{r}^{(0)}_{0,1} \\ 0 & 0 & 0 & 1 \end{bmatrix} \cdots \begin{bmatrix} & \mathbf{A}_{i-1,i} & & \mathbf{r}^{(i-1)}_{i-1,i} \\ 0 & 0 & 0 & 1 \end{bmatrix}.$$

Multiplying the components yields

$$\mathbf{H}_{0,i} = \begin{bmatrix} & \mathbf{A}_{0,1}\cdots\mathbf{A}_{i-1,i} & & \mathbf{r}_{0,1}^{(0)} + \cdots + \mathbf{A}_{0,i-1}\mathbf{r}_{i-1,i}^{(i-1)} \\ 0 & 0 & 0 & 1 \end{bmatrix},$$

where we recognize

$$\mathbf{H}_{0,i} = \begin{bmatrix} & \mathbf{A}_{0,i} & & \mathbf{r}_{0,i}^{(0)} \\ 0 & 0 & 0 & 1 \end{bmatrix}. \tag{3.9}$$

The homogeneous transformation matrix $\mathbf{H}_{0,i}$ includes the orientation of link i, which is captured in the rotation matrix $\mathbf{A}_{0,i}$, and position of the center of the i-th joint, described by the vector $\mathbf{r}_{0,i}^{(0)}$.

Similar to the previous example, we can rewrite the expression $\mathbf{H}_{0,i}$ in a recursive form

$$\mathbf{H}_{0,i} = \mathbf{H}_{0,i-1}\mathbf{H}_{i-1,i}, \quad i = 2, 3, \ldots, n, \tag{3.10}$$

in order to be able to calculate the position and orientation of the mechanism's links sequentially,

$$\begin{aligned} \mathbf{H}_{0,2} &= \mathbf{H}_{0,1}\mathbf{H}_{1,2}, \\ \mathbf{H}_{0,3} &= \mathbf{H}_{0,2}\mathbf{H}_{2,3}, \\ \mathbf{H}_{0,4} &= \mathbf{H}_{0,3}\mathbf{H}_{3,4}, \\ &\vdots \end{aligned}$$

The matrix $\mathbf{H}_{0,n}$ encompasses the orientation of the end-effector link and the position of the joint center n. The position of the mechanism's end-point can be obtained as follows

$$\begin{bmatrix} \mathbf{r}_{0,n+1}^{(0)} \\ 1 \end{bmatrix} = \mathbf{H}_{0,i} \begin{bmatrix} \mathbf{r}_{n,n+1}^{(n)} \\ 1 \end{bmatrix} = \begin{bmatrix} \mathbf{r}_{0,n}^{(0)} + \mathbf{A}_{0,n}\mathbf{r}_{n,n+1}^{(n)} \\ 1 \end{bmatrix}.$$

Another possibility is to assign an additional coordinate frame $\mathbf{x}_{n+1}, \mathbf{y}_{n+1}, \mathbf{z}_{n+1}$, to the mechanism's end-point. This frame would be parallel to the frame $\mathbf{x}_n, \mathbf{y}_n, \mathbf{z}_n$. The matrix

$$\mathbf{H}_{0,n+1} = \mathbf{H}_{0,n}\mathbf{H}_{n,n+1}, \tag{3.11}$$

where

$$\mathbf{H}_{n,n+1} = \begin{bmatrix} & \mathbf{I} & & \mathbf{r}_{n,n+1}^{(n)} \\ 0 & 0 & 0 & 1 \end{bmatrix}, \tag{3.12}$$

now includes the position of the mechanism's reference end-point, as we have

$$\mathbf{H}_{0,n+1} = \begin{bmatrix} & \mathbf{A}_{0,n} & & \mathbf{r}_{0,n+1}^{(0)} \\ 0 & 0 & 0 & 1 \end{bmatrix}. \tag{3.13}$$

Let us consider the following recursive equation

$$\mathbf{H}_{i-1,n+1} = \mathbf{H}_{i-1,i}\mathbf{H}_{i,n+1}, \quad i = n, n-1, \ldots, 1, \tag{3.14}$$

which proceeds in the opposite direction

$$\begin{aligned}
\mathbf{H}_{n-1,n+1} &= \mathbf{H}_{n-1,n}\mathbf{H}_{n,n+1},\\
\mathbf{H}_{n-2,n+1} &= \mathbf{H}_{n-2,n-1}\mathbf{H}_{n-1,n+1},\\
\mathbf{H}_{n-3,n+1} &= \mathbf{H}_{n-3,n-2}\mathbf{H}_{n-2,n+1},\\
&\vdots
\end{aligned}$$

It is not difficult to realize that

$$\mathbf{H}_{i-1,n+1} = \begin{bmatrix} & \mathbf{A}_{i-1,n} & & \mathbf{r}^{(i-1)}_{i-1,n+1} \\ 0 & 0 & 0 & 1 \end{bmatrix}. \tag{3.15}$$

The matrix includes the orientation of link $i-1$ with respect to link n, described by the rotation matrix $\mathbf{A}_{i-1,n}$, and the position of the reference end-point of the mechanism with respect to the joint center $i-1$, defined by the vector $\mathbf{r}^{(i-1)}_{i-1,n+1}$, as shown in Fig. 3.3.

Let us recap the rotation matrices and position vectors which determine the pose of a mechanism:

$\mathbf{A}_{i-1,i}(\theta_i)$: rotation matrix between links $i-1$ and i. It is calculated by formula (3.1).

$\mathbf{A}_{0,i}(\theta_1,\ldots,\theta_i)$: rotation matrix between links 0 and i. It is calculated by recursive formula (3.2), working in steps from the robot's base to its end-point. Prior to this the rotation matrices must be calculated (3.1).

$\mathbf{A}_{i-1,n}(\theta_i,\ldots,\theta_n)$: rotation matrix between links $i-1$ and n. It is obtained by recursive expression (3.3), working in steps from the end-point to the base of the robot mechanism. Prior to this the rotation matrices must be calculated (3.1).

$\mathbf{r}^{(i-1)}_{i-1,i}(d_i)$: position vector of joint center i with respect to joint center $i-1$. It is defined by (3.4).

$\mathbf{r}^{(0)}_{0,i}(\theta_1,\ldots,\theta_{i-1},d_1,\ldots,d_i)$: position vector of joint center i with respect to the base. It is calculated with recursive formula (3.5), working in steps from the base to the robot mechanism's end-point. Prior to this the rotation matrices (3.2) and position vectors must be calculated (3.4).

$\mathbf{r}^{(i)}_{i-1,n+1}(\theta_i,\ldots,\theta_n,d_i,\ldots,d_n)$: position vector of the robot's end-point with respect to joint center $i-1$. It is obtained by recursive expression (3.7), working in steps from the robot's end-point towards the base of the mechanism. Prior to that the rotation matrices (3.1) and position vectors (3.4) must be obtained.

We should be aware of which of the above-mentioned mathematical expressions requires fewer arithmetic operations. In general it is advantageous to use the recursive form of the equations in such a way that we calculate the link orientations sequentially, from the base to the end-point of the mechanism, while the position is calculated in the opposite direction. It is not advisable to use the homogeneous transformations, as they require useless multiplications by zero and one.

As an example, consider finding the pose of each link of the four degree mechanism from Fig. 2.18. The coordinate frames and pertinent kinematic parameters are shown in Fig. 2.20. Let us first calculate the orientations of the mechanism's links sequentially, from the base towards the robot's end-point, while using the following two abbreviations $\sin\theta_i = s_i$ and $\cos\theta_i = c_i$. The orientation of the first link is given by the rotation matrix

$$\mathbf{A}_{0,1} = \begin{bmatrix} c_1 & -s_1 & 0 \\ s_1 & c_1 & 0 \\ 0 & 0 & 1 \end{bmatrix}.$$

The orientation of the second link is described by the following matrix

$$\mathbf{A}_{0,2} = \mathbf{A}_{0,1}\mathbf{A}_{1,2} = \begin{bmatrix} c_1 & -s_1c_2 & s_1s_2 \\ s_1 & c_1c_2 & -c_1s_2 \\ 0 & s_2 & c_2 \end{bmatrix},$$

which is valid also for the third link, because the rotational variable is $\theta_3 = 0$

$$\mathbf{A}_{0,3} = \mathbf{A}_{0,2}.$$

The orientation of the fourth link is included in the following rotation matrix

$$\mathbf{A}_{0,4} = \mathbf{A}_{0,3}\mathbf{A}_{3,4} = \begin{bmatrix} c_1c_4 - s_1c_2s_4 & -c_1s_4 - s_1c_2c_4 & s_1s_2 \\ s_1c_4 + c_1c_2s_4 & -s_1s_4 + c_1c_2c_4 & -c_1s_2 \\ s_2s_4 & s_2c_4 & c_2 \end{bmatrix}.$$

The orientation of the end-effector link can be described by the three Euler angles. In accordance with the definition of the Euler angles given in Sect. 1.5.3, they can be calculated with the elements of the rotation matrix $\mathbf{A}_{0,4}$ as follows

$$\theta = \arccos c_2 = \theta_2,$$
$$\psi = \arctan\frac{s_1s_2}{c_1s_2} = \theta_1,$$
$$\phi = \arctan\frac{s_2s_4}{s_2c_4} = \theta_4.$$

The simple expressions obtained are not surprising as the arrangement of the rotations in the mechanism is the same as with the Euler angles defined in Sect. 1.5.3. If using the YPR angles or some other representation of orientation, the equations would not be that simple. Unlike in this example, in general, the orientation angles cannot be explicitly described by the rotational variables of a mechanism.

As the rotation matrices $\mathbf{A}_{0,1}$, $\mathbf{A}_{0,2}$, $\mathbf{A}_{0,3}$, and $\mathbf{A}_{0,4}$ are already known, we will use them to calculate the positions of the joint centers, together with the position of

Table 3.1 Vectors between the centers of neighboring joints of the mechanism from Fig. 2.20

i	1	2	3	4	5
$\mathbf{r}_{i-1,i}^{(i-1)}$	0	0	0	0	0
	0	l_1	l_2+d_3	l_3	l_4
	h_0	h_1	0	$-h_3$	0

the robot's end-point. The position vectors between the centers of the neighboring joints (3.4) are gathered in Table 3.1.

The position of the first joint center is given by the vector $\mathbf{r}_{0,1}^{(0)}$ in Table 3.1, while the position of the center of the second joint is given by the vector

$$\mathbf{r}_{0,2}^{(0)} = \mathbf{r}_{0,1}^{(0)} + \mathbf{A}_{0,1}\mathbf{r}_{1,2}^{(1)} = \mathbf{r}_{0,1}^{(0)} + \begin{bmatrix} -s_1 l_1 \\ c_1 l_1 \\ h_1 \end{bmatrix}.$$

The center of the third joint is given by

$$\mathbf{r}_{0,3}^{(0)} = \mathbf{r}_{0,2}^{(0)} + \mathbf{A}_{0,2}\mathbf{r}_{2,3}^{(2)} = \mathbf{r}_{0,2}^{(0)} + \begin{bmatrix} -s_1 c_2 (l_2+d_3) \\ c_1 c_2 (l_2+d_3) \\ s_2 (l_2+d_3) \end{bmatrix},$$

while the position of the fourth joint center is

$$\mathbf{r}_{0,4}^{(0)} = \mathbf{r}_{0,3}^{(0)} + \mathbf{A}_{0,3}\mathbf{r}_{3,4}^{(3)} = \mathbf{r}_{0,3}^{(0)} + \begin{bmatrix} -s_1 c_2 l_3 - s_1 s_2 h_3 \\ c_1 c_2 l_3 + c_1 s_2 h_3 \\ s_2 l_3 - c_2 h_3 \end{bmatrix}.$$

The following vector

$$\mathbf{r}_{0,5}^{(0)} = \mathbf{r}_{0,4}^{(0)} + \mathbf{A}_{0,4}\mathbf{r}_{4,5}^{(4)} = \mathbf{r}_{0,4}^{(0)} + l_4 \begin{bmatrix} -c_1 s_4 - s_1 c_2 c_4 \\ -s_1 s_4 + c_1 c_2 c_4 \\ s_2 c_4 \end{bmatrix}$$

describes the position of the robot mechanism's end-point, i.e. the reference point on the robot's gripper.

3.1.2 Angular and Translational Velocities of a Mechanism

The angular velocity matrix, which defines the relative angular velocity between links $i-1$ and i, is defined as the product of a rotation matrix and its time derivative, (1.104).

$$\boldsymbol{\Omega}_{i-1,i} = \dot{\mathbf{A}}_{i-1,i}\mathbf{A}_{i-1,i}^{\mathrm{T}}.$$

Recall formula (1.57)

$$\dot{\mathbf{A}}_{i-1,i} = \dot{\theta}_i \frac{\partial \mathbf{A}_{i-1,i}}{\partial \theta_i} = \dot{\theta}_i \mathbf{e}_i^{(i-1)} \otimes \mathbf{A}_{i-1,i},$$

so that we have

$$\boldsymbol{\Omega}_{i-1,i} = \dot{\theta}_i \mathbf{e}_i^{(i-1)} \otimes \mathbf{I} = \dot{\theta}_i \boldsymbol{\Delta}_{i-1,i}.$$

The matrix $\boldsymbol{\Delta}_{i-1,i}$ is well known to us (1.49), together with its properties, which were described in Sect. 1.3.1.

The angular velocity is defined as the vector invariant of the matrix $\boldsymbol{\Omega}_{i-1,i}$, i.e.

$$\boldsymbol{\omega}_{i-1,i}^{(i-1)} = \operatorname{vect}(\boldsymbol{\Omega}_{i-1,i}) = \dot{\theta}_i \operatorname{vect}\bigl(\mathbf{e}_i^{(i-1)} \otimes \mathbf{I}\bigr) = \dot{\theta}_i \operatorname{vect}(\boldsymbol{\Delta}_{i-1,i}).$$

It follows from the properties of the matrix $\boldsymbol{\Delta}_{i-1,i}$, that the angular velocity between the links $i-1$ and i is a consequence of the time variation of the rotational variable

$$\boldsymbol{\omega}_{i-1,i}^{(i-1)} = \dot{\theta}_i \mathbf{e}_i^{(i-1)}, \tag{3.16}$$

which is valid for all $i = 1, 2, \ldots, n$. Now we can realize

$$\dot{\mathbf{A}}_{i-1,i} = \boldsymbol{\omega}_{i-1,i}^{(i-1)} \otimes \mathbf{A}_{i-1,i}. \tag{3.17}$$

The angular velocity between links i and j, where $j > i$, is the vector sum of the angular velocities among the intermediate kinematic pairs. This statement cannot be considered self explanatory and should be proven. It follows from the definition

$$\begin{aligned}
\boldsymbol{\omega}_{i,j}^{(i)} &= \operatorname{vect}(\boldsymbol{\Omega}_{i,j}) = \operatorname{vect}\bigl(\dot{\mathbf{A}}_{i,j}\mathbf{A}_{i,j}^{\mathrm{T}}\bigr) \\
&= \operatorname{vect}\bigl(\dot{\mathbf{A}}_{i,i+1}\mathbf{A}_{i,i+1}^{\mathrm{T}} + \mathbf{A}_{i,i+1}\dot{\mathbf{A}}_{i+1,i+2}\mathbf{A}_{i+1,i+2}^{\mathrm{T}}\mathbf{A}_{i,i+1}^{\mathrm{T}} \\
&\quad + \cdots + \mathbf{A}_{i,j-1}\dot{\mathbf{A}}_{j-1,j}\mathbf{A}_{j-1,j}^{\mathrm{T}}\mathbf{A}_{i,j-1}^{\mathrm{T}}\bigr).
\end{aligned}$$

By considering the characteristic properties of the operation $\otimes$ and the vector invariant, we obtain

$$\begin{aligned}
\boldsymbol{\omega}_{i,j}^{(i)} &= \operatorname{vect}\bigl(\boldsymbol{\omega}_{i,i+1}^{(i)} \otimes \mathbf{I} + \mathbf{A}_{i,i+1}\bigl(\boldsymbol{\omega}_{i+1,i+2}^{(i+1)} \otimes \mathbf{I}\bigr)\mathbf{A}_{i,i+1}^{\mathrm{T}} \\
&\quad + \cdots + \mathbf{A}_{i,j-1}\bigl(\boldsymbol{\omega}_{j-1,j}^{(j-1)} \otimes \mathbf{I}\bigr)\mathbf{A}_{i,j-1}^{\mathrm{T}}\bigr) \\
&= \operatorname{vect}\bigl(\boldsymbol{\omega}_{i,i+1}^{(i)} \otimes \mathbf{I} + \boldsymbol{\omega}_{i+1,i+2}^{(i)} \otimes \mathbf{I} + \cdots + \boldsymbol{\omega}_{j-1,j}^{(i)} \otimes \mathbf{I}\bigr)
\end{aligned}$$

and finally

$$\boldsymbol{\omega}_{i,j}^{(i)} = \boldsymbol{\omega}_{i,i+1}^{(i)} + \boldsymbol{\omega}_{i+1,i+2}^{(i)} + \cdots + \boldsymbol{\omega}_{j-1,j}^{(i)}.$$

The angular velocity of link i with respect to the base, which is described by the Cartesian vector $\boldsymbol{\omega}_{0,i}$, is defined by the following sum

$$\boldsymbol{\omega}_{0,i}^{(0)} = \boldsymbol{\omega}_{0,1}^{(0)} + \mathbf{A}_{0,1}\boldsymbol{\omega}_{1,2}^{(1)} + \cdots + \mathbf{A}_{0,i-1}\boldsymbol{\omega}_{i-1,i}^{(i-1)}.$$

This equation can be solved sequentially

$$\begin{aligned}
\boldsymbol{\omega}_{0,2}^{(0)} &= \boldsymbol{\omega}_{0,1}^{(0)} + \mathbf{A}_{0,1}\boldsymbol{\omega}_{1,2}^{(1)},\\
\boldsymbol{\omega}_{0,3}^{(0)} &= \boldsymbol{\omega}_{0,2}^{(0)} + \mathbf{A}_{0,2}\boldsymbol{\omega}_{2,3}^{(2)},\\
\boldsymbol{\omega}_{0,4}^{(0)} &= \boldsymbol{\omega}_{0,3}^{(0)} + \mathbf{A}_{0,3}\boldsymbol{\omega}_{3,4}^{(3)},\\
&\vdots
\end{aligned}$$

which can be written in the more compact form

$$\boldsymbol{\omega}_{0,i}^{(0)} = \boldsymbol{\omega}_{0,i-1}^{(0)} + \mathbf{A}_{0,i-1}\boldsymbol{\omega}_{i-1,i}^{(i-1)}, \quad i = 2, 3, \ldots, n. \tag{3.18}$$

In each step we sequentially calculate the angular velocity of a link of the mechanism, starting from the base and working towards the end-point. The angular velocity at the end-point is $\boldsymbol{\omega}_{0,n}^{(0)}$.

The above equation can be rewritten in such a way that the rotation matrices are exposed. It follows that

$$\boldsymbol{\omega}_{0,n}^{(0)} = \boldsymbol{\omega}_{0,1}^{(0)} + \mathbf{A}_{0,1}\big(\boldsymbol{\omega}_{1,2}^{(1)} + \mathbf{A}_{1,2}\big(\boldsymbol{\omega}_{2,3}^{(2)} + \cdots + \mathbf{A}_{n-2,n-1}\boldsymbol{\omega}_{n-1,n}^{(n-1)}\big)\ldots\big).$$

In the above equation, only the rotation matrices between neighboring coordinate frames $\mathbf{A}_{0,1}, \mathbf{A}_{1,2}, \ldots, \mathbf{A}_{n-1,n}$, are involved. It is therefore not necessary to calculate the orientations of the links before calculating their angular velocities. In the following equations the calculations go sequentially from the end-point to the base of the mechanism

$$\begin{aligned}
\boldsymbol{\omega}_{n-2,n}^{(n-2)} &= \boldsymbol{\omega}_{n-2,n-1}^{(n-2)} + \mathbf{A}_{n-2,n-1}\boldsymbol{\omega}_{n-1,n}^{(n-1)},\\
\boldsymbol{\omega}_{n-3,n}^{(n-3)} &= \boldsymbol{\omega}_{n-3,n-2}^{(n-3)} + \mathbf{A}_{n-3,n-2}\boldsymbol{\omega}_{n-2,n}^{(n-2)},\\
\boldsymbol{\omega}_{n-4,n}^{(n-4)} &= \boldsymbol{\omega}_{n-4,n-3}^{(n-4)} + \mathbf{A}_{n-4,n-3}\boldsymbol{\omega}_{n-3,n}^{(n-3)},\\
&\vdots
\end{aligned}$$

which can be written as follows

$$\boldsymbol{\omega}_{i-1,n}^{(i-1)} = \boldsymbol{\omega}_{i-1,i}^{(i-1)} + \mathbf{A}_{i-1,i}\boldsymbol{\omega}_{i,n}^{(i)}, \quad i = n-1, n-2, \ldots, 1. \tag{3.19}$$

When the orientation of the mechanism is given by Euler angles or YPR angles, we need the relation between the angular velocity and the time derivatives of the orientation angles. The transformations between them were developed in Sects. 1.6.3 and 1.6.4.

Now we have the more difficult task of developing the expressions for the translational velocities of the mechanism. We start with the vector connecting the joint centers $i-1$ and i, i.e. vector $\mathbf{r}_{i-1,i}$ in Fig. 3.4. Its time derivative is the translational velocity between these two points. Mathematically this can be expressed as follows

$$\mathbf{v}_{i-1,i}^{(i-1)} = \dot{\mathbf{r}}_{i-1,i}^{(i-1)} = \dot{d}_i \mathbf{e}_i^{(i-1)}. \tag{3.20}$$

The equation is valid for $i = 1, 2, \ldots, n$, while $\mathbf{v}_{n,n+1}^{(n)} = 0$. All these statements and equations are valid, because link vectors do not change and because their time derivatives are zero. In general the translational velocity between the joint centers i and j, where $j > i$, is equal to the sum of the time derivatives

$$\mathbf{v}_{i,j}^{(i)} = \dot{\mathbf{r}}_{i,i+1}^{(i)} + \dot{\mathbf{r}}_{i+1,i+2}^{(i)} + \cdots + \dot{\mathbf{r}}_{j-1,j}^{(i)}.$$

In practical calculations this formula is useless, as the essence of the problem is to calculate the very time derivatives which occur in the formula. Let us take a step further and consider the translational velocity of the center of the joint i with respect to the reference frame. At first glance things look complex

$$\begin{aligned}\mathbf{v}_{0,i}^{(0)} = \dot{\mathbf{r}}_{0,i}^{(0)} &= \dot{\mathbf{r}}_{0,1}^{(0)} + \dot{\mathbf{A}}_{0,1}\mathbf{r}_{1,2}^{(1)} + \mathbf{A}_{0,1}\dot{\mathbf{r}}_{1,2}^{(1)} + \dot{\mathbf{A}}_{0,2}\mathbf{r}_{2,3}^{(2)} \\ &+ \mathbf{A}_{0,2}\dot{\mathbf{r}}_{2,3}^{(2)} + \cdots + \dot{\mathbf{A}}_{0,i-1}\mathbf{r}_{i-1,i}^{(i-1)} + \mathbf{A}_{0,i-1}\dot{\mathbf{r}}_{i-1,i}^{(i-1)},\end{aligned}$$

because of the time derivatives of the rotation matrices

$$\begin{aligned}\dot{\mathbf{A}}_{0,2} &= \dot{\mathbf{A}}_{0,1}\mathbf{A}_{1,2} + \mathbf{A}_{0,1}\dot{\mathbf{A}}_{1,2}, \\ \dot{\mathbf{A}}_{0,3} &= \dot{\mathbf{A}}_{0,1}\mathbf{A}_{1,2}\mathbf{A}_{2,3} + \mathbf{A}_{0,1}\dot{\mathbf{A}}_{1,2}\mathbf{A}_{2,3} + \mathbf{A}_{0,1}\mathbf{A}_{1,2}\dot{\mathbf{A}}_{2,3}, \\ &\vdots\end{aligned}$$

The situation however is not hopeless. Recall that

$$\dot{\mathbf{A}}_{0,k} = \boldsymbol{\omega}_{0,k}^{(0)} \otimes \mathbf{A}_{0,k},$$

where the angular velocity $\boldsymbol{\omega}_{0,k}^{(0)}$ is the sum of angular velocities of all the intermediate joints

$$\boldsymbol{\omega}_{0,k}^{(0)} = \boldsymbol{\omega}_{0,1}^{(0)} + \mathbf{A}_{0,1}\boldsymbol{\omega}_{1,2}^{(1)} + \cdots + \mathbf{A}_{0,k-1}\boldsymbol{\omega}_{k-1,k}^{(k-1)}.$$

According to the rules of the operation $\otimes$, we have

$$\dot{\mathbf{A}}_{0,k}\mathbf{r}_{k-1,i}^{(k-1)} = \boldsymbol{\omega}_{0,k}^{(0)} \times \mathbf{A}_{0,k}\mathbf{r}_{k-1,i}^{(k-1)}.$$

It follows that

$$\begin{aligned}\mathbf{v}_{0,i}^{(0)} &= \mathbf{v}_{0,1}^{(0)} + \boldsymbol{\omega}_{0,1}^{(0)} \times \mathbf{A}_{0,1}\mathbf{r}_{1,2}^{(1)} + \mathbf{A}_{0,1}\mathbf{v}_{1,2}^{(1)} + \boldsymbol{\omega}_{0,2}^{(0)} \times \mathbf{A}_{0,2}\mathbf{r}_{2,3}^{(2)} \\ &+ \mathbf{A}_{0,2}\mathbf{v}_{2,3}^{(2)} + \cdots + \boldsymbol{\omega}_{0,i-1}^{(0)} \times \mathbf{A}_{0,i-1}\mathbf{r}_{i-1,i}^{(i-1)} + \mathbf{A}_{0,i-1}\mathbf{v}_{i-1,i}^{(i-1)}.\end{aligned}$$

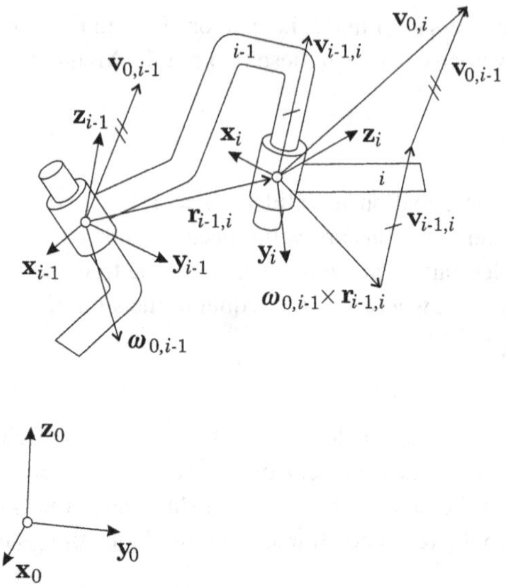

Fig. 3.4 Translational velocity vector of the joint center i with respect to the base of mechanism

This equation can be solved in steps, i.e. sequentially

$$\mathbf{v}_{0,2}^{(0)} = \mathbf{v}_{0,1}^{(0)} + \boldsymbol{\omega}_{0,1}^{(0)} \times \mathbf{A}_{0,1}\mathbf{r}_{1,2}^{(1)} + \mathbf{A}_{0,1}\mathbf{v}_{1,2}^{(1)},$$

$$\mathbf{v}_{0,3}^{(0)} = \mathbf{v}_{0,2}^{(0)} + \boldsymbol{\omega}_{0,2}^{(0)} \times \mathbf{A}_{0,2}\mathbf{r}_{2,3}^{(2)} + \mathbf{A}_{0,2}\mathbf{v}_{2,3}^{(2)},$$

$$\mathbf{v}_{0,4}^{(0)} = \mathbf{v}_{0,3}^{(0)} + \boldsymbol{\omega}_{0,3}^{(0)} \times \mathbf{A}_{0,3}\mathbf{r}_{3,4}^{(3)} + \mathbf{A}_{0,3}\mathbf{v}_{3,4}^{(3)},$$

$$\vdots$$

and expressed in the following form

$$\mathbf{v}_{0,i}^{(0)} = \mathbf{v}_{0,i-1}^{(0)} + \boldsymbol{\omega}_{0,i-1}^{(0)} \times \mathbf{A}_{0,i-1}\mathbf{r}_{i-1,i}^{(i-1)} + \mathbf{A}_{0,i-1}\mathbf{v}_{i-1,i}^{(i-1)}, \quad i = 2, 3, \ldots, n. \tag{3.21}$$

The terms in the above equation have geometric meaning, shown in Fig. 3.4. We recognize the vectors. They are related to the translational velocity of the point, which rotates and translates for the case when rotation comes before translation. This was mathematically expressed with (1.108) and presented in Fig. 1.19. The vector $\mathbf{v}_{0,i-1}$ represents the translational velocity of joint center $i-1$ with respect to the base of the mechanism, while $\boldsymbol{\omega}_{0,i-1}$ is the angular velocity of link $i-1$ with respect to the base. The cross product $\boldsymbol{\omega}_{0,i-1} \times \mathbf{r}_{i-1,i}$ runs tangentially to the axis of rotation of link $i-1$. The radius of rotation is represented by the component of the vector $\mathbf{r}_{i-1,i}$, which is perpendicular to this axis. The velocity $\mathbf{v}_{i-1,i}$ results from the translational coordinate of the i-th joint.

Consider now the velocity of an arbitrary point P on link i with respect to the base of the mechanism, as shown in Fig. 3.2. We have

$$\dot{\mathbf{p}}_{0,i}^{(0)} = \dot{\mathbf{r}}_{0,i}^{(0)} + \dot{\mathbf{A}}_{0,i}\mathbf{p}_{i}^{(i)}.$$

As in the previous example we arrive at the following result

$$\dot{\mathbf{p}}_{0,i}^{(0)} = \mathbf{v}_{0,i}^{(0)} + \boldsymbol{\omega}_{0,i}^{(0)} \times \mathbf{A}_{0,i}\mathbf{p}_i^{(i)}. \tag{3.22}$$

The translational velocity of the joint at the end of the mechanism, i.e. joint n, is determined by the vector $\mathbf{v}_{0,n}$, while the translational velocity of the reference end-point of the mechanism is $\mathbf{v}_{0,n+1}$. They are related by the following equation

$$\mathbf{v}_{0,n+1}^{(0)} = \mathbf{v}_{0,n}^{(0)} + \boldsymbol{\omega}_{0,n}^{(0)} \times \mathbf{A}_{0,n}\mathbf{r}_{n,n+1}^{(n)}. \tag{3.23}$$

It is convenient to calculate the position of the robot's end-point $\mathbf{r}_{0,n+1}$ in sequential steps, working from the end-point towards the base. The same is true for the translational velocities. From the expressions for the position (3.7) it follows that

$$\begin{aligned}
\dot{\mathbf{r}}_{n-1,n+1}^{(n-1)} &= \dot{\mathbf{r}}_{n-1,n}^{(n-1)} + \dot{\mathbf{A}}_{n-1,n}\mathbf{r}_{n,n+1}^{(n)} + \mathbf{A}_{n-1,n}\dot{\mathbf{r}}_{n,n+1}^{(n)},\\
\dot{\mathbf{r}}_{n-2,n+1}^{(n-2)} &= \dot{\mathbf{r}}_{n-2,n-1}^{(n-2)} + \dot{\mathbf{A}}_{n-2,n-1}\mathbf{r}_{n-1,n+1}^{(n-1)} + \mathbf{A}_{n-2,n-1}\dot{\mathbf{r}}_{n-1,n+1}^{(n-1)},\\
\dot{\mathbf{r}}_{n-3,n+1}^{(n-3)} &= \dot{\mathbf{r}}_{n-3,n-2}^{(n-3)} + \dot{\mathbf{A}}_{n-3,n-2}\mathbf{r}_{n-2,n+1}^{(n-2)} + \mathbf{A}_{n-3,n-2}\dot{\mathbf{r}}_{n-2,n+1}^{(n-2)},\\
&\vdots
\end{aligned}$$

Taking into account that the rotation vector is unchanged after multiplication with the rotation matrix $\mathbf{A}_{k-1,k}\mathbf{e}_k^{(k-1)} = \mathbf{e}_k^{(k-1)}$, it follows that

$$\mathbf{A}_{k-1,k}\boldsymbol{\omega}_{k-1,k}^{(k-1)} = \boldsymbol{\omega}_{k-1,k}^{(k-1)},$$

which gives

$$\begin{aligned}
\mathbf{v}_{n-1,n+1}^{(n-1)} &= \mathbf{v}_{n-1,n}^{(n-1)} + \mathbf{A}_{n-1,n}\left(\mathbf{v}_{n,n+1}^{(n)} + \boldsymbol{\omega}_{n-1,n}^{(n-1)} \times \mathbf{r}_{n,n+1}^{(n)}\right),\\
\mathbf{v}_{n-2,n+1}^{(n-2)} &= \mathbf{v}_{n-2,n-1}^{(n-2)} + \mathbf{A}_{n-2,n-1}\left(\mathbf{v}_{n-1,n+1}^{(n)} + \boldsymbol{\omega}_{n-2,n-1}^{(n-2)} \times \mathbf{r}_{n-1,n+1}^{(n-1)}\right),\\
\mathbf{v}_{n-3,n+1}^{(n-3)} &= \mathbf{v}_{n-3,n-2}^{(n-3)} + \mathbf{A}_{n-3,n-2}\left(\mathbf{v}_{n-2,n+1}^{(n)} + \boldsymbol{\omega}_{n-3,n-2}^{(n-3)} \times \mathbf{r}_{n-2,n+1}^{(n-2)}\right),\\
&\vdots
\end{aligned}$$

Here, it is not difficult to see the rule

$$\begin{aligned}
&\mathbf{v}_{i-1,n+1}^{(i-1)} = \mathbf{v}_{i-1,i}^{(i-1)} + \mathbf{A}_{i-1,i}\left(\mathbf{v}_{i,n+1}^{(i)} + \boldsymbol{\omega}_{i-1,i}^{(i-1)} \times \mathbf{r}_{i,n+1}^{(i)}\right),\\
&i = n, n-1, \ldots, 1.
\end{aligned} \tag{3.24}$$

The terms in the above equations have geometric meaning, presented in Fig. 3.5. The vector $\mathbf{v}_{i,n+1}$ is the translational velocity of the mechanism's end-point with respect to the joint center i , while $\boldsymbol{\omega}_{i-1,i}$ is the angular velocity of link i with respect to link $i-1$ resulting from the rotational coordinate of joint i. The cross product

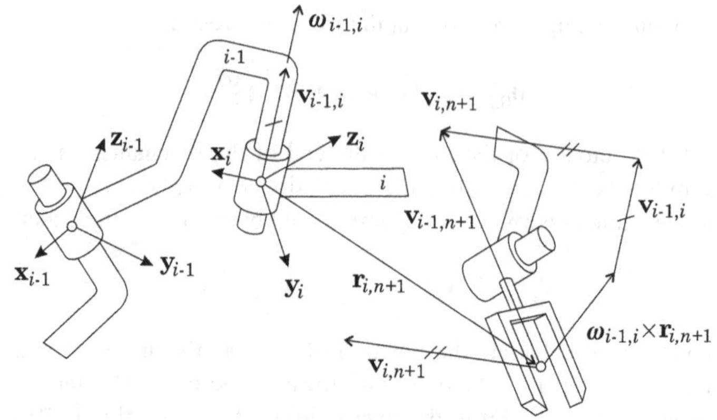

Fig. 3.5 Translational velocity vector of the mechanism's end-point with respect to the joint center $i-1$

$\omega_{i-1,i} \times \mathbf{r}_{i,n+1}$ has a tangential direction with respect to the joint vector $\mathbf{e}_i$, while the rotation radius is the component of the vector $\mathbf{r}_{i,n+1}$, which is perpendicular to $\mathbf{e}_i$. The velocity $\mathbf{v}_{i-1,i}$ results from the translational coordinate of the i-th joint.

When calculating position and orientation of a mechanism, we observed that homogeneous transformation matrices represent an efficient and straightforward mathematical tool, however when calculating velocities and accelerations the situation is more complex. Let us study one of the possible ways of using homogeneous transformations for calculating the translational and angular velocities of a mechanism. The approach is based on time derivatives of homogeneous transformation matrices. Let us assume that the homogeneous transformation matrices for individual kinematic pairs of mechanism are given by (2.12), therefore we have

$$\dot{\mathbf{H}}_{i-1,i} = \begin{bmatrix} & \dot{\mathbf{A}}_{i-1,i} & & \mathbf{v}_{i-1,i}^{(i-1)} \\ 0 & 0 & 0 & 0 \end{bmatrix} \tag{3.25}$$

for each $i = 1, 2, \ldots, n$, while the derivative of the end-effector matrix is

$$\dot{\mathbf{H}}_{n,n+1} = \mathbf{0}, \tag{3.26}$$

as it is described by (3.12).

Let us first observe the time derivative of the homogeneous transformation matrix $\mathbf{H}_{0,i}$. In accordance with (3.9) we have

$$\dot{\mathbf{H}}_{0,i} = \begin{bmatrix} & \dot{\mathbf{A}}_{0,i} & & \mathbf{v}_{0,i}^{(0)} \\ 0 & 0 & 0 & 0 \end{bmatrix}. \tag{3.27}$$

The recursive equation, working from the base towards the end-point of the mechanism is given by

$$\dot{\mathbf{H}}_{0,i} = \dot{\mathbf{H}}_{0,i-1}\mathbf{H}_{i-1,i} + \mathbf{H}_{0,i-1}\dot{\mathbf{H}}_{i-1,i}, \quad i = 2, 3, \ldots, n. \tag{3.28}$$

How to use the above equation becomes clear after writing out a few of the first steps

$$\begin{aligned}
\dot{\mathbf{H}}_{0,2} &= \dot{\mathbf{H}}_{0,1}\mathbf{H}_{1,2} + \mathbf{H}_{0,1}\dot{\mathbf{H}}_{1,2}, \\
\dot{\mathbf{H}}_{0,3} &= \dot{\mathbf{H}}_{0,2}\mathbf{H}_{2,3} + \mathbf{H}_{0,2}\dot{\mathbf{H}}_{2,3}, \\
\dot{\mathbf{H}}_{0,4} &= \dot{\mathbf{H}}_{0,3}\mathbf{H}_{3,4} + \mathbf{H}_{0,3}\dot{\mathbf{H}}_{3,4}, \\
&\vdots
\end{aligned}$$

The calculation results in a set of translational velocities $\mathbf{v}_{0,2}^{(0)}, \mathbf{v}_{0,3}^{(0)}, \ldots, \mathbf{v}_{0,n+1}^{(0)}$ and time derivatives of the rotation matrices $\dot{\mathbf{A}}_{0,1}, \dot{\mathbf{A}}_{0,2}, \ldots, \dot{\mathbf{A}}_{0,n}$, which allows us to also calculate the angular velocities $\boldsymbol{\omega}_{0,1}^{(0)}, \boldsymbol{\omega}_{0,2}^{(0)}, \ldots, \boldsymbol{\omega}_{0,n}^{(0)}$, as

$$\boldsymbol{\omega}_{0,i}^{(0)} = \operatorname{vect}\big(\dot{\mathbf{A}}_{0,i}\mathbf{A}_{0,i}^{\mathrm{T}}\big), \quad i = 2, 3, \ldots, n. \tag{3.29}$$

According to (3.15), the time derivative of the homogeneous transformation matrix $\mathbf{H}_{i-1,n+1}$ is given by

$$\dot{\mathbf{H}}_{i-1,n+1} = \begin{bmatrix} & \dot{\mathbf{A}}_{i-1,n} & & \mathbf{v}_{i-1,n+1}^{(i-1)} \\ 0 & 0 & 0 & 0 \end{bmatrix}. \tag{3.30}$$

The recursive equation, working from the end-point towards the base, is

$$\dot{\mathbf{H}}_{i-1,n+1} = \dot{\mathbf{H}}_{i-1,i}\mathbf{H}_{i,n+1} + \mathbf{H}_{i-1,i}\dot{\mathbf{H}}_{i,n+1}, \quad i = n, n-1, \ldots, 1. \tag{3.31}$$

It is calculated in sequential steps

$$\begin{aligned}
\dot{\mathbf{H}}_{n-1,n+1} &= \dot{\mathbf{H}}_{n-1,n}\mathbf{H}_{n,n+1}, \\
\dot{\mathbf{H}}_{n-2,n+1} &= \dot{\mathbf{H}}_{n-2,n-1}\mathbf{H}_{n-1,n+1} + \mathbf{H}_{n-2,n-1}\dot{\mathbf{H}}_{n-1,n+1}, \\
\dot{\mathbf{H}}_{i-3,n+1} &= \dot{\mathbf{H}}_{i-3,n-2}\mathbf{H}_{n-2,n+1} + \mathbf{H}_{n-3,n-2}\dot{\mathbf{H}}_{n-2,n+1}, \\
&\vdots
\end{aligned}$$

The result is a set of translational velocities $\mathbf{v}_{n-1,n+1}^{(n-1)}, \mathbf{v}_{n-2,n+1}^{(n-2)}, \ldots, \mathbf{v}_{0,n+1}^{(0)}$ and time derivatives of the rotation matrices $\dot{\mathbf{A}}_{n-2,n}, \dot{\mathbf{A}}_{n-3,n}, \ldots, \dot{\mathbf{A}}_{0,n}$, from which we can obtain the angular velocities $\boldsymbol{\omega}_{n-2,n}^{(n-2)}, \boldsymbol{\omega}_{n-3,n}^{(n-3)}, \ldots, \boldsymbol{\omega}_{0,n}^{(0)}$, as

$$\boldsymbol{\omega}_{i-1,n}^{(i-1)} = \operatorname{vect}\big(\dot{\mathbf{A}}_{i-1,n}\mathbf{A}_{i-1,n}^{\mathrm{T}}\big), \quad i = n-1, n-2, \ldots, 1. \tag{3.32}$$

Let us recap the variables which define the angular and translational velocities of the mechanism:

$\boldsymbol{\omega}_{i-1,i}^{(i-1)}(\dot{\theta}_i)$: angular velocity vector between links $i-1$ and i. It is calculated by formula (3.16).

$\boldsymbol{\omega}_{0,i}^{(0)}(\theta_1, \ldots, \theta_{i-1}, \dot{\theta}_1, \ldots, \dot{\theta}_i)$: angular velocity vector between links 0 and i. It is calculated by the use of recursive formula (3.18), working in steps from the base to the robot's end-point. Prior to this, the rotation matrices (3.2) and angular velocities (3.16) must be calculated.

$\boldsymbol{\omega}_{i-1,n}^{(i-1)}(\theta_i, \ldots, \theta_{n-1}, \dot{\theta}_i, \ldots, \dot{\theta}_n)$: angular velocity vector between links $i-1$ and n. It is calculated by the recursive expression (3.19), working in steps from the end-point towards the base of the mechanism. Prior to this, the rotation matrices (3.1) and angular velocities (3.16) must be calculated.

$\mathbf{v}_{i-1,i}^{(i-1)}(\dot{d}_i)$: translational velocity vector of the center of joint i with respect to the joint center $i-1$. It is calculated by formula (3.20).

$\mathbf{v}_{0,i}^{(0)}(\theta_1, \ldots, \theta_{i-1}, d_2, \ldots, d_i, \dot{\theta}_1, \ldots, \dot{\theta}_{i-1}, \dot{d}_1, \ldots, \dot{d}_i)$: translational velocity vector of the joint center i with respect to the base. It is calculated by recursive formula (3.21), working in steps from the base towards the end-point of the mechanism. Prior to this, the rotation matrices (3.2), position vectors (3.4), angular velocities (3.18), and translational velocities (3.20) must be calculated.

$\mathbf{v}_{i-1,n+1}^{(i)}(\theta_i, \ldots, \theta_n, d_{i+1}, \ldots, d_n, \dot{\theta}_i, \ldots, \dot{\theta}_n, \dot{d}_i, \ldots, \dot{d}_n)$: translational velocity vector of the robot's end-point with respect to the joint center $i-1$. It is calculated by recursive formula (3.24), working in steps from the end-point to the base of the mechanism. Beforehand the rotation matrices (3.1), position vectors (3.7), angular velocities (3.16) and translational velocities (3.20) should be calculated.

Let us consider the example of the four degree of freedom mechanism whose kinematic parameters are shown in Fig. 2.20. Positions and orientations of the mechanism's links were already calculated in the preceding section. With the use of the expressions developed in this section we will calculate the angular and translational velocities of the mechanism. First we find the angular velocities between the neighboring links of the mechanism (3.16) and the translational velocities between the centers of the neighboring joints (3.20) and write them into Table 3.2.

The angular velocity of the first link is described by the vector $\boldsymbol{\omega}_{0,1}^{(0)}$, which is presented in Table 3.2, while the following vector belongs to the angular velocity of the second link

$$\boldsymbol{\omega}_{0,2}^{(0)} = \boldsymbol{\omega}_{0,1}^{(0)} + \mathbf{A}_{0,1}\boldsymbol{\omega}_{1,2}^{(1)} = \boldsymbol{\omega}_{0,1}^{(0)} + \begin{bmatrix} c_1\dot{\theta}_2 \\ s_1\dot{\theta}_2 \\ 0 \end{bmatrix}.$$

Since the joint between links 2 and 3 has only a translation variable, these links must have the same angular velocity,

$$\boldsymbol{\omega}_{0,3}^{(0)} = \boldsymbol{\omega}_{0,2}^{(0)},$$

while the end-effector link rotates with the following angular velocity

$$\boldsymbol{\omega}_{0,4}^{(0)} = \boldsymbol{\omega}_{0,3}^{(0)} + \mathbf{A}_{0,3}\boldsymbol{\omega}_{3,4}^{(3)} = \boldsymbol{\omega}_{0,3}^{(0)} + \begin{bmatrix} s_1 s_2\dot{\theta}_4 \\ -c_1 s_2\dot{\theta}_4 \\ c_2\dot{\theta}_4 \end{bmatrix}.$$

Table 3.2 Angular velocities between neighboring links and translational velocities between the centers of neighboring joints of the mechanism from Fig. 2.20

i	1	2	3	4
$\boldsymbol{\omega}_{i-1,i}^{(i-1)}$	0	$\dot{\theta}_2$	0	0
	0	0	0	0
	$\dot{\theta}_1$	0	0	$\dot{\theta}_4$

i	1	2	3	4	5
$\mathbf{v}_{i-1,i}^{(i-1)}$	0	0	0	0	0
	0	0	$\dot{d}_3$	0	0
	0	0	0	0	0

Now we find the relations between the angular velocity of the end-effector link and the time derivatives of the Euler angles. We have already shown in (1.115) that the relation is linear

$$\boldsymbol{\omega}_{0,4}^{(0)} = \mathbf{W} \begin{bmatrix} \dot{\psi} \\ \dot{\theta} \\ \dot{\phi} \end{bmatrix}.$$

In our case we have

$$\mathbf{W} = \begin{bmatrix} 0 & c_1 & s_1 s_2 \\ 0 & s_1 & -c_1 s_2 \\ 1 & 0 & c_2 \end{bmatrix}.$$

The translational velocity of the center of the first joint is given by the vector $\mathbf{v}_{0,1}^{(0)}$ in Table 3.2, while the translational velocity of the center of the second joint is

$$\mathbf{v}_{0,2}^{(0)} = \mathbf{v}_{0,1}^{(0)} + \boldsymbol{\omega}_{0,1}^{(0)} \times \mathbf{A}_{0,1}\mathbf{r}_{1,2}^{(1)} + \mathbf{A}_{0,1}\mathbf{v}_{1,2}^{(1)} = \mathbf{v}_{0,1}^{(0)} + l_1 \begin{bmatrix} -c_1\dot{\theta}_1 \\ -s_1\dot{\theta}_1 \\ 0 \end{bmatrix}.$$

The velocity of the center of the third joint is

$$\begin{aligned} \mathbf{v}_{0,3}^{(0)} &= \mathbf{v}_{0,2}^{(0)} + \boldsymbol{\omega}_{0,2}^{(0)} \times \mathbf{A}_{0,2}\mathbf{r}_{2,3}^{(2)} + \mathbf{A}_{0,2}\mathbf{v}_{2,3}^{(2)} \\ &= \mathbf{v}_{0,2}^{(0)} + \begin{bmatrix} -c_1 c_2 (l_2 + d_3)\dot{\theta}_1 + s_1 s_2 (l_2 + d_3)\dot{\theta}_2 - s_1 c_2 \dot{d}_3 \\ -s_1 c_2 (l_2 + d_3)\dot{\theta}_1 - c_1 s_2 (l_2 + d_3)\dot{\theta}_2 + c_1 c_2 \dot{d}_3 \\ c_2 (l_2 + d_3)\dot{\theta}_2 + s_2 \dot{d}_3 \end{bmatrix}, \end{aligned}$$

while the velocity of the fourth joint center is

$$\begin{aligned} \mathbf{v}_{0,4}^{(0)} &= \mathbf{v}_{0,3}^{(0)} + \boldsymbol{\omega}_{0,3}^{(0)} \times \mathbf{A}_{0,3}\mathbf{r}_{3,4}^{(3)} + \mathbf{A}_{0,3}\mathbf{v}_{3,4}^{(3)} \\ &= \mathbf{v}_{0,3}^{(0)} + \begin{bmatrix} -(c_1 c_2 l_3 + c_1 s_2 h_3)\dot{\theta}_1 + (s_1 s_2 l_3 - s_1 c_2 h_3)\dot{\theta}_2 \\ -(s_1 c_2 l_3 + s_1 s_2 h_3)\dot{\theta}_1 - (c_1 s_2 l_3 - c_1 c_2 h_3)\dot{\theta}_2 \\ (c_2 l_3 + s_2 h_3)\dot{\theta}_2 \end{bmatrix}. \end{aligned}$$

Finally, we calculate the translational velocity of the robot's end-point. Taking into account that $\mathbf{v}_{4,5}^{(4)} = 0$, we obtain

$$\mathbf{v}_{0,5}^{(0)} = \mathbf{v}_{0,4}^{(0)} + \boldsymbol{\omega}_{0,4}^{(0)} \times \mathbf{A}_{0,4}\mathbf{r}_{4,5}^{(4)}$$

$$= \mathbf{v}_{0,4}^{(0)} + l_4 \begin{bmatrix} (s_1s_4 - c_1c_2c_4)\dot{\theta}_1 + s_1s_2c_4\dot{\theta}_2 - (c_1c_4 - s_1c_2s_4)\dot{\theta}_4 \\ -(c_1s_4 + s_1c_2c_4)\dot{\theta}_1 - c_1s_2c_4\dot{\theta}_2 - (s_1c_4 + c_1c_2s_4)\dot{\theta}_4 \\ c_2c_4\dot{\theta}_2 - s_2s_4\dot{\theta}_4 \end{bmatrix}.$$

In the above equations we have intentionally exposed the time derivatives of translational and rotational coordinates.

3.1.3 Angular and Translational Accelerations of a Mechanism

Having developed formulas for computing the orientations and positions of a mechanism's links, together with their angular and translational velocities, all that remains to be dealt with are the link's angular and translational accelerations. Higher time derivatives will not be considered, although they have practical significance, specifically in designing control systems for robot mechanisms.

The so called angular acceleration matrix between links $i-1$ and i is defined in (1.123), by use of the angular velocity matrix

$$\boldsymbol{\Psi}_{i-1,i} = \dot{\boldsymbol{\Omega}}_{i-1,i} + \boldsymbol{\Omega}_{i-1,i}^2.$$

The time derivative of the angular velocity matrix can be written as

$$\dot{\boldsymbol{\Omega}}_{i-1,i} = \ddot{\theta}_i \mathbf{e}_i^{(i-1)} \otimes \mathbf{I} = \ddot{\theta}_i \boldsymbol{\Delta}_{i-1,i},$$

and the square of angular velocity matrix can be simplified in the following way

$$\boldsymbol{\Omega}_{i-1,i}^2 = \dot{\theta}_i^2 \left(\mathbf{e}_i^{(i-1)} \otimes \mathbf{I}\right)^2 = \dot{\theta}_i^2 \boldsymbol{\Delta}_{i-1,i}^2 = \dot{\theta}_i^2 (\mathbf{I} - \boldsymbol{\Lambda}_{i-1,i}).$$

We already know the properties of the matrices $\boldsymbol{\Delta}_{i-1,i}$ and $\boldsymbol{\Lambda}_{i-1,i}$, as they were introduced in Sect. 1.3.1.

The angular acceleration is defined as the vector invariant of the matrix $\boldsymbol{\Psi}_{i-1,i}$. As the vector invariant of the symmetrical part of the matrix equals zero, the angular acceleration belongs only to the non-symmetrical part, i.e. the time derivative of the angular velocity matrix

$$\mathbf{u}_{i-1,i}^{(i-1)} = \text{vect}(\dot{\boldsymbol{\Omega}}_{i-1,i}) = \ddot{\theta}_i \,\text{vect}\left(\mathbf{e}_i^{(i-1)} \otimes \mathbf{I}\right) = \ddot{\theta}_i \,\text{vect}(\boldsymbol{\Delta}_{i-1,i}).$$

As the vector invariant of the matrix $\boldsymbol{\Delta}_{i-1,i}$ is equal to the rotational vector $\mathbf{e}_i^{(i-1)}$, the angular acceleration between links $i-1$ and i is

$$\mathbf{u}_{i-1,i}^{(i-1)} = \ddot{\theta}_i \mathbf{e}_i^{(i-1)}. \tag{3.33}$$

This relation holds for all $i = 1, 2, \ldots, n$. By adding the relation between the angular acceleration and the second time derivative of the rotation matrix

$$\ddot{\mathbf{A}}_{i-1,i} = \frac{d}{dt}\dot{\mathbf{A}}_{i-1,i} = \frac{d}{dt}\left(\boldsymbol{\omega}_{i-1,i}^{(i-1)} \otimes \mathbf{A}_{i-1,i}\right),$$

we obtain

$$\ddot{\mathbf{A}}_{i-1,i} = \mathbf{u}_{i-1,i}^{(i-1)} \otimes \mathbf{A}_{i-1,i} + \boldsymbol{\omega}_{i-1,i}^{(i-1)} \otimes \left(\boldsymbol{\omega}_{i-1,i}^{(i-1)} \otimes \mathbf{A}_{i-1,i}\right). \tag{3.34}$$

Thus the angular acceleration between links i and j, where $j > i$, is represented by the sum of time derivatives

$$\mathbf{u}_{i,j}^{(i)} = \dot{\boldsymbol{\omega}}_{i,i+1}^{(i)} + \dot{\boldsymbol{\omega}}_{i+1,i+2}^{(i)} + \cdots + \dot{\boldsymbol{\omega}}_{j-1,j}^{(i)}.$$

This relation can be used when calculating the angular acceleration of link i with respect to the fixed reference frame

$$\begin{aligned}\mathbf{u}_{0,i}^{(0)} = \dot{\boldsymbol{\omega}}_{0,i}^{(0)} &= \dot{\boldsymbol{\omega}}_{0,1}^{(0)} + \dot{\mathbf{A}}_{0,1}\boldsymbol{\omega}_{1,2}^{(1)} + \mathbf{A}_{0,1}\dot{\boldsymbol{\omega}}_{1,2}^{(1)} + \dot{\mathbf{A}}_{0,2}\boldsymbol{\omega}_{2,3}^{(2)} \\ &+ \mathbf{A}_{0,2}\dot{\boldsymbol{\omega}}_{2,3}^{(2)} + \cdots + \dot{\mathbf{A}}_{0,i-1}\boldsymbol{\omega}_{i-1,i}^{(i-1)} + \mathbf{A}_{0,i-1}\dot{\boldsymbol{\omega}}_{i-1,i}^{(i-1)}.\end{aligned}$$

From here we obtain

$$\begin{aligned}\mathbf{u}_{0,i}^{(0)} &= \mathbf{u}_{0,1}^{(0)} + \boldsymbol{\omega}_{0,1}^{(0)} \times \mathbf{A}_{0,1}\boldsymbol{\omega}_{1,2}^{(1)} + \mathbf{A}_{0,1}\mathbf{u}_{1,2}^{(1)} + \boldsymbol{\omega}_{0,2}^{(0)} \times \mathbf{A}_{0,2}\boldsymbol{\omega}_{2,3}^{(2)} \\ &+ \mathbf{A}_{0,2}\mathbf{u}_{2,3}^{(2)} + \cdots + \boldsymbol{\omega}_{0,i-1}^{(0)} \times \mathbf{A}_{0,i-1}\boldsymbol{\omega}_{i-1,i}^{(i-1)} + \mathbf{A}_{0,i-1}\mathbf{u}_{i-1,i}^{(i-1)}.\end{aligned}$$

This can be reformulated into the following sequential steps

$$\begin{aligned}\mathbf{u}_{0,2}^{(0)} &= \mathbf{u}_{0,1}^{(0)} + \boldsymbol{\omega}_{0,1}^{(0)} \times \mathbf{A}_{0,1}\boldsymbol{\omega}_{1,2}^{(1)} + \mathbf{A}_{0,1}\mathbf{u}_{1,2}^{(1)}, \\ \mathbf{u}_{0,3}^{(0)} &= \mathbf{u}_{0,2}^{(0)} + \boldsymbol{\omega}_{0,2}^{(0)} \times \mathbf{A}_{0,2}\boldsymbol{\omega}_{2,3}^{(2)} + \mathbf{A}_{0,2}\mathbf{u}_{2,3}^{(2)}, \\ \mathbf{u}_{0,4}^{(0)} &= \mathbf{u}_{0,3}^{(0)} + \boldsymbol{\omega}_{0,3}^{(0)} \times \mathbf{A}_{0,3}\boldsymbol{\omega}_{3,4}^{(3)} + \mathbf{A}_{0,3}\mathbf{u}_{3,4}^{(3)}, \\ &\vdots\end{aligned}$$

which can be combined into the following recursive equation

$$\begin{aligned}&\mathbf{u}_{0,i}^{(0)} = \mathbf{u}_{0,i-1}^{(0)} + \boldsymbol{\omega}_{0,i-1}^{(0)} \times \mathbf{A}_{0,i-1}\boldsymbol{\omega}_{i-1,i}^{(i-1)} + \mathbf{A}_{0,i-1}\mathbf{u}_{i-1,i}^{(i-1)}, \\ &i = 2, 3, \ldots, n.\end{aligned} \tag{3.35}$$

The calculation proceeds from the base towards the robot's end-point. Sequentially, we can calculate the angular accelerations of all the links of a mechanism. As we see, the angular acceleration of link i is product of angular accelerations, contributed by the rotational variables in the joints $1, 2, \ldots, i$, and angular velocities contributed by the same rotational variables. The angular acceleration of the end-effector link is

$\mathbf{u}_{0,n}^{(0)}$. When the orientation of the end-effector is given by the orientation angles, the angular acceleration can be expressed in terms of the first and second time derivatives of the orientation angles, as shown in Sect. 1.7.

The time derivatives of the angular velocities, working in the opposite direction (3.19), are as follows

$$
\begin{aligned}
\dot{\boldsymbol{\omega}}_{n-2,n}^{(n-2)} &= \dot{\boldsymbol{\omega}}_{n-2,n-1}^{(n-2)} + \dot{\mathbf{A}}_{n-2,n-1}\boldsymbol{\omega}_{n-1,n}^{(n-1)} + \mathbf{A}_{n-2,n-1}\dot{\boldsymbol{\omega}}_{n-1,n}^{(n-1)},\\
\dot{\boldsymbol{\omega}}_{n-3,n}^{(n-3)} &= \dot{\boldsymbol{\omega}}_{n-3,n-2}^{(n-3)} + \dot{\mathbf{A}}_{n-3,n-2}\boldsymbol{\omega}_{n-2,n}^{(n-2)} + \mathbf{A}_{n-3,n-2}\dot{\boldsymbol{\omega}}_{n-2,n}^{(n-2)},\\
\dot{\boldsymbol{\omega}}_{n-4,n}^{(n-4)} &= \dot{\boldsymbol{\omega}}_{n-4,n-3}^{(n-4)} + \dot{\mathbf{A}}_{n-4,n-3}\boldsymbol{\omega}_{n-3,n}^{(n-3)} + \mathbf{A}_{n-4,n-3}\dot{\boldsymbol{\omega}}_{n-3,n}^{(n-3)},\\
&\vdots
\end{aligned}
$$

which can be rewritten in the following form

$$
\begin{aligned}
\mathbf{u}_{n-2,n}^{(n-2)} &= \mathbf{u}_{n-2,n-1}^{(n-2)} + \mathbf{A}_{n-2,n-1}\big(\mathbf{u}_{n-1,n}^{(n)} + \boldsymbol{\omega}_{n-2,n-1}^{(n-2)} \times \boldsymbol{\omega}_{n-1,n}^{(n-1)}\big),\\
\mathbf{u}_{n-3,n}^{(n-3)} &= \mathbf{u}_{n-3,n-2}^{(n-3)} + \mathbf{A}_{n-3,n-2}\big(\mathbf{u}_{n-2,n}^{(n)} + \boldsymbol{\omega}_{n-3,n-2}^{(n-3)} \times \boldsymbol{\omega}_{n-2,n}^{(n-2)}\big),\\
\mathbf{u}_{n-4,n}^{(n-4)} &= \mathbf{u}_{n-4,n-3}^{(n-4)} + \mathbf{A}_{n-4,n-3}\big(\mathbf{u}_{n-3,n}^{(n)} + \boldsymbol{\omega}_{n-4,n-3}^{(n-4)} \times \boldsymbol{\omega}_{n-3,n}^{(n-3)}\big),\\
&\vdots
\end{aligned}
$$

Finally, we combine all above the equations into the following compressed recursive formula

$$
\begin{aligned}
&\mathbf{u}_{i-1,n}^{(i-1)} = \mathbf{u}_{i-1,i}^{(i-1)} + \mathbf{A}_{i-1,i}\big(\mathbf{u}_{i,n}^{(i)} + \boldsymbol{\omega}_{i-1,i}^{(i-1)} \times \boldsymbol{\omega}_{i,n}^{(i)}\big),\\
&i = n-1, n-2, \ldots, 1,
\end{aligned}
\tag{3.36}
$$

which can be solved in sequential steps, working from the robot's end-point towards its base.

We can also calculate the translational accelerations as time derivatives of the translational velocities of a mechanism. Let us first determine the translational acceleration between the centers of two neighboring joints $i-1$ and i. We proceed from (3.20). The result of time differentiation is as expected. The translational acceleration between the neighboring joint centers is a consequence of the acceleration of the intermediate translational coordinate

$$
\mathbf{a}_{i-1,i}^{(i-1)} = \dot{\mathbf{v}}_{i-1,i}^{(i-1)} = \ddot{d}_i \mathbf{e}_i^{(i-1)}. \tag{3.37}
$$

This equation is valid for all $i = 1, 2, \ldots, n$, while $\mathbf{a}_{n,n+1}^{(n)} = 0$. The translational acceleration between the joint centers i and j, where $j > i$, equals the sum of time derivatives of the translational velocities

$$
\mathbf{a}_{i,j}^{(i)} = \dot{\mathbf{v}}_{i,i+1}^{(i)} + \dot{\mathbf{v}}_{i+1,i+2}^{(i)} + \cdots + \dot{\mathbf{v}}_{j-1,j}^{(i)}.
$$

Such a formula is not very useful, as we do not know the time derivatives of the velocities. Instead of developing a general solution, we direct our attention to calculating the translational acceleration of the joint center i with respect to the mechanism's base. The calculation becomes complicated, as

$$\begin{aligned}\mathbf{a}_{0,i}^{(0)} = \dot{\mathbf{v}}_{0,i}^{(0)} &= \mathbf{a}_{0,1}^{(0)} + \ddot{\mathbf{A}}_{0,1}\mathbf{r}_{1,2}^{(1)} \\ &+ 2\dot{\mathbf{A}}_{0,1}\mathbf{v}_{1,2}^{(1)} + \mathbf{A}_{0,1}\mathbf{a}_{1,2}^{(1)} + \ddot{\mathbf{A}}_{0,2}\mathbf{r}_{2,3}^{(2)} + 2\dot{\mathbf{A}}_{0,2}\mathbf{v}_{2,3}^{(2)} + \mathbf{A}_{0,2}\mathbf{a}_{2,3}^{(2)} \\ &+ \cdots + \ddot{\mathbf{A}}_{0,i-1}\mathbf{r}_{i-1,i}^{(i-1)} + 2\dot{\mathbf{A}}_{0,i-1}\mathbf{v}_{i-1,i}^{(i-1)} + \mathbf{A}_{0,i-1}\mathbf{a}_{i-1,i}^{(i-1)}.\end{aligned}$$

We know the second time derivative of the rotation matrix depends on both angular acceleration and angular velocity

$$\ddot{\mathbf{A}}_{0,k} = \mathbf{u}_{0,k}^{(0)} \otimes \mathbf{A}_{0,k} + \boldsymbol{\omega}_{0,k}^{(0)} \otimes \left(\boldsymbol{\omega}_{0,k}^{(0)} \otimes \mathbf{A}_{0,k}\right).$$

This equation, together with the first time derivative of the rotation matrix $\dot{\mathbf{A}}_{0,k} = \boldsymbol{\omega}_{0,k}^{(0)} \otimes \mathbf{A}_{0,k}$ is inserted into the preceding equation

$$\begin{aligned}\mathbf{a}_{0,i}^{(0)} &= \mathbf{a}_{0,1}^{(0)} + \mathbf{u}_{0,1}^{(0)} \times \mathbf{A}_{0,1}\mathbf{r}_{1,2}^{(1)} + \boldsymbol{\omega}_{0,1}^{(0)} \times \left(\boldsymbol{\omega}_{0,1}^{(0)} \times \mathbf{A}_{0,1}\mathbf{r}_{1,2}^{(1)}\right) \\ &+ 2\boldsymbol{\omega}_{0,1}^{(0)} \times \mathbf{A}_{0,1}\mathbf{v}_{1,2}^{(1)} + \mathbf{A}_{0,1}\mathbf{a}_{1,2}^{(1)} \\ &+ \mathbf{u}_{0,2}^{(0)} \times \mathbf{A}_{0,2}\mathbf{r}_{2,3}^{(2)} + \boldsymbol{\omega}_{0,2}^{(0)} \times \left(\boldsymbol{\omega}_{0,2}^{(0)} \times \mathbf{A}_{0,2}\mathbf{r}_{2,3}^{(2)}\right) \\ &+ 2\boldsymbol{\omega}_{0,2}^{(0)} \times \mathbf{A}_{0,2}\mathbf{v}_{2,3}^{(2)} + \mathbf{A}_{0,2}\mathbf{a}_{2,3}^{(2)} + \cdots \\ &+ \mathbf{u}_{0,i-1}^{(0)} \times \mathbf{A}_{0,i-1}\mathbf{r}_{i-1,i}^{(i-1)} + \boldsymbol{\omega}_{0,i-1}^{(0)} \times \left(\boldsymbol{\omega}_{0,i-1}^{(0)} \times \mathbf{A}_{0,i-1}\mathbf{r}_{i-1,i}^{(i-1)}\right) \\ &+ 2\boldsymbol{\omega}_{0,i-1}^{(0)} \times \mathbf{A}_{0,i-1}\mathbf{v}_{i-1,i}^{(i-1)} + \mathbf{A}_{0,i-1}\mathbf{a}_{i-1,i}^{(i-1)}.\end{aligned}$$

This long expression can be written in terms of sequential steps, working from the base towards the robot's end-point, in the following way

$$\begin{aligned}\mathbf{a}_{0,2}^{(0)} &= \mathbf{a}_{0,1}^{(0)} + \mathbf{u}_{0,1}^{(0)} \times \mathbf{A}_{0,1}\mathbf{r}_{1,2}^{(1)} + \boldsymbol{\omega}_{0,1}^{(0)} \times \left(\boldsymbol{\omega}_{0,1}^{(0)} \times \mathbf{A}_{0,1}\mathbf{r}_{1,2}^{(1)}\right) \\ &\quad + 2\boldsymbol{\omega}_{0,1}^{(0)} \times \mathbf{A}_{0,1}\mathbf{v}_{1,2}^{(1)} + \mathbf{A}_{0,1}\mathbf{a}_{1,2}^{(1)}, \\ \mathbf{a}_{0,3}^{(0)} &= \mathbf{a}_{0,2}^{(0)} + \mathbf{u}_{0,2}^{(0)} \times \mathbf{A}_{0,2}\mathbf{r}_{2,3}^{(2)} + \boldsymbol{\omega}_{0,2}^{(0)} \times \left(\boldsymbol{\omega}_{0,2}^{(0)} \times \mathbf{A}_{0,2}\mathbf{r}_{2,3}^{(2)}\right) \\ &\quad + 2\boldsymbol{\omega}_{0,2}^{(0)} \times \mathbf{A}_{0,2}\mathbf{v}_{2,3}^{(2)} + \mathbf{A}_{0,2}\mathbf{a}_{2,3}^{(2)}, \\ \mathbf{a}_{0,4}^{(0)} &= \mathbf{a}_{0,3}^{(0)} + \mathbf{u}_{0,3}^{(0)} \times \mathbf{A}_{0,3}\mathbf{r}_{3,4}^{(3)} + \boldsymbol{\omega}_{0,3}^{(0)} \times \left(\boldsymbol{\omega}_{0,3}^{(0)} \times \mathbf{A}_{0,3}\mathbf{r}_{3,4}^{(3)}\right) \\ &\quad + 2\boldsymbol{\omega}_{0,3}^{(0)} \times \mathbf{A}_{0,3}\mathbf{v}_{3,4}^{(3)} + \mathbf{A}_{0,3}\mathbf{a}_{3,4}^{(3)}, \\ &\vdots\end{aligned}$$

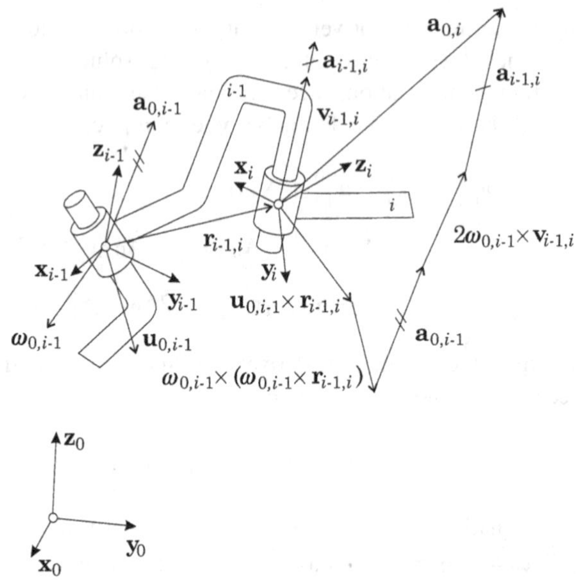

Fig. 3.6 Vector of translational acceleration of the joint center i with respect to the base of mechanism

The sequence of the steps is developed in such a way that the result of the preceding step represents the input data for the next step. The steps can be combined into the following compressed recursive equation

$$\mathbf{a}_{0,i}^{(0)} = \mathbf{a}_{0,i-1}^{(0)} + \mathbf{u}_{0,i-1}^{(0)} \times \mathbf{A}_{0,i-1}\mathbf{r}_{i-1,i}^{(i-1)} + \boldsymbol{\omega}_{0,i-1}^{(0)} \times \left(\boldsymbol{\omega}_{0,i-1}^{(0)} \times \mathbf{A}_{0,i-1}\mathbf{r}_{i-1,i}^{(i-1)}\right)$$
$$+ 2\boldsymbol{\omega}_{0,i-1}^{(0)} \times \mathbf{A}_{0,i-1}\mathbf{v}_{i-1,i}^{(i-1)} + \mathbf{A}_{0,i-1}\mathbf{a}_{i-1,i}^{(i-1)}, \quad i = 2, 3, \ldots, n. \tag{3.38}$$

The vectors in the above equation are shown in Fig. 3.6. Here, $\mathbf{a}_{0,i-1}$ represents the acceleration of the joint center $i-1$ with respect to the base of the mechanism, while $\mathbf{a}_{i-1,i}$ is the acceleration contributed by the translational coordinate of the joint i. We also recognize the tangential acceleration $\mathbf{u}_{0,i-1} \times \mathbf{r}_{i-1,i}$, encompassing the angular acceleration of link $i-1$ with respect to the basis ($\mathbf{u}_{0,i-1}$), radial acceleration $\boldsymbol{\omega}_{0,i-1} \times \boldsymbol{\omega}_{0,i-1} \times \mathbf{r}_{i-1,i}$, including the angular velocity of link $i-1$ with respect to the base ($\boldsymbol{\omega}_{0,i-1}$), and the Coriolis acceleration $2\boldsymbol{\omega}_{0,i-1} \times \mathbf{v}_{i-1,i}$, combining the angular velocity of link $i-1$ with respect to the base and the translational velocity $\mathbf{v}_{i-1,i}$, contributed by the translational coordinate of the i-th joint.

Consider the translational acceleration of an arbitrary point P on link i with respect to the base, as shown in Fig. 3.2. By differentiating the translational velocity, we obtain

$$\ddot{\mathbf{p}}_{0,i}^{(0)} = \ddot{\mathbf{r}}_{0,i}^{(0)} + \ddot{\mathbf{A}}_{0,i}\mathbf{p}_i^{(i)}.$$

In this equation we substitute for $\ddot{\mathbf{A}}_{0,i}$ and obtain the following

$$\ddot{\mathbf{p}}_{0,i}^{(0)} = \mathbf{a}_{0,i}^{(0)} + \mathbf{u}_{0,i}^{(0)} \times \mathbf{A}_{0,i}\mathbf{p}_i^{(i)} + \boldsymbol{\omega}_{0,i}^{(0)} \times \left(\boldsymbol{\omega}_{0,i}^{(0)} \times \mathbf{A}_{0,i}\mathbf{p}_i^{(i)}\right). \tag{3.39}$$

The translational acceleration of the center of the uppermost joint is determined by the vector $\mathbf{a}_{0,n}$, while the translational acceleration of the robot's end-point is given by $\mathbf{a}_{0,n+1}$, and is calculated as follows

$$\mathbf{a}_{0,n+1}^{(0)} = \mathbf{a}_{0,n}^{(0)} + \mathbf{u}_{0,n}^{(0)} \times \mathbf{A}_{0,n}\mathbf{r}_{n,n+1}^{(n)} + \boldsymbol{\omega}_{0,n}^{(0)} \times \left(\boldsymbol{\omega}_{0,n}^{(0)} \times \mathbf{A}_{0,n}\mathbf{r}_{n,n+1}^{(n)}\right). \tag{3.40}$$

We need to calculate the translational accelerations of the centers of particular joints with respect to the robot's end-point. They are conveniently found from the time derivatives of the translational velocities. By doing this sequentially the following expressions are obtained

$$\begin{aligned}
\mathbf{a}_{n-1,n+1}^{(n-1)} &= \mathbf{a}_{n-1,n}^{(n-1)} + \mathbf{A}_{n-1,n}\big(\mathbf{a}_{n,n+1}^{(n)} + \mathbf{u}_{n-1,n}^{(n-1)} \times \mathbf{r}_{n,n+1}^{(n)} \\
&\quad + \boldsymbol{\omega}_{n-1,n}^{(n-1)} \times \left(\boldsymbol{\omega}_{n-1,n}^{(n-1)} \times \mathbf{r}_{n,n+1}^{(n)}\right) + 2\boldsymbol{\omega}_{n-1,n}^{(n-1)} \times \mathbf{v}_{n,n+1}^{(n)}\big), \\
\mathbf{a}_{n-2,n+1}^{(n-2)} &= \mathbf{a}_{n-2,n-1}^{(n-2)} + \mathbf{A}_{n-2,n-1}\big(\mathbf{a}_{n-1,n+1}^{(n-1)} + \mathbf{u}_{n-2,n-1}^{(n-2)} \times \mathbf{r}_{n-1,n+1}^{(n-1)} \\
&\quad + \boldsymbol{\omega}_{n-2,n-1}^{(n-2)} \times \left(\boldsymbol{\omega}_{n-2,n-1}^{(n-2)} \times \mathbf{r}_{n-1,n+1}^{(n-1)}\right) + 2\boldsymbol{\omega}_{n-2,n-1}^{(n-2)} \times \mathbf{v}_{n-1,n+1}^{(n-1)}\big), \\
\mathbf{a}_{n-3,n+1}^{(n-3)} &= \mathbf{a}_{n-3,n-2}^{(n-3)} + \mathbf{A}_{n-3,n-2}\big(\mathbf{a}_{n-2,n+1}^{(n-2)} + \mathbf{u}_{n-3,n-2}^{(n-3)} \times \mathbf{r}_{n-2,n+1}^{(n-2)} \\
&\quad + \boldsymbol{\omega}_{n-3,n-2}^{(n-3)} \times \left(\boldsymbol{\omega}_{n-3,n-2}^{(n-3)} \times \mathbf{r}_{n-2,n+1}^{(n-2)}\right) + 2\boldsymbol{\omega}_{n-3,n-2}^{(n-3)} \times \mathbf{v}_{n-2,n+1}^{(n-2)}\big), \\
&\vdots
\end{aligned}$$

From the above, it is not difficult to see the following equation

$$\begin{aligned}
\mathbf{a}_{i-1,n+1}^{(i-1)} &= \mathbf{a}_{i-1,i}^{(i-1)} + \mathbf{A}_{i-1,i}\big(\mathbf{a}_{i,n+1}^{(i)} + \mathbf{u}_{i-1,i}^{(i-1)} \times \mathbf{r}_{i,n+1}^{(i)} \\
&\quad + \boldsymbol{\omega}_{i-1,i}^{(i-1)} \times \left(\boldsymbol{\omega}_{i-1,i}^{(i-1)} \times \mathbf{r}_{i,n+1}^{(i)}\right) + 2\boldsymbol{\omega}_{i-1,i}^{(i-1)} \times \mathbf{v}_{i,n+1}^{(i)}\big), \\
i &= n, n-1, \ldots, 1.
\end{aligned} \tag{3.41}$$

The Cartesian vectors in this equation are shown in Fig. 3.7. Here, $\mathbf{a}_{i,n+1}$ is the acceleration from the joint center i to the end-point and $\mathbf{a}_{i-1,i}$ is a component of acceleration contributed by the translational coordinate of the joint i. Between joint i and the robot's end-point there is also the tangential acceleration $\mathbf{u}_{i-1,i} \times \mathbf{r}_{i,n+1}$, radial acceleration $\boldsymbol{\omega}_{i-1,i} \times \boldsymbol{\omega}_{i-1,i} \times \mathbf{r}_{i,n+1}$, and Coriolis acceleration $2\boldsymbol{\omega}_{i-1,i} \times \mathbf{v}_{i,n+1}$.

Similar to our discussion of mechanism velocities, we shall again use the homogeneous transformation matrix (2.12) and now calculate its second time derivative

$$\ddot{\mathbf{H}}_{i-1,i} = \begin{bmatrix} & \ddot{\mathbf{A}}_{i-1,i} & & \mathbf{a}_{i-1,i}^{(i-1)} \\ 0 & 0 & 0 & 0 \end{bmatrix} \tag{3.42}$$

for each $i = 1, 2, \ldots, n$, while

$$\ddot{\mathbf{H}}_{n,n+1} = \mathbf{0}, \tag{3.43}$$

according to (3.12).

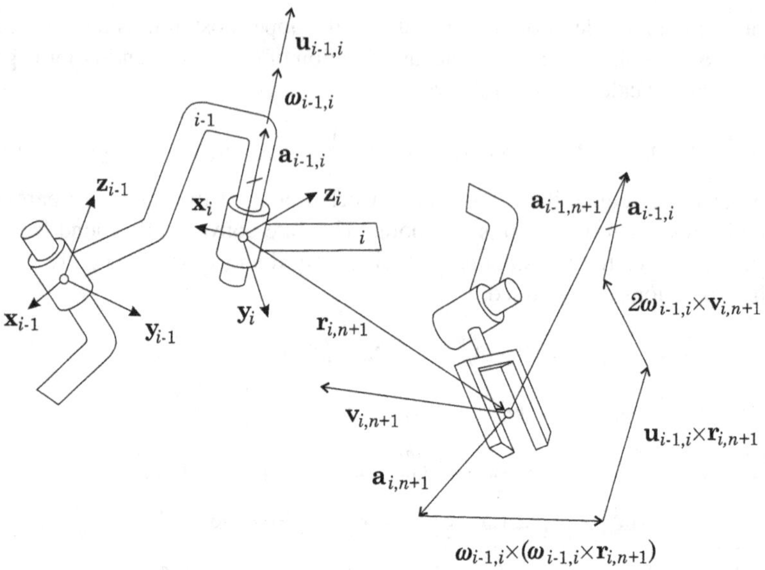

Fig. 3.7 Vector of translational acceleration of the robot's end-point with respect to joint center $i-1$

In a similar way we also write the time derivative of the homogeneous transformation matrix $\mathbf{H}_{0,i}$. In accordance with (3.9) this is given by

$$\ddot{\mathbf{H}}_{0,i} = \begin{bmatrix} & \ddot{\mathbf{A}}_{0,i} & & \mathbf{a}_{0,i}^{(0)} \\ 0 & 0 & 0 & 0 \end{bmatrix}. \tag{3.44}$$

Now the recursive equation, working from the base towards the robot's end-point, has the following form

$$\ddot{\mathbf{H}}_{0,i} = \ddot{\mathbf{H}}_{0,i-1}\mathbf{H}_{i-1,i} + 2\dot{\mathbf{H}}_{0,i-1}\dot{\mathbf{H}}_{i-1,i} + \mathbf{H}_{0,i-1}\ddot{\mathbf{H}}_{i-1,i}, \quad i = 2, 3, \ldots, n. \tag{3.45}$$

This equation is applied in such a way that the result of the current step represents the input for the next step

$$\begin{aligned}
\ddot{\mathbf{H}}_{0,2} &= \ddot{\mathbf{H}}_{0,1}\mathbf{H}_{1,2} + 2\dot{\mathbf{H}}_{0,1}\dot{\mathbf{H}}_{1,2} + \mathbf{H}_{0,1}\ddot{\mathbf{H}}_{1,2},\\
\ddot{\mathbf{H}}_{0,3} &= \ddot{\mathbf{H}}_{0,2}\mathbf{H}_{2,3} + 2\dot{\mathbf{H}}_{0,2}\dot{\mathbf{H}}_{2,3} + \mathbf{H}_{0,2}\ddot{\mathbf{H}}_{2,3},\\
\ddot{\mathbf{H}}_{0,4} &= \ddot{\mathbf{H}}_{0,3}\mathbf{H}_{3,4} + 2\dot{\mathbf{H}}_{0,3}\dot{\mathbf{H}}_{3,4} + \mathbf{H}_{0,3}\ddot{\mathbf{H}}_{3,4},\\
&\vdots
\end{aligned}$$

Here, we shall assume that the homogeneous transformation matrices $\mathbf{H}_{0,1}$, $\mathbf{H}_{0,2}$, $\ldots$, $\mathbf{H}_{0,i}$ were calculated in advance together with their time derivatives. The results of these calculations can be presented in terms of the translational accelerations $\mathbf{a}_{0,2}^{(0)}$, $\mathbf{a}_{0,3}^{(0)}, \ldots, \mathbf{a}_{0,n+1}^{(0)}$ and the second time derivatives of the rotation matrices $\mathbf{A}_{0,1}$, $\mathbf{A}_{0,2}$, $\ldots$, $\mathbf{A}_{0,n}$. The angular accelerations are obtained by the use of the vector invariant $\mathbf{u}_{0,i}^{(0)} = \mathrm{vect}(\boldsymbol{\Psi}_{0,i})$. As the vector invariant of the symmetrical part of the matrix equals zero, it is sufficient to calculate $\mathbf{u}_{0,i}^{(0)} = \mathrm{vect}(\dot{\boldsymbol{\Omega}}_{0,i})$. This last expression can be further simplified, as

$$\dot{\boldsymbol{\Omega}}_{0,i} = \frac{d}{dt}\left(\dot{\mathbf{A}}_{0,i}\mathbf{A}_{0,i}^{\mathrm{T}}\right) = \ddot{\mathbf{A}}_{0,i}\mathbf{A}_{0,i}^{\mathrm{T}} + \dot{\mathbf{A}}_{0,i}\dot{\mathbf{A}}_{0,i}^{\mathrm{T}}$$

and the symmetrical part can be omitted. It remains to calculate

$$\mathbf{u}_{0,i}^{(0)} = \mathrm{vect}\left(\ddot{\mathbf{A}}_{0,i}\mathbf{A}_{0,i}^{\mathrm{T}}\right), \quad i = 2, 3, \ldots, n. \tag{3.46}$$

From this, the angular accelerations of the mechanism $\mathbf{u}_{0,1}^{(0)}$, $\mathbf{u}_{0,2}^{(0)}$, $\ldots$, $\mathbf{u}_{0,n}^{(0)}$ can be obtained.

According to (3.15), the time derivative of the homogeneous transformation matrix $\mathbf{H}_{i-1,n+1}$ has the following form

$$\ddot{\mathbf{H}}_{i-1,n+1} = \begin{bmatrix} & \ddot{\mathbf{A}}_{i-1,n} & & \mathbf{a}_{i-1,n+1}^{(i-1)} \\ 0 & 0 & 0 & 0 \end{bmatrix}. \tag{3.47}$$

The recursive equation, working from the end-point to the base, is given by

$$\begin{aligned} &\ddot{\mathbf{H}}_{i-1,n+1} = \ddot{\mathbf{H}}_{i-1,i}\mathbf{H}_{i,n+1} + 2\dot{\mathbf{H}}_{i-1,i}\dot{\mathbf{H}}_{i,n+1} + \mathbf{H}_{i-1,i}\ddot{\mathbf{H}}_{i,n+1}, \\ &\quad i = n, n-1, \ldots, 1. \end{aligned} \tag{3.48}$$

It is calculated in the following sequence of steps

$$\begin{aligned} \ddot{\mathbf{H}}_{n-1,n+1} &= \ddot{\mathbf{H}}_{n-1,n}\mathbf{H}_{n,n+1}, \\ \ddot{\mathbf{H}}_{n-2,n+1} &= \ddot{\mathbf{H}}_{n-2,n-1}\mathbf{H}_{n-1,n+1} + 2\dot{\mathbf{H}}_{n-2,n-1}\dot{\mathbf{H}}_{n-1,n+1} \\ &\quad + \mathbf{H}_{n-2,n-1}\ddot{\mathbf{H}}_{n-1,n+1}, \\ \ddot{\mathbf{H}}_{i-3,n+1} &= \ddot{\mathbf{H}}_{i-3,n-2}\mathbf{H}_{n-2,n+1} + 2\dot{\mathbf{H}}_{i-3,n-2}\dot{\mathbf{H}}_{n-2,n+1} \\ &\quad + \mathbf{H}_{n-3,n-2}\ddot{\mathbf{H}}_{n-2,n+1}, \\ &\ \ \vdots \end{aligned}$$

The result consists of a series of translational accelerations $\mathbf{a}_{n-1,n+1}^{(n-1)}$, $\mathbf{a}_{n-2,n+1}^{(n-2)}$, $\ldots$, $\mathbf{a}_{0,n+1}^{(0)}$ and second derivatives of rotation matrices $\mathbf{A}_{n-2,n}$, $\mathbf{A}_{n-3,n}$, $\ldots$, $\mathbf{A}_{0,n}$, which by additional manipulation yields the angular accelerations $\mathbf{u}_{n-2,n}^{(n-2)}$, $\mathbf{u}_{n-3,n}^{(n-3)}$, $\ldots$, $\mathbf{u}_{0,n}^{(0)}$, since we know that

$$\mathbf{u}_{i-1,n}^{(i-1)} = \mathrm{vect}\left(\ddot{\mathbf{A}}_{i-1,n}\mathbf{A}_{i-1,n}^{\mathrm{T}}\right), \quad i = n-1, n-2, \ldots, 1. \tag{3.49}$$

Before considering an example, we recap which variables determine the angular and translational accelerations of a mechanism:

$\mathbf{u}_{i-1,i}^{(i-1)}(\ddot{\theta}_i)$: angular acceleration vector between links $i-1$ in i. It is calculated by formula (3.33).

$\mathbf{u}_{0,i}^{(0)}(\theta_1,\ldots,\theta_{i-1},\dot{\theta}_1,\ldots,\dot{\theta}_i,\ddot{\theta}_1,\ldots,\ddot{\theta}_i)$: angular acceleration vector between links 0 and i. It is calculated by recursive formula (3.35), working in steps from the base towards the robot's end-point. Prior to this, the rotation matrices (3.2), angular velocities (3.16) and (3.18), and angular accelerations (3.33), must be calculated

$\mathbf{u}_{i-1,n}^{(i-1)}(\theta_i,\ldots,\theta_{n-1},\dot{\theta}_i,\ldots,\dot{\theta}_n,\ddot{\theta}_i,\ldots,\ddot{\theta}_n)$: angular acceleration vector between links $i-1$ and n. It is calculated by the use of recursive formula (3.36), working in steps from the robot's end-point to the base of the mechanism. Prior to this, the rotation matrices (3.1), angular velocities (3.16) and (3.19), and angular accelerations (3.33), must be calculated.

$\mathbf{a}_{i-1,i}^{(i-1)}(\ddot{d}_i)$: translational acceleration vector of joint center i with respect to the joint center $i-1$. It is calculated with expression (3.37).

$\mathbf{a}_{0,i}^{(0)}(\theta_1,\ldots,\theta_{i-1},d_2,\ldots,d_i,\dot{\theta}_1,\ldots,\dot{\theta}_{i-1},\dot{d}_1,\ldots,\dot{d}_i,\ddot{\theta}_1,\ldots,\ddot{\theta}_{i-1},\ddot{d}_1,\ldots,\ddot{d}_i)$: vector of translational acceleration of the joint center i with respect to the base. It is calculated by recursive formula (3.38), working in steps from the base towards the end-point of the mechanism. Prior to this, the rotation matrices (3.2), position vectors (3.4), angular velocities (3.18), translational velocities (3.20), angular accelerations (3.35) and translational accelerations (3.37) must be calculated.

$\mathbf{a}_{i-1,n+1}^{(i)}(\theta_i,\ldots,\theta_n,d_{i+1},\ldots,d_n,\dot{\theta}_i,\ldots,\dot{\theta}_n,\dot{d}_{i+1},\ldots,\dot{d}_n,\ddot{\theta}_i,\ldots,\ddot{\theta}_n,\ddot{d}_i,\ldots,\ddot{d}_n)$: translational acceleration vector of the robot's end-point with respect to the joint center $i-1$. It is calculated by recursive formula (3.41), working in steps from the robot's end-point towards the base of the mechanism. Prior to this, the rotation matrices (3.1), position vectors (3.7), angular velocities (3.16), translational velocities (3.24), angular accelerations (3.33) and translational accelerations (3.37) must be calculated.

Finally, we shall calculate the angular and translational accelerations of the four degree of freedom mechanism shown in Fig. 2.20. We are continuing with this example for which the positions and orientations of the mechanism's links, together with their translational and angular velocities, were calculated in the examples of the two preceding sections. The angular accelerations between neighboring links (3.33) and translational accelerations between the neighboring joint centers (3.37) are given in Table 3.3.

The angular acceleration of the first link $\mathbf{u}_{0,1}^{(0)}$ is given in Table 3.3, while the angular acceleration of the second link is given by

$$\mathbf{u}_{0,2}^{(0)}=\mathbf{u}_{0,1}^{(0)}+\boldsymbol{\omega}_{0,1}^{(0)}\times\mathbf{A}_{0,1}\boldsymbol{\omega}_{1,2}^{(1)}+\mathbf{A}_{0,1}\mathbf{u}_{1,2}^{(1)}=\mathbf{u}_{0,1}^{(0)}+\begin{bmatrix}-s_1\dot{\theta}_1\dot{\theta}_2+c_1\ddot{\theta}_2\\ c_1\dot{\theta}_1\dot{\theta}_2+s_1\ddot{\theta}_2\\ 0\end{bmatrix}.$$

Table 3.3 Angular accelerations between neighboring links and translational accelerations between neighboring joint centers of the mechanism from Fig. 2.20

i	1	2	3	4
$\mathbf{u}_{i-1,i}^{(i-1)}$	0	$\ddot{\theta}_2$	0	0
	0	0	0	0
	$\ddot{\theta}_1$	0	0	$\ddot{\theta}_4$

i	1	2	3	4	5
$\mathbf{a}_{i-1,i}^{(i-1)}$	0	0	0	0	0
	0	0	$\ddot{d}_3$	0	0
	0	0	0	0	0

As the joint between links 2 and 3 has only a translation variable, these links must have the same angular acceleration

$$\mathbf{u}_{0,3}^{(0)} = \mathbf{u}_{0,2}^{(0)},$$

while the end-effector rotates with the angular acceleration

$$\begin{aligned}\mathbf{u}_{0,4}^{(0)} &= \mathbf{u}_{0,3}^{(0)} + \boldsymbol{\omega}_{0,3}^{(0)} \times \mathbf{A}_{0,3}\boldsymbol{\omega}_{3,4}^{(3)} + \mathbf{A}_{0,3}\mathbf{u}_{3,4}^{(3)} \\ &= \mathbf{u}_{0,3}^{(0)} + \begin{bmatrix} c_1 s_2 \dot{\theta}_1 \dot{\theta}_4 + s_1 c_2 \dot{\theta}_2 \dot{\theta}_4 + s_1 s_2 \ddot{\theta}_4 \\ s_1 s_2 \dot{\theta}_1 \dot{\theta}_4 - c_1 c_2 \dot{\theta}_2 \dot{\theta}_4 - c_1 s_2 \ddot{\theta}_4 \\ -s_2 \dot{\theta}_2 \dot{\theta}_4 + c_2 \ddot{\theta}_4 \end{bmatrix}.\end{aligned}$$

The angular acceleration of the end-effector is related to the time derivatives of the Euler angles, as was shown in Sect. 1.7.3. We therefore have

$$\mathbf{u}_{0,4}^{(0)} = \mathbf{W}_0 \begin{bmatrix} \dot{\psi}\dot{\theta} \\ \dot{\psi}\dot{\phi} \\ \dot{\theta}\dot{\phi} \end{bmatrix} + \mathbf{W} \begin{bmatrix} \ddot{\psi} \\ \ddot{\theta} \\ \ddot{\phi} \end{bmatrix}.$$

For this example, the $\mathbf{W}$ matrix was already calculated in the preceding section, so now we only need to find the matrix $\mathbf{W}_0$. It has the following form

$$\mathbf{W}_0 = \begin{bmatrix} -s_1 & c_1 s_2 & s_1 c_2 \\ c_1 & s_1 s_2 & -c_1 c_2 \\ 0 & 0 & -s_2 \end{bmatrix}.$$

The translational acceleration of the center of the first joint is given by vector $\mathbf{a}_{0,1}^{(0)}$ in Table 3.3, while the translational acceleration of the center of the second joint equals

$$\begin{aligned}\mathbf{a}_{0,2}^{(0)} = \mathbf{a}_{0,1}^{(0)} &+ \mathbf{u}_{0,1}^{(0)} \times \mathbf{A}_{0,1}\mathbf{r}_{1,2}^{(1)} + \boldsymbol{\omega}_{0,1}^{(0)} \times \left(\boldsymbol{\omega}_{0,1}^{(0)} \times \mathbf{A}_{0,1}\mathbf{r}_{1,2}^{(1)}\right) \\ &+ 2\boldsymbol{\omega}_{0,1}^{(0)} \times \mathbf{A}_{0,1}\mathbf{v}_{1,2}^{(1)} + \mathbf{A}_{0,1}\mathbf{a}_{1,2}^{(1)}\end{aligned}$$

$$
= \mathbf{a}_{0,1}^{(0)} + l_1 \begin{bmatrix} s_1\dot{\theta}_1^2 - c_1\ddot{\theta}_1 \\ -c_1\dot{\theta}_1^2 - s_1\ddot{\theta}_1 \\ 0 \end{bmatrix}.
$$

The translational acceleration of the center of the third joint is given by vector

$$
\begin{aligned}
\mathbf{a}_{0,3}^{(0)} &= \mathbf{a}_{0,2}^{(0)} + \mathbf{u}_{0,2}^{(0)} \times \mathbf{A}_{0,2}\mathbf{r}_{2,3}^{(1)} + \boldsymbol{\omega}_{0,2}^{(0)} \times \left(\boldsymbol{\omega}_{0,2}^{(0)} \times \mathbf{A}_{0,2}\mathbf{r}_{2,3}^{(2)}\right) \\
&\quad + 2\boldsymbol{\omega}_{0,2}^{(0)} \times \mathbf{A}_{0,2}\mathbf{v}_{2,3}^{(2)} + \mathbf{A}_{0,2}\mathbf{a}_{2,3}^{(2)} \\
&= \mathbf{a}_{0,2}^{(0)} + \begin{bmatrix} s_1c_2(l_2+d_3)\dot{\theta}_1^2 + 2c_1s_2(l_2+d_3)\dot{\theta}_1\dot{\theta}_2 - 2c_1c_2\dot{\theta}_1\dot{d}_3 \\ \quad + s_1c_2(l_2+d_3)\dot{\theta}_2^2 + 2s_1s_2\dot{\theta}_2\dot{d}_3 \\ \quad - c_1c_2(l_2+d_3)\ddot{\theta}_1 + s_1s_2(l_2+d_3)\ddot{\theta}_2 - s_1c_2\ddot{d}_3 \\ -c_1c_2(l_2+d_3)\dot{\theta}_1^2 + 2s_1s_2(l_2+d_3)\dot{\theta}_1\dot{\theta}_2 - 2s_1c_2\dot{\theta}_1\dot{d}_3 \\ \quad - c_1c_2(l_2+d_3)\dot{\theta}_2^2 - 2c_1s_2\dot{\theta}_2\dot{d}_3 \\ \quad - s_1c_2(l_2+d_3)\ddot{\theta}_1 - c_1s_2(l_2+d_3)\ddot{\theta}_2 + c_1c_2\ddot{d}_3 \\ -s_2(l_2+d_3)\dot{\theta}_2^2 + 2c_2\dot{\theta}_2\dot{d}_3 + c_2(l_2+d_3)\ddot{\theta}_2 + s_2\ddot{d}_3 \end{bmatrix}.
\end{aligned}
$$

The expressions are becoming increasingly complex and expand over several lines, in spite of the use of recursive equations. Typically, in these equations, the joint coordinates appear as arguments of trigonometric functions while their velocities are in quadratic forms. We also note that the accelerations of joint variables are linearly connected with the accelerations of the mechanism.

The translational acceleration of the fourth joint center is

$$
\begin{aligned}
\mathbf{a}_{0,4}^{(0)} &= \mathbf{a}_{0,3}^{(0)} + \mathbf{u}_{0,3}^{(0)} \times \mathbf{A}_{0,3}\mathbf{r}_{3,4}^{(3)} + \boldsymbol{\omega}_{0,3}^{(0)} \times \left(\boldsymbol{\omega}_{0,3}^{(0)} \times \mathbf{A}_{0,3}\mathbf{r}_{3,4}^{(3)}\right) \\
&\quad + 2\boldsymbol{\omega}_{0,3}^{(0)} \times \mathbf{A}_{0,3}\mathbf{v}_{3,4}^{(3)} + \mathbf{A}_{0,3}\mathbf{a}_{3,4}^{(3)} \\
&= \mathbf{a}_{0,3}^{(0)} + \begin{bmatrix} (s_1c_2l_3 + s_1s_2h_3)\dot{\theta}_1^2 + 2(c_1s_2l_3 - c_1c_2h_3)\dot{\theta}_1\dot{\theta}_2 \\ \quad + (s_1c_2l_3 + s_1s_2h_3)\dot{\theta}_2^2 - (c_1c_2l_3 + c_1s_2h_3)\ddot{\theta}_1 \\ \quad + (s_1s_2l_3 - s_1c_2h_3)\ddot{\theta}_2 \\ -(c_1c_2l_3 + c_1s_2h_3)\dot{\theta}_1^2 + 2(s_1s_2l_3 - s_1c_2h_3)\dot{\theta}_1\dot{\theta}_2 \\ \quad - (c_1c_2l_3 + c_1s_2h_3)\dot{\theta}_2^2 - (s_1c_2l_3 + s_1s_2h_3)\ddot{\theta}_1 \\ \quad - (c_1s_2l_3 - c_1c_2h_3)\ddot{\theta}_2 \\ -(s_2l_3 - c_2h_3)\dot{\theta}_2^2 + (c_2l_3 + s_2h_3)\ddot{\theta}_2 \end{bmatrix}.
\end{aligned}
$$

The translational acceleration of the mechanism's end-point is calculated by taking into account that $\mathbf{v}_{4,5}^{(4)} = 0$ and $\mathbf{a}_{4,5}^{(4)} = 0$. We obtain

$$
\mathbf{a}_{0,5}^{(0)} = \mathbf{a}_{0,4}^{(0)} + \mathbf{u}_{0,4}^{(0)} \times \mathbf{A}_{0,4}\mathbf{r}_{4,5}^{(4)} + \boldsymbol{\omega}_{0,4}^{(0)} \times \left(\boldsymbol{\omega}_{0,4}^{(0)} \times \mathbf{A}_{0,4}\mathbf{r}_{4,5}^{(4)}\right)
$$

$$
= \mathbf{a}_{0,4}^{(0)} + l_4 \begin{bmatrix} (c_1s_4 + s_1c_2c_4)\dot{\theta}_1^2 + 2c_1s_2c_4\dot{\theta}_1\dot{\theta}_2 + 2(s_1c_4 + c_1c_2s_4)\dot{\theta}_1\dot{\theta}_4 \\ + s_1c_2c_4\dot{\theta}_2^2 - 2s_1s_2s_4\dot{\theta}_2\dot{\theta}_4 + (c_1s_4 + s_1c_2c_4)\dot{\theta}_4^2 \\ + (s_1s_4 - c_1c_2c_4)\ddot{\theta}_1 + s_1s_2c_4\ddot{\theta}_2 - (c_1c_4 - s_1c_2s_4)\ddot{\theta}_4 \\ (s_1s_4 - c_1c_2c_4)\dot{\theta}_1^2 + 2s_1s_2c_4\dot{\theta}_1\dot{\theta}_2 - 2(c_1c_4 - s_1c_2s_4)\dot{\theta}_1\dot{\theta}_4 \\ - c_1c_2c_4\dot{\theta}_2^2 + 2c_1s_2s_4\dot{\theta}_2\dot{\theta}_4 + (s_1s_4 - c_1c_2c_4)\dot{\theta}_4^2 \\ - (c_1s_4 + s_1c_2c_4)\ddot{\theta}_1 - c_1s_2c_4\ddot{\theta}_2 - (s_1c_4 + c_1c_2s_4)\ddot{\theta}_4 \\ -s_2c_4\dot{\theta}_2^2 - 2c_2s_4\dot{\theta}_2\dot{\theta}_4 - s_2c_4\dot{\theta}_4^2 + c_2c_4\ddot{\theta}_2 - s_2s_4\ddot{\theta}_4 \end{bmatrix}.
$$

Although the four degree of freedom mechanism we have considered in the three examples in this chapter is quite simple, we can notice that the expressions for its pose in space, and the corresponding velocities and accelerations, are rather complex. By increasing the number of degrees of freedom, in particular the rotational ones, the kinematic formulas become increasingly more complex and the probability of making an error increases as well. The developed recursive formulas are very convenient for computer programming and numeric computations. The benefits and disadvantages in the numeric kinematic analysis of mechanisms will be considered later.

3.2 Direct Kinematics

In the preceding discussions of serial mechanisms, we have assumed that the i-th kinematic pair has both a translational variable d_i and a rotational variable θ_i. From this point forward we shall adopt a convention where each kinematic pair is a single degree of freedom joint

$$f_i = 1, \quad i = 1, 2, \ldots, n$$

and therefore each can be either translational or rotational, but not both. This means that going forward, we shall consider only serial mechanisms where the number of degrees of freedom is the same as number of joints

$$F = \sum_{i=1}^{n} f_i = n.$$

In these mechanisms only one of the coordinates d_i or θ_i is variable, while the other always equals zero. The variable of the i-th joint is called the internal or the generalized coordinate q_i and is defined as follows

$$q_i = \begin{cases} \theta_i & (d_i = 0), \\ d_i & (\theta_i = 0), \end{cases} \quad i = 1, 2, \ldots, n. \tag{3.50}$$

The assumption of only one joint variable in (3.50) does not invalidate the kinematic equations developed to this point, but only serves to simplify them.

3.2.1 Equations of Direct Kinematics

The internal coordinates of the mechanism are $q_1, q_2, \ldots, q_n$. The following vector

$$\mathbf{q} = \begin{bmatrix} q_1 \\ q_2 \\ \vdots \\ q_n \end{bmatrix}$$

is called the internal coordinates vector. It is important to recognize that the values of the internal coordinates uniquely define the position and orientation of all the links of serial mechanism. Thus, a unique pose of the mechanism corresponds to a particular internal coordinates vector. The pose is called the configuration of the mechanism.

The following vector is called the external coordinates vector

$$\mathbf{p} = \begin{bmatrix} p_1 \\ p_2 \\ \vdots \\ p_m \end{bmatrix}.$$

The external coordinates are $p_1, p_2, \ldots, p_m$. They can be selected in different ways, usually they are defined in connection to a particular task. Let us focus on the case when the external coordinates describe the position and orientation of the robot end-effector. Let λ denote the number of coordinates describing the pose of an object in space or in the plane. In general we are not necessarily interested in all of these external coordinates, therefore, in general we have $m \leq \lambda$. To start with, we consider the example where

$$m = \lambda.$$

When considering the pose of an object in space, we have $m = 6$, where three coordinates belong to the position of a mechanism's end-point, while three coordinates describe the orientation of the last link of the mechanism.

From a kinematic point of view, the relation between the external and internal coordinates is of utmost importance. The external coordinates can be expressed as functions of the internal coordinates by using the formulas we have developed to this point. In general, the relation between the external and internal coordinates is represented by a system of nonlinear algebraic equations

$$\mathbf{p} = \mathbf{p}_{\mathrm{q}}, \tag{3.51}$$

where $\mathbf{p}$ is the external coordinates vector and $\mathbf{p}_{\mathrm{q}}$ is a vector function of the internal coordinates $\mathbf{q}$, representing the kinematic equations for position and orientation of the robot end-effector. When the coordinates are rotational, they appear in function

$\mathbf{p}_\mathrm{q}$ as arguments of sines and cosines. When the internal coordinates are translational, they appear explicitly in the function $\mathbf{p}_\mathrm{q}$, as we have seen in preceding section. When the external coordinates correspond to the position and orientation of the robot end-effector, the above system of equations should be considered as two subsystems, each with three equations

$$\begin{bmatrix} \mathbf{r} \\ \mathbf{g} \end{bmatrix} = \begin{bmatrix} \mathbf{r}_\mathrm{q} \\ \mathbf{g}_\mathrm{q} \end{bmatrix},$$

where $\mathbf{p} = (\mathbf{r}^\mathrm{T}, \mathbf{g}^\mathrm{T})^\mathrm{T}$ and $\mathbf{p}_\mathrm{q} = (\mathbf{r}_\mathrm{q}^\mathrm{T}, \mathbf{g}_\mathrm{q}^\mathrm{T})^\mathrm{T}$. The vector function

$$\mathbf{r}_\mathrm{q} = \mathbf{r}_{0,n+1}^{(0)}$$

is known and can be calculated with (3.5) or (3.7), while the vector function

$$\mathbf{g}_\mathrm{q} = \mathbf{g}_{0,n}^{(0)}$$

depends on three orientation parameters, i.e. Euler angle or YPR orientation angles, or whatever triple of orientation parameters we choose to use. These orientation parameters can always be considered as belonging to the rotation matrix $\mathbf{A}_{0,n}$, which can be calculated by the use of (3.2) or (3.3).

The so called direct kinematics problem can be formulated as follows

$$\mathbf{q} \rightarrow \mathbf{p}.$$

In the direct kinematics problem, the internal coordinates $q_1, q_2, \ldots, q_n$ are known and the external coordinates $p_1, p_2, \ldots, p_m$ are to be computed. It is, therefore, necessary to solve the system of algebraic equations (3.51), where the external coordinates are unknowns and the internal coordinates are known. The system of equations (3.51) are called the mechanism's equations of direct kinematics. In serial mechanisms, the direct kinematics problem is solved quite simply, without any particularities. The transformation from internal into external coordinates is single valued and can be expressed in a closed-form.

3.2.2 Jacobian Matrix

The kinematics of a mechanism may also consider velocities. To this end, we find the time derivative of (3.51)

$$\dot{\mathbf{p}} = \frac{d}{dt}\mathbf{p}_\mathrm{q} = \left\{ \frac{\partial \mathbf{p}_\mathrm{q}}{\partial \mathbf{q}} \right\} \dot{\mathbf{q}}.$$

The matrix within the braces contains the partial derivatives of the external coordinates with respect to the internal coordinates. It is referred to as the Jacobian matrix and is denoted by $\mathbf{J}$ and in general its dimension is $m \times n$.

$$\mathbf{J} = \begin{bmatrix} \frac{\partial p_{q1}}{\partial q_1} & \frac{\partial p_{q1}}{\partial q_2} & \cdots & \frac{\partial p_{q1}}{\partial q_n} \\ \frac{\partial p_{q2}}{\partial q_1} & \frac{\partial p_{q2}}{\partial q_2} & \cdots & \frac{\partial p_{q2}}{\partial q_n} \\ \vdots & \vdots & & \vdots \\ \frac{\partial p_{qm}}{\partial q_1} & \frac{\partial p_{qm}}{\partial q_2} & \cdots & \frac{\partial p_{qm}}{\partial q_n} \end{bmatrix}. \tag{3.52}$$

In mathematics the Jacobian is frequently used within iterative methods which solve systems of nonlinear equations. The system of equations

$$\dot{\mathbf{p}} = \mathbf{J}\dot{\mathbf{q}}, \tag{3.53}$$

describes the linear relationship between the velocities of the external and internal coordinates. It represents the mechanism's direct kinematics in Jacobian or differential form. In fact the Jacobian maps the differentials of the internal coordinates into the differentials of the external coordinates

$$d\mathbf{p} = \mathbf{J}d\mathbf{q}. \tag{3.54}$$

The kinematic equations in the differential form can be seen as the first-order terms in the Taylor series which approximates the system of kinematic equations in (3.51). They are used in both the design and control of robots [23, 90].

When the external coordinates are defined as position and orientation of the robot end-effector, the velocities of the external coordinates are split into two parts

$$\begin{bmatrix} \mathbf{v} \\ \dot{\mathbf{g}} \end{bmatrix} = \begin{bmatrix} \mathbf{v}_{\mathrm{q}} \\ \dot{\mathbf{g}}_{\mathrm{q}} \end{bmatrix}, \tag{3.55}$$

where $\dot{\mathbf{p}} = (\mathbf{v}^{\mathrm{T}}, \dot{\mathbf{g}}^{\mathrm{T}})^{\mathrm{T}}$ and $\dot{\mathbf{p}}_{\mathrm{q}} = (\mathbf{v}_{\mathrm{q}}^{\mathrm{T}}, \dot{\mathbf{g}}_{\mathrm{q}}^{\mathrm{T}})^{\mathrm{T}}$. Here, the vector $\mathbf{v} = \dot{\mathbf{r}}$ represents the velocities of the translational part of the external coordinates, while the vectorial function

$$\mathbf{v}_{\mathrm{q}} = \dot{\mathbf{r}}_{\mathrm{q}} = \dot{\mathbf{r}}_{0,n+1}^{(0)} = \mathbf{v}_{0,n+1}^{(0)}$$

is the translational velocity of the robot's end-point. The vector $\dot{\mathbf{g}}$ encompasses the first time derivative of the orientational part of the external coordinates and the vectorial function

$$\dot{\mathbf{g}}_{\mathrm{q}} = \dot{\mathbf{g}}_{0,n}^{(0)}$$

represents the velocities of the three orientational parameters of the robot end link. The Jacobian matrix has two submatrices

$$\mathbf{J} = \begin{bmatrix} \mathbf{J}_{\mathrm{v}} \\ \mathbf{J}_{\dot{\mathrm{g}}} \end{bmatrix}, \tag{3.56}$$

where

$$\mathbf{J}_{\mathrm{v}} = \begin{bmatrix} \frac{\partial \mathbf{r}_{\mathrm{q}}}{\partial q_1} & \frac{\partial \mathbf{r}_{\mathrm{q}}}{\partial q_2} & \cdots & \frac{\partial \mathbf{r}_{\mathrm{q}}}{\partial q_n} \end{bmatrix},$$

$$\mathbf{J}_{\dot{\mathrm{g}}} = \begin{bmatrix} \frac{\partial \mathbf{g}_{\mathrm{q}}}{\partial q_1} & \frac{\partial \mathbf{g}_{\mathrm{q}}}{\partial q_2} & \cdots & \frac{\partial \mathbf{g}_{\mathrm{q}}}{\partial q_n} \end{bmatrix}.$$

Both dimensions are $3 \times n$.

Recall that the first time derivatives of the orientational parameters of the robot end-effector are related to the angular velocity of this link. When using either Euler or YPR orientational angles, this relation is determined with matrix $\mathbf{W}$, as follows

$$\boldsymbol{\omega}_{0,n}^{(0)} = \mathbf{W}\dot{\mathbf{g}}_{\mathrm{q}}.$$

The matrix $\mathbf{W}$ has been introduced in Sects. 1.6.3 and 1.6.4 for Euler and YPR angles. When the equation

$$\dot{\mathbf{g}}_{\mathrm{q}} = \mathbf{J}_{\dot{\mathrm{g}}}\dot{\mathbf{q}}$$

is multiplied by the matrix $\mathbf{W}$, we obtain

$$\boldsymbol{\omega}_{0,n}^{(0)} = \mathbf{W}\mathbf{J}_{\dot{\mathrm{g}}}\dot{\mathbf{q}}.$$

We denote

$$\mathbf{J}_{\omega} = \mathbf{W}\mathbf{J}_{\dot{\mathrm{g}}},$$

from where

$$\boldsymbol{\omega}_{0,n}^{(0)} = \mathbf{J}_{\omega}\dot{\mathbf{q}}.$$

In practice, instead of (3.55), the following equations are used

$$\begin{bmatrix} \mathbf{v} \\ \boldsymbol{\omega} \end{bmatrix} = \begin{bmatrix} \mathbf{v}_{\mathrm{q}} \\ \boldsymbol{\omega}_{\mathrm{q}} \end{bmatrix}, \tag{3.57}$$

where

$$\boldsymbol{\omega} = \mathbf{W}\dot{\mathbf{g}}$$

and

$$\boldsymbol{\omega}_{\mathrm{q}} = \boldsymbol{\omega}_{0,n}^{(0)}.$$

The following Jacobian matrix corresponds to the above equations

$$\mathbf{J} = \begin{bmatrix} \mathbf{J}_{\mathrm{v}} \\ \mathbf{J}_{\omega} \end{bmatrix}. \tag{3.58}$$

It describes the translational and angular velocity of the robot end-effector.

Based on the kinematic equations developed in the preceding sections, we shall derive formulas to determine the Jacobian matrix $\mathbf{J}$, by separately considering the submatrices $\mathbf{J}_\mathrm{v}$ and $\mathbf{J}_\omega$. Calculating submatrix $\mathbf{J}_\mathrm{v}$ comes directly from the expressions for translational velocity (3.24). To begin, we denote the columns of the matrix in the following way

$$\mathbf{J}_\mathrm{v} = \begin{bmatrix} \mathbf{J}_{\mathrm{v}1} & \mathbf{J}_{\mathrm{v}2} & \dots & \mathbf{J}_{\mathrm{v}n} \end{bmatrix}.$$

When the time derivatives of the rotational joint coordinates $\dot{\theta}_1, \dot{\theta}_2, \dots, \dot{\theta}_n$ are exposed in the equations describing the translational velocities (3.24), the following equations are obtained

$$\begin{aligned}
\mathbf{J}_{\mathrm{v}1} &= \mathbf{A}_{0,1}\big(\mathbf{e}_1^{(0)} \times \mathbf{r}_{1,n+1}^{(1)}\big),\\
\mathbf{J}_{\mathrm{v}2} &= \mathbf{A}_{0,2}\big(\mathbf{e}_2^{(1)} \times \mathbf{r}_{2,n+1}^{(2)}\big),\\
\mathbf{J}_{\mathrm{v}3} &= \mathbf{A}_{0,3}\big(\mathbf{e}_3^{(2)} \times \mathbf{r}_{3,n+1}^{(3)}\big),\\
&\ \ \vdots
\end{aligned}$$

A general expression for the i-th column of the submatrix $\mathbf{J}_\mathrm{v}$, when the internal coordinate is rotational $q_i = \theta_i$, is the following

$$\mathbf{J}_{\mathrm{v}i} = \mathbf{A}_{0,i}\big(\mathbf{e}_i^{(i-1)} \times \mathbf{r}_{i,n+1}^{(i)}\big), \quad i = 1, 2, \dots, n. \tag{3.59}$$

There is less involved with exposing the velocities of the translational coordinates $\dot{d}_1, \dot{d}_2, \dots, \dot{d}_n$, as there is

$$\begin{aligned}
\mathbf{J}_{\mathrm{v}1} &= \mathbf{A}_{0,1}\mathbf{e}_1^{(0)},\\
\mathbf{J}_{\mathrm{v}2} &= \mathbf{A}_{0,2}\mathbf{e}_2^{(1)},\\
\mathbf{J}_{\mathrm{v}3} &= \mathbf{A}_{0,3}\mathbf{e}_3^{(2)},\\
&\ \ \vdots
\end{aligned}$$

The formula for the i-th column of the matrix $\mathbf{J}_\mathrm{v}$, when the internal coordinate $q_i = d_i$, is therefore

$$\mathbf{J}_{\mathrm{v}i} = \mathbf{A}_{0,i}\mathbf{e}_i^{(i-1)}, \quad i = 1, 2, \dots, n. \tag{3.60}$$

We shall base calculating submatrix $\mathbf{J}_\omega$ on the expressions for the angular velocity of the robot end-effector (3.18). To begin, we denote the columns of the matrix as

$$\mathbf{J}_\omega = \begin{bmatrix} \mathbf{J}_{\omega 1} & \mathbf{J}_{\omega 2} & \dots & \mathbf{J}_{\omega n} \end{bmatrix}.$$

When exposing the rotational joint coordinates $\dot{\theta}_1, \dot{\theta}_2, \ldots, \dot{\theta}_n$ in the equations for angular velocity (3.18), we obtain

$$\mathbf{J}_{\omega 1} = \mathbf{A}_{0,1}\mathbf{e}_1^{(0)},$$
$$\mathbf{J}_{\omega 2} = \mathbf{A}_{0,2}\mathbf{e}_2^{(1)},$$
$$\mathbf{J}_{\omega 3} = \mathbf{A}_{0,3}\mathbf{e}_3^{(2)},$$
$$\vdots$$

The formula for the i-th column of the submatrix $\mathbf{J}_\omega$, when the internal coordinate is rotational $q_i = \theta_i$, is therefore

$$\mathbf{J}_{\omega i} = \mathbf{A}_{0,i}\mathbf{e}_i^{(i-1)}, \quad i = 1, 2, \ldots, n. \tag{3.61}$$

We know that the angular velocities of the mechanism are not dependent on the velocities of the translational joint coordinates $\dot{d}_1, \dot{d}_2, \ldots, \dot{d}_n$, therefore we have the following expression when the internal coordinate is $q_i = d_i$

$$\mathbf{J}_{\omega i} = \mathbf{0}, \quad i = 1, 2, \ldots, n. \tag{3.62}$$

The Jacobian matrix of a mechanism is obtained by calculating its submatrices $\mathbf{J}_\mathrm{v}$ and $\mathbf{J}_\omega$:

$\mathbf{J}_\mathrm{v}(\theta_1, \ldots, \theta_n, d_2, \ldots, d_n)$: a $3 \times n$ matrix belonging to the translational velocity of the robot's end-point. It is calculated column by column for $i = 1, 2, \ldots, n$ with the formula (3.59), when the internal coordinate is $q_i = \theta_i$, and with formula (3.60), when the internal coordinate is $q_i = d_i$. In order to make these calculations, the rotation matrices (3.1) and (3.2), and position vectors (3.7) must be known;

$\mathbf{J}_\omega(\theta_1, \ldots, \theta_{n-1})$: a $3 \times n$ matrix belonging to the angular velocity of the robot's end-point. It is calculated column by column for $i = 1, 2, \ldots, n$ with the formula (3.61), when the internal coordinate is $q_i = \theta_i$, and with formula (3.62), when the internal coordinate is $q_i = d_i$. In order to make these calculations, the rotation matrices (3.2) must be known.

When computation time is important, the Jacobian matrix must be calculated with as few operations as possible [44]. Therefore, formula (3.59) should be used after the position vectors (3.7) are known. If the position of the mechanism had been calculated by the use of (3.5), it is preferred to use the relation (3.8) and in case of $q_i = \theta_i$ to use the following expression

$$\mathbf{J}_{\mathrm{v}i} = \left(\mathbf{A}_{0,i}\mathbf{e}_i^{(i-1)}\right) \times \left(\mathbf{r}_{0,n+1}^{(0)} - \mathbf{r}_{0,i}^{(0)}\right), \quad i = 1, 2, \ldots, n.$$

When $q_i = \theta_i$, it is advantageous to first determine the columns of the submatrix $\mathbf{J}_\omega$ and thereafter use them as an intermediate result

$$\mathbf{J}_{\mathrm{v}i} = \mathbf{J}_{\omega i} \times \left(\mathbf{r}_{0,n+1}^{(0)} - \mathbf{r}_{0,i}^{(0)}\right), \quad i = 1, 2, \ldots, n. \tag{3.63}$$

We shall use this approach when examining the mechanism, whose kinematic parameters are shown in Fig. 2.20. The columns of the matrix $\mathbf{J}_\omega$ are derived without any calculation by simply rewriting the corresponding columns of the rotation matrices

$$\mathbf{J}_{\omega 1} = \begin{bmatrix} 0 \\ 0 \\ 1 \end{bmatrix}, \qquad \mathbf{J}_{\omega 2} = \begin{bmatrix} c_1 \\ s_1 \\ 0 \end{bmatrix}, \qquad \mathbf{J}_{\omega 3} = \begin{bmatrix} 0 \\ 0 \\ 0 \end{bmatrix}, \qquad \mathbf{J}_{\omega 4} = \begin{bmatrix} s_1 s_2 \\ -c_1 s_2 \\ c_2 \end{bmatrix}.$$

The matrix $\mathbf{J}_\mathrm{v}$ is obtained by calculating the vector products

$$\mathbf{J}_{\mathrm{v}1} = \mathbf{J}_{\omega 1} \times \left(\mathbf{r}_{0,5}^{(0)} - \mathbf{r}_{0,1}^{(0)}\right),$$
$$\mathbf{J}_{\mathrm{v}2} = \mathbf{J}_{\omega 2} \times \left(\mathbf{r}_{0,5}^{(0)} - \mathbf{r}_{0,2}^{(0)}\right),$$
$$\mathbf{J}_{\mathrm{v}4} = \mathbf{J}_{\omega 4} \times \left(\mathbf{r}_{0,5}^{(0)} - \mathbf{r}_{0,4}^{(0)}\right),$$

as we already know the vectors $\mathbf{r}_{0,1}^{(0)}$, $\mathbf{r}_{0,2}^{(0)}$, $\mathbf{r}_{0,4}^{(0)}$ in $\mathbf{r}_{0,5}^{(0)}$. The third column does not have to be calculated, it is hidden in the rotation matrix $\mathbf{A}_{0,3}$

$$\mathbf{J}_{\mathrm{v}3} = \begin{bmatrix} -s_1 c_2 \\ c_1 c_2 \\ s_2 \end{bmatrix}.$$

3.2.3 Hessian Matrix

The Hessian matrix is predominantly used to obtain numerical solutions to systems of nonlinear equations. In robotics however, it is much less utilized than the Jacobian matrix. The Hessian matrix is obtained by time differentiation of (3.53)

$$\ddot{\mathbf{p}} = \dot{\mathbf{J}}\dot{\mathbf{q}} + \mathbf{J}\ddot{\mathbf{q}}.$$

Let us expand the time derivative of Jacobian matrix using the chain rule from calculus

$$\ddot{\mathbf{p}} = \left(\frac{\partial \mathbf{J}}{\partial q_1}\dot{q}_1 + \frac{\partial \mathbf{J}}{\partial q_2}\dot{q}_2 + \cdots + \frac{\partial \mathbf{J}}{\partial q_n}\dot{q}_n\right)\dot{\mathbf{q}} + \mathbf{J}\ddot{\mathbf{q}},$$

which can be rewritten in the following way

$$\ddot{\mathbf{p}} = \dot{\mathbf{q}}^\mathrm{T}\mathbf{H}\dot{\mathbf{q}} + \mathbf{J}\ddot{\mathbf{q}}. \tag{3.64}$$

where $\mathbf{H}$ is the Hessian matrix. The Hessian is three-dimensional, that is, it is a $1 \times n$ column vector, whose elements are the $m \times n$ matrices which are the partial derivatives of the Jacobian matrix. Thus the Hessian is of dimension $n \times (m \times n)$.

$$\mathbf{H} = \begin{bmatrix} \frac{\partial \mathbf{J}}{\partial q_1} \\ \frac{\partial \mathbf{J}}{\partial q_2} \\ \vdots \\ \frac{\partial \mathbf{J}}{\partial q_n} \end{bmatrix}.$$

The product $\dot{\mathbf{q}}^T\mathbf{H}$ is obtained in such a way, that particular components of the vector $\dot{\mathbf{q}}^T$ are multiplied by the entire contiguous submatrix of the matrix $\mathbf{H}$.

The system of equations (3.64) are called the mechanism's direct kinematic equations in Hessian form. Here, the Jacobian matrix directly defines the linear relationship between the second time derivatives of the external and internal coordinates, while the Hessian is part of the quadratic form of the first derivatives of the internal coordinates and encompasses all the Coriolis and radial components of accelerations.

Now it remains to derive the elements of the Hessian matrix, specifically for $\mathbf{J}_v$ and $\mathbf{J}_\omega$. Two indices must be used. Let the index $i = 1, 2, \ldots, n$ belong to the columns of the matrices $\mathbf{J}_v$ and $\mathbf{J}_\omega$, while index $j = 1, 2, \ldots, n$ should denote the internal coordinate with respect to which the matrix is being differentiated.

Let us first find the derivatives of the matrix $\mathbf{J}_v$. When $q_i = \theta_i$ we have the following

$$\begin{aligned} \frac{\partial \mathbf{J}_{vi}}{\partial \theta_j} &= \left(\mathbf{A}_{0,j}\mathbf{e}_j^{(j-1)}\right) \times \mathbf{A}_{0,i}\left(\mathbf{e}_i^{(i-1)} \times \mathbf{r}_{i,n+1}^{(i)}\right), \quad j \le i, \\ \frac{\partial \mathbf{J}_{vi}}{\partial \theta_j} &= \left(\mathbf{A}_{0,i}\mathbf{e}_i^{(i-1)}\right) \times \mathbf{A}_{0,j}\left(\mathbf{e}_j^{(j-1)} \times \mathbf{r}_{j,n+1}^{(j)}\right), \quad j > i \end{aligned} \tag{3.65}$$

and

$$\begin{aligned} \frac{\partial \mathbf{J}_{vi}}{\partial d_j} &= \mathbf{0}, \quad j \le i, \\ \frac{\partial \mathbf{J}_{vi}}{\partial d_j} &= \left(\mathbf{A}_{0,i}\mathbf{e}_i^{(i-1)}\right) \times \left(\mathbf{A}_{0,j}\mathbf{e}_j^{(j-1)}\right), \quad j > i. \end{aligned} \tag{3.66}$$

When $q_i = d_i$ we have

$$\begin{aligned} \frac{\partial \mathbf{J}_{vi}}{\partial \theta_j} &= \left(\mathbf{A}_{0,j}\mathbf{e}_j^{(j-1)}\right) \times \left(\mathbf{A}_{0,i}\mathbf{e}_i^{(i-1)}\right), \quad j < i, \\ \frac{\partial \mathbf{J}_{vi}}{\partial \theta_j} &= \mathbf{0}, \quad j \ge i \end{aligned} \tag{3.67}$$

and

$$\frac{\partial \mathbf{J}_{vi}}{\partial d_j} = \mathbf{0}, \quad \forall j. \tag{3.68}$$

The derivatives of the matrix $\mathbf{J}_\omega$ are simpler. When $q_i = \theta_i$ we have

$$\begin{aligned} \frac{\partial \mathbf{J}_{\omega i}}{\partial \theta_j} &= \left(\mathbf{A}_{0,j}\mathbf{e}_j^{(j-1)}\right) \times \left(\mathbf{A}_{0,i}\mathbf{e}_i^{(i-1)}\right), \quad j < i, \\ \frac{\partial \mathbf{J}_{\omega i}}{\partial \theta_j} &= \mathbf{0}, \qquad\qquad\qquad\qquad\qquad\qquad j \geq i \end{aligned} \tag{3.69}$$

and

$$\frac{\partial \mathbf{J}_{\omega i}}{\partial d_j} = \mathbf{0}, \quad \forall j. \tag{3.70}$$

When $q_i = d_i$, we have

$$\frac{\partial \mathbf{J}_{\omega i}}{\partial \theta_j} = \mathbf{0}, \quad \forall j \tag{3.71}$$

and

$$\frac{\partial \mathbf{J}_{\omega i}}{\partial d_j} = \mathbf{0}, \quad \forall j. \tag{3.72}$$

The kinematic equations we have developed for position, orientation, velocities, accelerations, Jacobian and Hessian matrices are valid for all serial mechanisms, independent of the number of degrees of freedom, the mechanism's kinematic parameters, or any other characteristics. When examining a particular mechanism, we can simplify the kinematic equations by eliminating all multiplications or additions with zero and multiplications by one. This can be incorporated directly into computer programs which automatically generate expressions with the minimal number of arithmetic operations. In the past, special computer programs were written in different computer languages based on various methods of symbolic programming. With the increase in computer capabilities the number of arithmetic operations required in kinematic equations has become less relevant.

3.3 Inverse Kinematics

A mechanism's inverse kinematics problem can be formulated as follows

$$\mathbf{p} \rightarrow \mathbf{q}.$$

In the inverse kinematics problem the external coordinates $p_1, p_2, \ldots, p_m$ are given and the internal coordinates $q_1, q_2, \ldots, q_n$ need to be computed. In this section we shall only consider such mechanisms where the number of internal coordinates is equal to the number of external coordinates,

$$n = m.$$

Later we shall study mechanisms, where $n > m$ or $n < m$. Similar to the direct kinematic equations, we write the inverse kinematic equations in the following way

$$\mathbf{q} = \mathbf{q}_{\mathrm{p}}, \tag{3.73}$$

where $\mathbf{q}$ is the internal coordinates vector and $\mathbf{q}_{\mathrm{p}}$ is a vector function of the external coordinates. This expression has only symbolic meaning. In general, in serial mechanisms it is not possible to explicitly write the internal coordinates as functions of external coordinates. Kinematic equations consist of trigonometric functions, which can be solved in a closed-form only in special cases. The complexity of the inverse kinematics problem increases when the mechanism has more rotational degrees of freedom and when the rotational axes are neither parallel, or perpendicular or intersecting. When a mechanism has only translational degrees of freedom, which rarely occurs, the inverse kinematics problem involves solving a system of linear equations.

To begin, we consider the 3R mechanism shown in Fig. 3.8. This is a planar mechanism with three internal coordinates q_1, q_2, and q_3 and the link lengths l_1, l_2, and l_3. The external coordinates are represented by the position of the reference point P on the gripper with the coordinates p_1 and p_2. The orientation of the gripper is described by the Euler angle ψ and is denoted as the variable p_3.

The reference frame $\mathbf{x}_0$, $\mathbf{y}_0$, $\mathbf{z}_0$ is placed in the center of the first joint. When the mechanism is in its initial pose, the links are completely extended in the direction of the $\mathbf{x}_0$ axis. In that configuration, the internal coordinates are taken to be zero, $q_1 = q_2 = q_3 = 0$ and the local coordinate frames are all parallel with $\mathbf{x}_0$, $\mathbf{y}_0$, $\mathbf{z}_0$. The local coordinate frames are placed onto the joint centers as shown in Fig. 3.8. The vector parameters of the mechanism are shown in Table 3.4.

First we determine the rotation matrices between the neighboring links. It is not necessary to calculate them, as all rotation axes $\mathbf{e}_1$, $\mathbf{e}_2$, and $\mathbf{e}_3$ are parallel to the vector $\mathbf{z}_0$. The three matrices can be written as follows

$$\mathbf{A}_{0,1} = \begin{bmatrix} c_1 & -s_1 & 0 \\ s_1 & c_1 & 0 \\ 0 & 0 & 1 \end{bmatrix},$$

$$\mathbf{A}_{1,2} = \begin{bmatrix} c_2 & -s_2 & 0 \\ s_2 & c_2 & 0 \\ 0 & 0 & 1 \end{bmatrix},$$

$$\mathbf{A}_{2,3} = \begin{bmatrix} c_3 & -s_3 & 0 \\ s_3 & c_3 & 0 \\ 0 & 0 & 1 \end{bmatrix}.$$

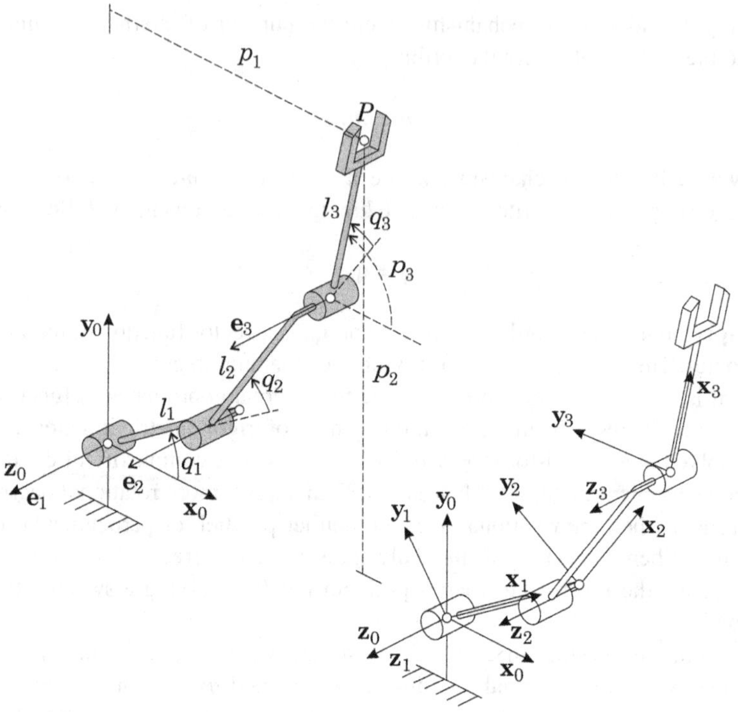

Fig. 3.8 Planar 3R mechanism

Table 3.4 Vector parameters of 3R mechanism

i	1	2	3
θ_i	q_1	q_2	q_3
d_i	0	0	0

i	1	2	3
$\mathbf{e}_i^{(i-1)}$	0	0	0
	0	0	0
	1	1	1

i	1	2	3	4
$\mathbf{b}_{i-1,i}^{(i-1)}$	0	0	0	0
	0	l_1	l_2	l_3
	0	0	0	0

Likewise, there is no real need to calculate the rotation matrices with respect to the base, as we are dealing with parallel rotations. Thus

$$\mathbf{A}_{0,2} = \begin{bmatrix} c_{12} & -s_{12} & 0 \\ s_{12} & c_{12} & 0 \\ 0 & 0 & 1 \end{bmatrix}, \qquad \mathbf{A}_{0,3} = \begin{bmatrix} c_{123} & -s_{123} & 0 \\ s_{123} & c_{123} & 0 \\ 0 & 0 & 1 \end{bmatrix}.$$

Here a shorthand notation is being used where, $s_{12} = \sin(q_1 + q_2)$, $c_{12} = \cos(q_1 + q_2)$, and $s_{123} = \sin(q_1 + q_2 + q_3)$, $c_{123} = \cos(q_1 + q_2 + q_3)$.

As the mechanism has no translational internal coordinates, we have

$$\mathbf{r}_{0,1}^{(0)} = \mathbf{b}_{0,1}^{(0)}, \qquad \mathbf{r}_{1,2}^{(1)} = \mathbf{b}_{1,2}^{(1)}, \qquad \mathbf{r}_{2,3}^{(2)} = \mathbf{b}_{2,3}^{(2)}, \qquad \mathbf{r}_{3,4}^{(3)} = \mathbf{b}_{3,4}^{(3)}.$$

The position of point P is obtained as follows

$$\mathbf{r}_{0,4}^{(0)} = \mathbf{A}_{0,1}\mathbf{r}_{1,2}^{(1)} + \mathbf{A}_{0,2}\mathbf{r}_{2,3}^{(2)} + \mathbf{A}_{0,3}\mathbf{r}_{3,4}^{(3)},$$

where we are only interested into the first two components of the vector, i.e. the coordinates in the $\mathbf{x}_0$ and $\mathbf{y}_0$ directions. The orientation of the gripper is obtained from the matrix $\mathbf{A}_{0,3}$ in accordance with the definition of the Euler angle ψ. As a result, we obtain the following direct kinematic equations

$$\begin{aligned} p_1 &= l_1 c_1 + l_2 c_{12} + l_3 c_{123}, \\ p_2 &= l_1 s_1 + l_2 s_{12} + l_3 s_{123}, \\ p_3 &= q_1 + q_2 + q_3. \end{aligned} \tag{3.74}$$

This system of equations corresponds to the general form (3.51). On the left side we have the external coordinates $\mathbf{p}$, while on the right side there is the vector function $\mathbf{p}_\mathrm{q}$, which depends on the internal coordinates $\mathbf{q}$. With the internal coordinates q_1, q_2, q_3 given, the external coordinates p_1, p_2, p_3 can be calculated by the use of (3.74). This is the solution to the direct kinematics problem.

Now consider the inverse kinematics problem, where the external coordinates p_1, p_2 and p_3 are given and we must compute the internal coordinates q_1, q_2 and q_3. We must use the system of equations (3.74), where, unfortunately, the internal coordinates appear as arguments of trigonometric functions. The solution can be found in different ways. Our solution will be based on the law of cosines.

As the coordinate p_3 is given, the sum of internal coordinates is known

$$q_1 + q_2 + q_3 = p_3,$$

therefore

$$s_{123} = \sin p_3,$$

and

$$c_{123} = \cos p_3.$$

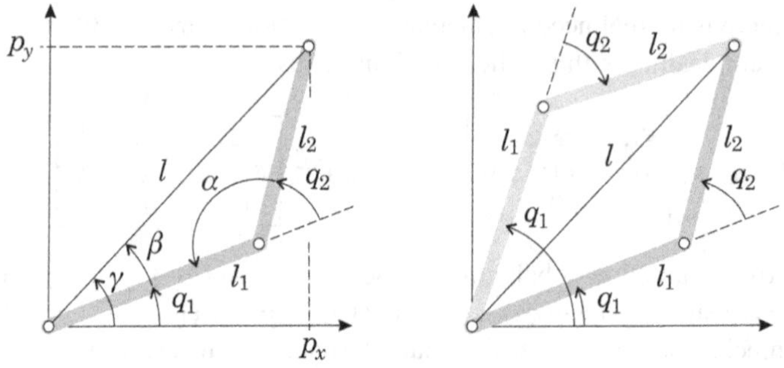

Fig. 3.9 Inclined triangle with sides l_1, l_2, and l

We define two parameters, p_{x} and p_{y}

$$p_{\mathrm{x}} = p_1 - l_3 \cos p_3,$$
$$p_{\mathrm{y}} = p_2 - l_3 \sin p_3,$$

which are knowns, since terms on the right hand sides are known. We then rewrite the first two equations in (3.74), placing the unknowns on the left side and the known parameters on the right

$$l_1 c_1 + l_2 c_{12} = p_{\mathrm{x}},$$
$$l_1 s_1 + l_2 s_{12} = p_{\mathrm{y}}.$$

The above equations have a geometric meaning. They describe the geometry of the inclined triangle in Fig. 3.9 with known sides l_1, l_2, and l, where

$$l^2 = p_{\mathrm{x}}^2 + p_{\mathrm{y}}^2.$$

It is clear that real solutions exist only inside a ring with radii $|l_1 - l_2|$ and $l_1 + l_2$. The following inequality holds

$$(l_1 - l_2)^2 \leq l^2 \leq (l_1 + l_2)^2. \tag{3.75}$$

The internal coordinate q_2 is found by applying the law of cosines

$$l^2 = l_1^2 + l_2^2 - 2 l_1 l_2 \cos\alpha,$$

so that we have

$$\alpha = \arccos\left(\frac{l_1^2 + l_2^2 - l^2}{2 l_1 l_2}\right),$$

and

$$q_2 = \pm(\pi - \alpha).$$

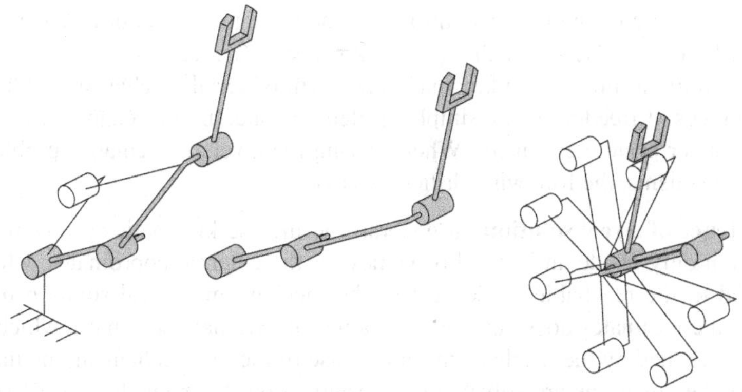

Fig. 3.10 3R mechanism has different number of configurations for the given values of external coordinates. Usually there exist two configurations. When the second joint is extended, only single solution exists. When the first and the third joint rotate around the same axis, there is infinite number of configurations

We obtain $q_2 = 0$, when $l = l_1 + l_2$ and $q_2 = \pm\pi$, when $l = |l_1 - l_2|$. From Fig. 3.9 we see that

$$\gamma = \arctan_2 \frac{p_y}{p_x}.$$

The law of cosines is used to calculate the angle β

$$l_2^2 = l_1^2 + l^2 - 2l_1 l \cos\beta,$$

so

$$\beta = \arccos\left(\frac{l_1^2 + l^2 - l_2^2}{2l_1 l}\right),$$

which holds only when $l \neq 0$. There are two solutions for the angle q_1. They depend on the selected solution for the angle q_2 as follows

$$\begin{aligned} q_2 > 0 \quad &\Longrightarrow \quad q_1 = \gamma - \beta, \\ q_2 < 0 \quad &\Longrightarrow \quad q_1 = \gamma + \beta. \end{aligned} \tag{3.76}$$

The solutions are shown in Figs. 3.9 and the left side of Fig. 3.10. The two extreme cases shown in Fig. 3.10 are important. When $q_2 = 0$, we find that $q_1 = \gamma$, and only one solution for internal coordinates exists. When $q_2 = \pm\pi$ we find that $q_1 = \gamma$. Also, when $l = 0$, which is only possible when the first two links have the same length ($l_1 = l_2$), the value of the internal coordinate q_1 is arbitrary and there exists an infinite number of solutions for the internal coordinates corresponding to a single value of the external coordinates.

Once the values of q_1 and q_2 are known, the last coordinate can be found from

$$q_3 = p_3 - q_1 - q_2.$$

Also, the periodic nature of the solutions for the three internal coordinates must be accounted for: $q_1 \pm 2k\pi$, $q_2 \pm 2k\pi$, $q_3 \pm 2k\pi$, $k = 1, 2, \ldots$.

The solution to the inverse kinematics problem of serial mechanisms with rotational degrees of freedom is not simple, as demonstrated by this example, which in fact is a rather basic mechanism. When solving the inverse kinematics problem of serial mechanisms, the following difficulties exist:

nonexistence of a real solution In general, the inverse kinematics problem has a solution only inside an interval of values for the external coordinates, which is related to the reachable workspace of the mechanism. A real solution for the internal coordinates does not exist for values of external coordinates which cannot be reached by the mechanism. In the case of the 3R mechanism, the interval where the equations are solvable is determined by the inequality (3.75). When this condition does not hold, the given external coordinates are out of the reach of the mechanism, hence no real solution for the internal coordinates exists.

multiple solutions In general, for values of external coordinates within the reachable workspace there exists multiple solutions for the internal coordinates. In general these solutions appear in pairs. The pose of a mechanism which corresponds to a selected combination of the internal coordinates is called the configuration of the mechanism. The number of configurations corresponding to given values of external coordinates depends on the kinematic structure of a mechanism. Theoretically, the largest possible number of configurations is 2^{k-1} for a mechanism with k rotations. For a mechanism with six rotations at most sixteen different configurations has been proven, which is the same as for a mechanism with five rotations and one translation. When a mechanism has four rotations and two translations, there exist eight different configurations. In the case of the 3R mechanism in our example, two configurations belong to the same external coordinates (3.76). They are shown in Figs. 3.9 and 3.10.

kinematic singularity For some values of the external coordinates the number of possible configurations of the mechanism is reduced. This can happen either for a single point in the external coordinates or for a continuous region of the external coordinates. These are the kinematic singularities of the mechanism. Usually this is a consequence of several solutions of the inverse kinematics problem merging into a single solution. For example, in the 3R mechanism of our example, we found that when the mechanism is fully extended by the second joint, i.e. when $q_2 = 0$, the variable q_1 has a single value and the mechanism also has only single configuration (Fig. 3.10). This is a kinematic singularity. However, there also exist kinematic singularities where the number of configurations is infinite. For example, in the 3R mechanism in our example we found that coordinate q_1 has an infinite number of solutions when $l = 0$, and this is also a kinematic singularity.

nonexistence of closed-form solutions Some systems of kinematic equations do not have closed-form solutions to the inverse kinematics problem. The exact solution cannot be obtained, even if for the given set of external coordinates, a real solution for the internal coordinates exists. In such cases the solution can only be found with numeric iterative approaches, which may not converge, and as well, may not find all the possible solutions.

periodic solutions There are an infinite number of equivalent periodic solutions for internal coordinates that are rotational. We must be cautious about them, above all when we use the internal coordinates as variables in robot control. In that situation, the internal coordinates expressed as functions of time must be continuous and cannot be allowed to skip from one period into another.

3.3.1 Algebraic Solutions to the Inverse Kinematics Problem

Much more than a single section can be written about how to algebraically solve a system of trigonometric equations. The fact remains that the inverse kinematics problem of a serial mechanism cannot be solved algebraically for the internal coordinates, explicitly expressed as functions of the external coordinates, when the mechanism has more than three rotations (except in special cases when the rotational axes are successively parallel or intersect at a common point). Regardless of the existence of an explicit algebraic solution, algebraic methods are useful in the analysis of mechanisms, e.g. when examining how many solutions to the inverse kinematic problem exist, or what the connections between solutions are, or how the solutions depend on the values of the kinematic parameters, and other matters. Algebraic solutions have several advantages when compared to numerical solutions. Nevertheless, the algebraic solutions can be so complex, that the solutions become so complicated that they lose their practical meaning.

Today, the inverse kinematics problem of mechanisms where the translational and rotational axes are consecutively either parallel, perpendicular or intersecting, are well known. However, when the neighboring axes are arbitrarily inclined with respect to one another, even seemingly simple mechanisms can have unpredictable characteristic properties which cannot be described by general postulates.

In practice, the most important inverse kinematics problem of serial mechanisms is for mechanisms with six degrees of freedom, which can be divided into a distinct three degree of freedom positioning mechanism and a distinct three degree of freedom orienting mechanism. This is the case in industrial robot manipulators. It is long well known that such six degree of freedom mechanisms have algebraic solutions to the inverse kinematics problem when the three rotational axes of the orienting mechanism intersect at a common point, regardless of the kinematic structure of the positional part of the mechanism [69]. Such three degree of freedom orienting mechanisms are called spherical wrists and the point of intersection of the three rotational axes is called the wrist center.

Let us assume we have a mechanism with $n = 6$ degrees of freedom, the positional mechanism with three coordinates q_1, q_2, and q_3, and a wrist (orienting mechanism) with three coordinates q_4, q_5, and q_6. When the external coordinates are given, the position of mechanism's end-point is known

$$\mathbf{r}_{0,7}^{(0)} = \mathbf{r}$$

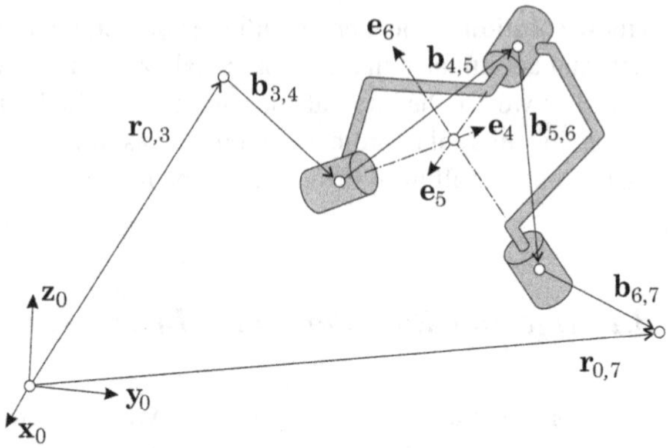

Fig. 3.11 Serial mechanism with six degrees of freedom and spherical wrist

together with the orientation of the end link, which is represented by the rotation matrix $\mathbf{A}$. Therefore we have

$$\mathbf{A}_{0,6} = \mathbf{A}.$$

In accordance with (3.8) we write

$$\mathbf{r}_{0,7}^{(0)} = \mathbf{r}_{0,3}^{(0)} + \mathbf{A}_{0,3}\mathbf{r}_{3,7}^{(3)}.$$

A mechanism has a spherical wrist when the coordinates $q_4 = \theta_4$, $q_5 = \theta_5$ and $q_6 = \theta_6$ are rotational ($d_4 = d_5 = d_6 = 0$) and the corresponding axes $\mathbf{e}_4$, $\mathbf{e}_5$ and $\mathbf{e}_6$ intersect at a common point, as shown in Fig. 3.11.

The vector $\mathbf{r}_{3,7}$, which includes no translational coordinates, is given as follows

$$\mathbf{r}_{3,7}^{(3)} = \mathbf{b}_{3,4}^{(3)} + \mathbf{A}_{3,4}\big(\mathbf{b}_{4,5}^{(4)} + \mathbf{A}_{4,5}\big(\mathbf{b}_{5,6}^{(5)} + \mathbf{A}_{5,6}\mathbf{b}_{6,7}^{(6)}\big)\big).$$

From Fig. 3.11 we observe that with a spherical wrist, the vectors $\mathbf{b}_{4,5}$, $\mathbf{e}_4$, and $\mathbf{e}_5$ are coplanar. The same is also true for the vectors $\mathbf{b}_{5,6}$, $\mathbf{e}_5$ and $\mathbf{e}_6$, which can be mathematically described as follows

$$\mathbf{b}_{4,5}^{(4)} = \upsilon_4\mathbf{A}_{3,4}^{\mathrm{T}}\mathbf{e}_4^{(3)} - \upsilon_5\mathbf{e}_5^{(4)}$$

and

$$\mathbf{b}_{5,6}^{(5)} = \upsilon_5\mathbf{A}_{4,5}^{\mathrm{T}}\mathbf{e}_5^{(4)} - \upsilon_6\mathbf{e}_6^{(5)}.$$

Here, υ_4, υ_5, and υ_6 are scalars of appropriate values. The above equations are inserted into the expression for $\mathbf{r}_{3,7}$, which yields

$$\mathbf{r}_{3,7}^{(3)} = \mathbf{b}_{3,4}^{(3)} + \upsilon_4\mathbf{e}_4^{(3)} + \mathbf{A}_{3,4}\mathbf{A}_{4,5}\mathbf{A}_{5,6}\big(-\upsilon_6\mathbf{e}_6^{(5)} + \mathbf{b}_{6,7}^{(6)}\big).$$

Here, we have taken into account that $\mathbf{e}_6^{(5)} = \mathbf{A}_{5,6}\mathbf{e}_6^{(5)}$. All these expressions must be included into the equation for $\mathbf{r}_{0,7}$, which gives

$$\mathbf{r}_{0,7}^{(0)} = \mathbf{r}_{0,3}^{(0)} + \mathbf{A}_{0,3}\big(\mathbf{b}_{3,4}^{(3)} + \upsilon_4\mathbf{e}_4^{(3)}\big) + \mathbf{A}_{0,6}\big(-\upsilon_6\mathbf{e}_6^{(5)} + \mathbf{b}_{6,7}^{(6)}\big).$$

The unknowns are transferred to the left, while the known parameters are on the right side

$$\mathbf{r}_{0,3}^{(0)} + \mathbf{A}_{0,3}\big(\mathbf{b}_{3,4}^{(3)} + \upsilon_4\mathbf{e}_4^{(3)}\big) = \mathbf{r} - \mathbf{A}\big(-\upsilon_6\mathbf{e}_6^{(5)} + \mathbf{b}_{6,7}^{(6)}\big).$$

On the left side we only have three unknowns, these are q_1, q_2, and q_3. This system of equations has algebraic solutions and the coordinates q_1, q_2, and q_3 can be found by several different methods. When these three coordinates are known, the rotation matrix $\mathbf{A}_{0,3}$ is determined by their values and a new system of equations with three variables is obtained

$$\mathbf{A}_{4,6} = \mathbf{A}_{0,3}^{\mathrm{T}}\mathbf{A}.$$

Here we have another triple of unknowns q_4, q_5, and q_6 which can be solved for algebraically. Using this method, known as wrist-partitioning, a six degree of freedom serial mechanism with a three degree of freedom spherical wrist, always has algebraic solutions to its inverse kinematics problem. This occurs because the kinematic equations of the mechanism can always be partitioned into two independent systems of equations, with three unknowns each.

Even when an algebraic solution to the inverse kinematics problem exists, searching for the algebraic solution can be rather difficult. There is no standard procedure for finding the algebraic solution in the most efficient way. When solving the system of kinematic equations, we make use of trigonometric identities in such a way that we gradually eliminate individual unknowns. The goal is to develop an equation which excludes all but one unknown

$$f_1(q_\alpha) = 0,$$

from which we compute this unknown q_α. After which we calculate the second one while using the value of the first one

$$f_2(q_\alpha, q_\beta) = 0,$$

and by use of first two values, we calculate the third unknown

$$f_3(q_\alpha, q_\beta, q_\gamma) = 0,$$

$$\vdots$$

and in the same way we continue to the end. The approach is called triangulation. Different triangulations can be adapted to a selected system of kinematic equations. The question remains which triangulation will lead to a final solution for all coordinates with the least amount of effort, or if it will lead to solutions at all [39].

Today, the most accepted and most general approach to solving inverse kinematics is by transformation of the system of trigonometric equations into a system of polynomial equations. Solving a system of trigonometric equations is thus transformed into searching for the roots of a system of polynomial equations in several variables [74]. The advantage of this transformation is that methods of symbolic mathematics for solving for the roots of polynomial equations are better developed than those for systems of trigonometric equations. Also, for a system of polynomial equations, it is easier to determine the number of real solutions and the number of solutions at infinity.

One of the possible transformations to a polynomial form of equations is realized by replacing the sines and cosines of the rotational coordinates using the tangent of the half angle formulas. If for each of the rotational coordinate q_i, we define the tangent of the half angle as

$$x_i = \tan\frac{q_i}{2},$$

the following expressions exist for the sines and cosines of q_i

$$\sin q_i = \frac{2x_i}{1+x_i^2}$$

and

$$\cos q_i = \frac{1-x_i^2}{1+x_i^2},$$

and by substituting the right hand sides in the trigonometric equations, they are transformed into a system of polynomial equations in terms of the variables x_i. As an example, consider the following trigonometric equation

$$a\sin q_i + b\cos q_i = c.$$

We might ask ourselves, how many solutions does it have and what are the solutions? By substituting the half angle formula, this equation is transformed into the polynomial

$$a\frac{2x_i}{1+x_i^2} + b\frac{1-x_i^2}{1+x_i^2} = c.$$

The equation is multiplied by $1+x_i^2$ and afterwards rearranged into a quadratic equation, from which the following two solutions are found

$$x_i = \frac{a \pm \sqrt{a^2+b^2-c^2}}{b+c},$$

so clearly there are at most 2 real solutions for x_i and in order for the solutions to be real the radicand must be positive. From here, the corresponding solutions to the trigonometric equations are

$$q_i = 2\arctan x_i.$$

For each value of x_i, the arctan function produces two solutions which differ by π. When these solutions are multiplied by 2, the corresponding solutions for q_i differ by 2π and are in fact the same solution. Thus each value of x_i produces one value of

q_i and since there are a total of two solutions for x_i there are a total of two solutions for q_i.

Another possibility is the following substitution

$$\sin q_i = y_i$$

and

$$\cos q_i = x_i.$$

This relation between the new variables must be also added as an equation

$$x_i^2 + y_i^2 = 1.$$

Solving the system of two equations for x_i and y_i gives

$$q_i = \arctan_2 \frac{y_i}{x_i}$$

or

$$q_i = 2 \arctan_2 \frac{y_i}{1 - x_i}.$$

The substitution for trigonometric functions in kinematic equations (such as given above) leads to a polynomial of n variables which can appear in quadratic forms. Thus, in the worst case, we end up with a $2n$-order polynomial. However, the peculiarities of a particular mechanism can decrease the polynomial's order. Once the polynomial has been defined, the method of solving it is considered to be well known, however a general method of solution does not exist. Additional troubles arise because polynomial equations may have some extraneous solutions [40]. It is also known that closed-form solutions are in general possible only for the roots of polynomials of the fifth or lower order and only in special cases for higher order polynomials. In the sixties, a method of Grübner bases was introduced. The method was aimed for symbolic solving polynomials with several variables. However, it is rather inefficient and in complex examples required a lot of calculations. It appears from the literature that the most efficient are the elimination methods, whose theoretical background originates from the first half of nineteenth century. In kinematics the most common is the method of dialytic elimination [73].

3.3.2 Numerical Solutions to the Inverse Kinematics Problem

A numerical solution to the inverse kinematics problem is used when the system of equations has no algebraic solution, or when we want to develop a general solver for all the possible kinematic equations which might arise, regardless of the special properties of a particular mechanism under consideration. Numerical approaches are well suited for computer programming, as the same program can be used for all the various structures of mechanisms. When solving either the original system of trigonometric equations or the transformed system of the polynomial equations, any numerical method can be applied, such as a secant method or a Monte Carlo method. The advantage of numerical methods is their simplicity.

Iterative gradient methods are most common in kinematics. Among them the most well-known is the Newton-Raphson method [90] and its numerous variations. With iterative methods some initial approximate value of internal coordinates $\mathbf{q}(0)$ must be selected. It is clear that the function $\mathbf{p}_\mathrm{q}(0)$ in general does not comply with the selected values of the external coordinates $\mathbf{p}$, therefore

$$\mathbf{p} - \mathbf{p}_\mathrm{q}(0) \neq \mathbf{0}.$$

The goal of the iterative methods is to create a sequence of values of internal coordinates $\mathbf{q}(\kappa)$, $\kappa = 1, 2, \ldots$ which through iterations get closer to the solution $\mathbf{q}$, so that the difference between the desired values of external coordinates $\mathbf{p}$ and the actual values $\mathbf{p}_\mathrm{q}(\kappa)$ approaches zero

$$\mathbf{p} - \mathbf{p}_\mathrm{q}(\kappa) \rightarrow \mathbf{0}.$$

The integer $\kappa = 1, 2, \ldots$ represents the number of iterations. In numerical solutions we start with a numerical guess of the solution and after a number of iterations, we finish with an approximate solution which is more accurate than the original guess. Only in rare cases does a numerical solution yield the exact solution $\mathbf{q}$, where

$$\mathbf{p} - \mathbf{p}_\mathrm{q} = \mathbf{0}$$

is obtained. The measure of the difference between the momentary and exact solution is given by the error, usually in the form of an Euclidean norm

$$\epsilon(\kappa) = \|\mathbf{p} - \mathbf{p}_\mathrm{q}(\kappa)\| = \sqrt{(\mathbf{p} - \mathbf{p}_\mathrm{q}(\kappa))^\mathrm{T}(\mathbf{p} - \mathbf{p}_\mathrm{q}(\kappa))}. \tag{3.77}$$

If the numerical solution is going to be used in place of the exact solution, then after a number of iterations κ, the error must be smaller than a prescribed maximum acceptable value ϵ

$$\epsilon(\kappa) \leq \epsilon, \tag{3.78}$$

where ϵ is a small positive scalar. The error in the solution obtained using an iterative method is a consequence of truncating the infinite process of calculation which would be required to obtain the exact solution. When the error is decreasing during the process of additional iterations, we say that the method is converging to a solution. It is to be expected that the rate of convergence of a method depends on the mathematical nature of the numerical method and on the system of equations under consideration. If the system of equations is ill defined, i.e. when a small change in data produces a large difference in results, there will be convergence problems.

With the systems of trigonometric kinematic equations we are dealing with, there exist areas of solutions where numerical methods converge well and other areas where almost all numerical methods fail. These areas of failure are at, or nearby, the kinematic singularities of the mechanism. Kinematic singularities will be studied in more detail later. For now it is sufficient to say that the kinematic singularities of a mechanism are related to the singularities of the mechanism's Jacobian matrix.

This observation is evident with gradient methods which are based on the equation of inverse kinematics in Jacobian form

$$\dot{\mathbf{q}} = \mathbf{J}^{-1}\dot{\mathbf{p}}. \tag{3.79}$$

The equation is only valid when the Jacobian matrix $\mathbf{J}$ is not singular. At certain values of external coordinates $\mathbf{q}$, the Jacobian matrix degenerates to a singular matrix and in that case it is evident from the above equation that an infinitesimally small displacement in external coordinates is transformed into an infinite displacement in internal coordinates, i.e.

$$d\mathbf{q} = \mathbf{J}^{-1}d\mathbf{p} \to \infty, \tag{3.80}$$

which means that the mechanism in this particular configuration cannot perform the desired displacement in external coordinates $d\mathbf{p}$. Kinematic singularities of a mechanism are a mathematical property of the system of kinematic equations, yet at the same time they are a physical property of the mechanism. Namely, when a mechanism is in a singular configuration, it cannot move in certain directions in the external coordinates. Kinematic singularities of serial mechanisms correspond to special poses of the mechanism. In this discussion of numerical methods we shall assume for the sake of simplicity that the Jacobian matrix does not degenerate to a singular matrix at any value of internal coordinates under consideration. These special cases will be studied later.

With the Newton-Raphson method we numerically integrate (3.79) or (3.80) using the Euler formula and the difference equation

$$\mathbf{q}(\kappa+1) = \mathbf{q}(\kappa) + \mathbf{J}(\kappa)^{-1}\big(\mathbf{p} - \mathbf{p}_\mathrm{q}(\kappa)\big), \quad \kappa = 0, 1, \ldots, \tag{3.81}$$

as it is not difficult to see that

$$d\mathbf{q}(\kappa) = \mathbf{J}(\kappa)^{-1}\big(\mathbf{p} - \mathbf{p}_\mathrm{q}(\kappa)\big) \tag{3.82}$$

and

$$d\mathbf{p}(\kappa) = \mathbf{p} - \mathbf{p}_\mathrm{q}(\kappa). \tag{3.83}$$

When the initial approximation is not adequate or the system of equations, which we are solving, is ill defined, this method can diverge. A better approach is realized by modifying the Newton-Raphson method and using a scalar weighting factor α which multiplies the amount of the descent

$$\mathbf{q}(\kappa+1) = \mathbf{q}(\kappa) + \alpha\mathbf{J}(\kappa)^{-1}\big(\mathbf{p} - \mathbf{p}_\mathrm{q}(\kappa)\big), \quad \kappa = 0, 1, \ldots. \tag{3.84}$$

This is called the ordinary gradient method or the method of steepest descent, where the scalar α is called the step size. The step size is estimated based on how the

Table 3.5 Desired values of external coordinates and initial approximative values of internal coordinates

i	1	2	3
p_i	1.50 m	1.60 m	30.00°
$q_i(0)$	10.00°	25.00°	−25.00°

method converges in the particular problem. When the step is smaller, the method converges slowly but reliably. By increasing α the opposite effect is achieved. When using the ordinary gradient method, solving the inverse kinematics problems uses the following steps:

Step 1 Set $\kappa = 0$, since this is the first iteration. The first approximate values of internal coordinates $\mathbf{q}(0)$ are selected for given values of external coordinates $\mathbf{p}$. We also choose the step size α and the maximum error ϵ. We then continue with the next step.

Step 2 Vector function $\mathbf{p}_{\mathrm{q}}(\kappa)$ is calculated for the current values $\mathbf{q}(\kappa)$. Since $\mathbf{p}$ is given the error $\epsilon(\kappa)$ can be obtained (3.77). If $\epsilon(\kappa) > \epsilon$, we continue the iterative process, when $\epsilon(\kappa) \leq \epsilon$, the iteration is stopped. The current values of internal coordinates $\mathbf{q}(\kappa)$ represent the approximate solution of inverse kinematics problem.

Step 3 We calculate the Jacobian matrix $\mathbf{J}(\kappa)$ together with its inverse $\mathbf{J}(\kappa)^{-1}$. The gradient $d\mathbf{q}(\kappa)$ can be obtained as given in (3.83).

Step 4 If the new value of κ exceeds the maximal permitted number of iterations, the process is concluded, as we were unable to obtain the solution in this case, i.e. the method is not converging. Otherwise we calculate a new approximative $\mathbf{q}(\kappa)$ (3.84), we increment the number of iterations κ by one and return to Step 2, making another loop/iteration.

For an example of a numerical solution to kinematic equations, we consider the 3R mechanism whose inverse kinematics problem was algebraically calculated in the preceding section. We start with the system of kinematic equations (3.74), where the values of the external coordinates p_1, p_2, and p_3 are given, and we have to calculate the corresponding values of internal coordinates q_1, q_2, and q_3. We shall assume that the links are of equal lengths, $l_1 = l_2 = l_3 = 1.0$ m. Table 3.5 includes the desired values of the external coordinates and the selected initial values of internal coordinates.

Convergence of the ordinary gradient method for the given data is shown by the error value $\epsilon(\kappa)$ as function of the number of iterations (Fig. 3.12). The coordinates p_1, p_2 in the error $\epsilon(\kappa)$ have units of meters, while the coordinate p_3 is in radians. Here it should be pointed out that combining of different units is problematic and must be carried out vary carefully. From the numerical point of view it is advantageous that the units are selected in such a way that the values of the error are of the same order of magnitude. In Fig. 3.12 we observe how the convergence of the method is changing with various step sizes α. When the step size is too large (in our example this is when $\alpha = 2.1$) the method diverges and does not bring us to

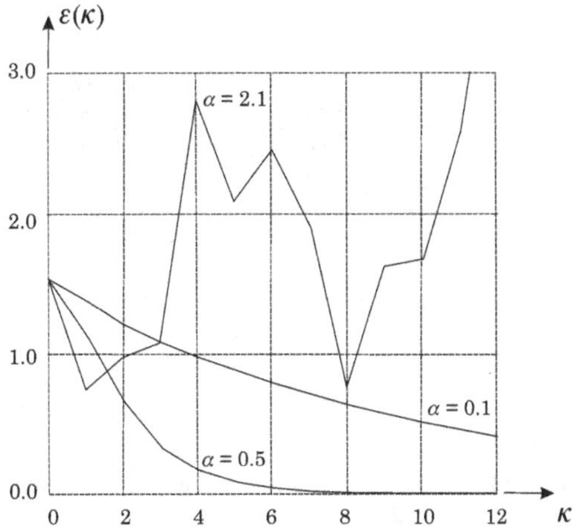

Fig. 3.12 Error $\epsilon(\kappa)$ as function of the number of iterations, when the step size is $\alpha = 0.1$, $\alpha = 0.5$, and $\alpha = 2.1$

a solution. When the step size is too short (when $\alpha = 0.1$) the method converges, however the number of iterations is unnecessarily high.

In Fig. 3.13 the mechanism configurations are shown, which belong to successive iterations using the different step sizes. We can observe that with $\alpha = 0.1$ and $\alpha = 0.5$ the mechanism converges towards the solution $q_1 = 9.45°$, $q_2 = 101.20°$, and $q_3 = -80.65°$. When $\alpha = 0.5$ the solution is found in only a few steps, with $\alpha = 0.1$ the process is slower, while with $\alpha = 2.1$ the method never converges and the solutions are dispersed all over the robot workspace, far from the actual solution.

Even with this simple example we realize the main disadvantages of a numerical solution of the inverse kinematics problem. First, there is the question of whether or not the method is converging for the given numerical data and how many steps are necessary. It is not surprising that for particular data we may require several thousand iterations to converge to a solution. The second risk is that numerical methods only lead to a single solution even though the kinematic equations probably have several. There is no general rule for how to reliably determine all solutions and for what amount of time would be required for solutions to be found.

By increasing the number of iterations in a numerical solution the computation time is increased and it is quite possible that the method does not yield the result in an acceptable amount of time. We can improve the process to some extent by selecting a good initial approximation. More efficiently, we can shorten the computation time by accelerating the convergence of the method in several ways, or by decreasing the number of arithmetic operations in particular steps. Many proposed improvements of the gradient methods are based on an observation which is also evident in our example. Namely, for given numerical data, it is possible to find a step size where the number of iterations is minimum. To this end, one may introduce a correction of the descent $d\mathbf{p}(\kappa)$ by taking into account the preceding iterations. This is the idea of the conjugate gradient method [45]

$$\mathbf{q}(\kappa + 1) = \mathbf{q}(\kappa) + \alpha \mathbf{J}(\kappa)^{-1}\mathbf{h}(\kappa),$$

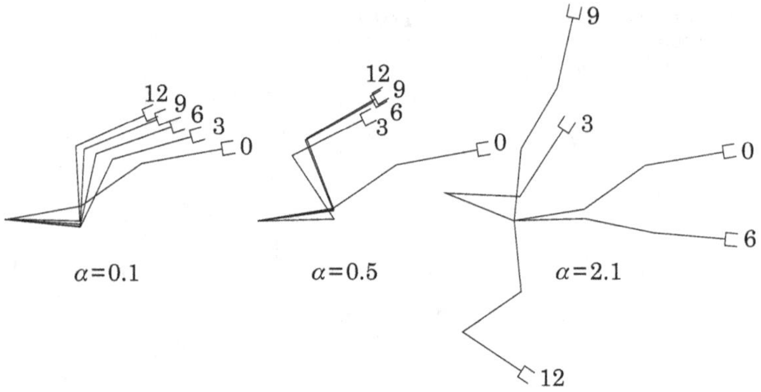

Fig. 3.13 Configurations of 3R mechanism at different iterations, when the step size is $\alpha = 0.1$, $\alpha = 0.5$, and $\alpha = 2.1$

$$\mathbf{h}(\kappa) = \mathbf{p} - \mathbf{p}_{\mathrm{q}}(\kappa) + \beta(\kappa - 1)\mathbf{h}(\kappa - 1),$$
$$\beta(\kappa - 1) = \frac{(\mathbf{p} - \mathbf{p}_{\mathrm{q}}(\kappa))^{\mathrm{T}}(\mathbf{p} - \mathbf{p}_{\mathrm{q}}(\kappa))}{(\mathbf{p} - \mathbf{p}_{\mathrm{q}}(\kappa - 1))^{\mathrm{T}}(\mathbf{p} - \mathbf{p}_{\mathrm{q}}(\kappa - 1))}, \quad \kappa = 1, 2, \ldots.$$

Here, the first iteration is the same as with the ordinary gradient method. The conjugate gradient method can, with favorable data, significantly accelerate the convergence as compared to the ordinary gradient method.

With a second order Newton-Raphson method, each iteration has two stages

$$\mathbf{q}(\kappa + 1)' = \mathbf{q}(\kappa) + \mathbf{J}(\kappa)^{-1}\big(\mathbf{p} - \mathbf{p}_{\mathrm{q}}(\kappa)\big),$$
$$\mathbf{q}(\kappa + 1) = \mathbf{q}(\kappa + 1)' + \mathbf{J}(\kappa + 1)'^{-1}\big(\mathbf{p} - \mathbf{p}_{\mathrm{q}}(\kappa + 1)'\big), \quad \kappa = 0, 1, \ldots.$$

Here, the error diminishes faster from one iteration to the next, however there is no savings in the overall number of arithmetic operations when compared to the first order Newton-Raphson method. The number of arithmetic operations is significantly smaller when using the following approach

$$\mathbf{q}(\kappa + 1)' = \mathbf{q}(\kappa) + \mathbf{J}(\kappa)^{-1}\big(\mathbf{p} - \mathbf{p}_{\mathrm{q}}(\kappa)\big),$$
$$\mathbf{q}(\kappa + 1) = \mathbf{q}(\kappa + 1)' + \mathbf{J}(\kappa)^{-1}\big(\mathbf{p} - \mathbf{p}_{\mathrm{q}}(\kappa + 1)'\big), \quad \kappa = 0, 1, \ldots,$$

where the Jacobian matrix and its inverse are calculated only once, while the convergence time does not significantly increase [24]. Note that most of the arithmetic operations are related to calculating the Jacobian matrix and its inverse.

In practice, the inverse of the Jacobian matrix does not play a major role when the system of equations is well defined and an approximate value is sufficient. Thus, with the gradient method the Jacobian matrix and its inverse can be calculated at each second iteration or even less frequently. Figure 3.14 corresponds to the ordinary gradient method applied to the 3R mechanism and for the data presented in

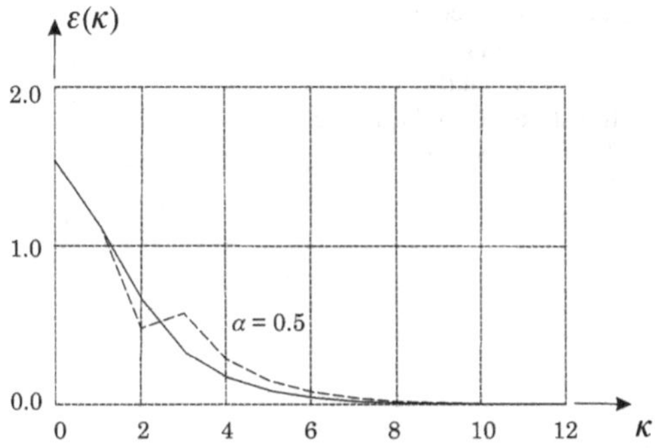

Fig. 3.14 Comparison of errors $\epsilon(\kappa)$ as functions of the number of iterations at step size $\alpha = 0.5$, when the Jacobian matrix is calculated at each step (*solid line*) and when it is calculated at every third step (*dashed line*)

Table 3.5. The figure shows the errors $\epsilon(\kappa)$ as functions of the number of iterations for the case when the Jacobian matrix is calculated at each iteration and when it is calculated at each third iteration. The selected step size is $\alpha = 0.5$. The latter approach is less stable during initial iterations when the values of internal coordinates are still far from the correct solution, while later it converges smoothly, as with the original method. It leads to the solution in approximately the same number of iterations, although the total number of arithmetic operations and the computation time are reduced about 50 %.

When the step size is sufficiently small, i.e. at small values of α, the following approximation

$$\mathbf{J}(\kappa)^{-1} \doteq \xi \mathbf{J}(\kappa)^{\mathrm{T}},$$

where ξ is a scalar, has been found to be quite efficient. It avoids time consuming matrix inversion, but negatively effects convergence. The approach can be considered in cases when we have a good initial approximation for the values of the internal coordinates [77].

To overcome the problems of gradient methods, the continuation method was developed in the sixties and was later applied to kinematics [87]. In place of the original system of equations, the continuation method begins with a substitute system of equations, which has a number of solutions equal to the number of solutions to the original system. The solutions to the substitute system are known. Through iterations, the parameters of the substitute system are gradually changed, morphing the substitute system into the original system. In the process, the solutions are tracked and at the point where the substitute system has completely morphed to the original system, you have all the solutions to the original system. Complex values of parameters can also be used. The difference between the various continuation methods is the manner in which the parameters of the substitute system are changed. Continuation methods are most useful when it is necessary to find all the solutions of the inverse kinematics problem and computation time is not critical. Continuation methods can be used in combination with different numerical and symbolic methods [73].

In conclusion, we state that there are no difficulties in finding solutions to the direct kinematics problem of serial mechanisms. However, solutions to the inverse

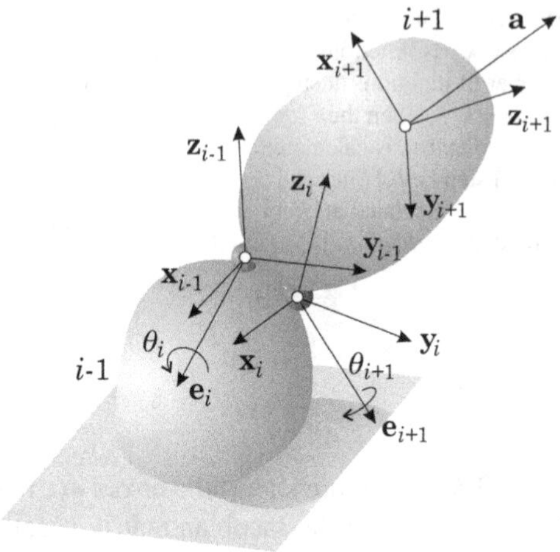

Fig. 3.15 Displacement of body $i+1$ with respect to body $i-1$, when the rotational axes $\mathbf{e}_i$ and $\mathbf{e}_{i+1}$ are fixed with respect to body $i-1$

kinematics problem is more difficult to obtain. If the solution can be found, the algebraic solution is preferred, however the algebraic solution exists only for special cases of mechanisms. For the general case, only numerical methods are available, which are in many aspects challenging, as either they do not lead to the solution or they find only a single solution and they require significantly longer computing time, to the point where their use may become unreasonable.

3.4 Serial Mechanisms with Fixed Rotational Axes

To this point we have examined mechanisms where links and joints occur in series. In this way a displacement in one joint provokes the displacement of all subsequent links and joints. In this section we shall briefly describe mechanisms where the joint axes are fixed with respect to the base or reference coordinate frame. One such a mechanism is the universal joint, realized by the use of a spherical ball and socket. An example is the mechanical computer mouse, where by rotating the ball we propel two perpendicular wheels rotating around two joint axes which are fixed with respect to the housing of the mouse.

In Fig. 3.15 the body $i+1$ is rotating around the fixed body $i-1$ with respect to the rotation vectors $\mathbf{e}_i$ and $\mathbf{e}_{i+1}$, as shown. Here, both rotation vectors are fixed to body $i-1$ and therefore do not move in the frame $\mathbf{x}_{i-1}$, $\mathbf{y}_{i-1}$, $\mathbf{z}_{i-1}$. It is our aim to calculate the values of a vector $\mathbf{a}$, which is attached to the body $i+1$, in the coordinate frame $\mathbf{x}_{i-1}$, $\mathbf{y}_{i-1}$, $\mathbf{z}_{i-1}$. This is mathematically expressed as follows

$$\mathbf{a}^{(i-1)} = \mathbf{A}_{i-1,i}\mathbf{A}_{i,i+1}\mathbf{a}^{(i+1)},$$

where $\mathbf{A}_{i-1,i}$ and $\mathbf{A}_{i,i+1}$ are rotation matrices describing the rotations around the vectors $\mathbf{e}_i$ and $\mathbf{e}_{i+1}$. Particular to this problem is that the vector $\mathbf{e}_i^{(i-1)}$, creating the matrix $\mathbf{A}_{i-1,i}$, is constant, while the vector $\mathbf{e}_{i+1}^{(i)}$, creating the matrix $\mathbf{A}_{i,i+1}$, is not.

The latter vector is fixed with respect to the frame $\mathbf{x}_{i-1}$, $\mathbf{y}_{i-1}$, $\mathbf{z}_{i-1}$, therefore its image $\mathbf{e}_{i+1}^{(i-1)}$ is constant. As there is

$$\mathbf{e}_{i+1}^{(i)} = \mathbf{A}_{i-1,i}^{\mathrm{T}} \mathbf{e}_{i+1}^{(i-1)},$$

we see the vector $\mathbf{e}_{i+1}^{(i)}$ depends on the angle θ_i and it follows that the matrix $\mathbf{A}_{i,i+1}$ is a function of the angles θ_i and θ_{i+1}. With this observation we could conclude our description, however a mechanism with fixed joints has an interesting mathematical particularity worth discussing.

Let us assume that the rotational vector $\mathbf{e}$, corresponding to the rotation matrix $\mathbf{A}$, changes in accordance with a rotation represented by the matrix $\mathbf{B}$. Let it be

$$\mathbf{e} = \mathbf{B}^{\mathrm{T}} \bar{\mathbf{e}},$$

where $\bar{\mathbf{e}}$ is a constant vector whose elements are known. Therefore, the vector $\bar{\mathbf{e}}$ is used instead of vector $\mathbf{e}$ when calculating the matrix $\mathbf{A}$. The matrix is written by the use of formula (1.48) as follows

$$\mathbf{A} = \boldsymbol{\Delta} \sin\theta + (\mathbf{I} - \boldsymbol{\Lambda}) \cos\theta + \boldsymbol{\Lambda},$$

where we calculate the matrices $\boldsymbol{\Lambda}$ and $\boldsymbol{\Delta}$ $\mathbf{B}^{\mathrm{T}}\bar{\mathbf{e}}$

$$\boldsymbol{\Lambda} = \mathbf{B}^{\mathrm{T}} \bar{\mathbf{e}} \bar{\mathbf{e}}^{\mathrm{T}} \mathbf{B},$$
$$\boldsymbol{\Delta} = \left(\mathbf{B}^{\mathrm{T}} \bar{\mathbf{e}}\right) \otimes \mathbf{I}.$$

Using the notation

$$\bar{\boldsymbol{\Lambda}} = \bar{\mathbf{e}} \bar{\mathbf{e}}^{\mathrm{T}},$$
$$\bar{\boldsymbol{\Delta}} = \bar{\mathbf{e}} \otimes \mathbf{I},$$

we immediately realize the relation

$$\boldsymbol{\Lambda} = \mathbf{B}^{\mathrm{T}} \bar{\boldsymbol{\Lambda}} \mathbf{B},$$

while some more skills are necessary to calculate the matrix $\boldsymbol{\Delta}$ using the rules (1.27)

$$\left(\mathbf{B}^{\mathrm{T}} \bar{\mathbf{e}}\right) \otimes \mathbf{I} = \mathbf{B}^{\mathrm{T}} \mathbf{B} \left(\left(\mathbf{B}^{\mathrm{T}} \bar{\mathbf{e}}\right) \otimes \mathbf{I}\right) = \mathbf{B}^{\mathrm{T}} \left(\left(\mathbf{B} \mathbf{B}^{\mathrm{T}} \bar{\mathbf{e}}\right) \otimes \mathbf{B}\right) = \mathbf{B}^{\mathrm{T}} (\bar{\mathbf{e}} \otimes \mathbf{I}) \mathbf{B},$$

where we can recognize

$$\boldsymbol{\Delta} = \mathbf{B}^{\mathrm{T}} \bar{\boldsymbol{\Delta}} \mathbf{B}.$$

While assuming

$$\bar{\mathbf{A}} = \bar{\boldsymbol{\Delta}} \sin\theta + (\mathbf{I} - \bar{\boldsymbol{\Lambda}}) \cos\theta + \bar{\boldsymbol{\Lambda}},$$

we write

$$\mathbf{A} = \mathbf{B}^{\mathrm{T}} \bar{\mathbf{A}} \mathbf{B},$$

where the matrix $\bar{\mathbf{A}}$ is created by the constant vector $\bar{\mathbf{e}}$ and is therefore independent of the transformation $\mathbf{B}$.

This short mathematical excursion was necessary to derive the following relation for the mechanism from Fig. 3.15

$$\mathbf{A}_{i,i+1} = \mathbf{A}_{i-1,i}^{\mathrm{T}} \bar{\mathbf{A}}_{i,i+1} \mathbf{A}_{i-1,i}, \tag{3.85}$$

where we create the rotation matrix $\bar{\mathbf{A}}_{i,i+1}$ with the constant vector $\mathbf{e}_{i+1}^{(i-1)}$, which is a parameter that is independent from the rotational angle θ_i. It holds

$$\mathbf{a}^{(i-1)} = \mathbf{A}_{i-1,i} \mathbf{A}_{i-1,i}^{\mathrm{T}} \bar{\mathbf{A}}_{i,i+1} \mathbf{A}_{i-1,i} \mathbf{a}^{(i+1)} = \bar{\mathbf{A}}_{i,i+1} \mathbf{A}_{i-1,i} \mathbf{a}^{(i+1)}. \tag{3.86}$$

It is surprising that in this formula, the matrices occur in the opposite order as observed in serial mechanisms, where the joints move along with the links.

It remains to generalize the formula for a serial mechanism where all rotational axes are fixed with respect to the reference coordinate frame. Let the mechanism have n rotational degrees of freedom, which are geometrically described by the vectors $\mathbf{e}_1, \mathbf{e}_2, \ldots, \mathbf{e}_n$, which are constant with respect to the reference frame. The rotation matrices $\mathbf{A}_{i-1,i}$, $i = 1, 2, \ldots, n$ are calculated in the usual way with formula (1.48), where we use the rotational vectors $\mathbf{e}_i^{(0)}$, $i = 1, 2, \ldots, n$ expressed in the frame $\mathbf{x}_0, \mathbf{y}_0, \mathbf{z}_0$. When applying vector parameters of the mechanism with the reference pose, where all frames are parallel to the reference frame, and $\mathbf{e}_i^{(0)} = \mathbf{e}_i^{(i-1)}$, $i = 1, 2, \ldots, n$, the calculation of rotation matrices $\mathbf{A}_{i-1,i}$, $i = 1, 2, \ldots, n$ for the mechanism with fixed rotational axes does not differ from the calculation introduced for the mechanisms where the joints are displaced together with the links. However, the rotation matrix transforming from the frame $\mathbf{x}_n, \mathbf{y}_n, \mathbf{z}_n$ into the frame $\mathbf{x}_0, \mathbf{y}_0, \mathbf{z}_0$ has the following unusual form

$$\mathbf{A}_{0,n} = \mathbf{A}_{n-1,n} \mathbf{A}_{n-2,n-1} \cdots \mathbf{A}_{0,1}. \tag{3.87}$$

Here, the intermediate rotation matrices occur in the opposite order to the that encountered in serial mechanisms where the joints move together with the links.

Chapter 4
Evaluation of Mechanisms

Abstract Among the various criteria used to represent and evaluate the functional properties of a mechanism we describe the reachable and the dexterous workspace expressed by their volume and compactness. We also described the kinematic flexibility associated with the number of inverse kinematics solutions, the manipulability and the kinematic index associated with the kinematic singularities. Attention is given to the associated computational aspects, in particular to the determination and visualization of robot workspaces, which usually requires an enormous number of numerical operations.

The functional properties of a mechanism, such as reachability or sensitivity to manufacturing tolerances, are determined by its kinematic structure. When describing functional properties of a mechanism, various mathematical criteria are used, among them are the reachable and dexterous workspace, the manipulability index and the kinematic index. In this chapter we will consider several of the more important criteria for evaluating the functionality of a mechanism and we will learn how to improve some properties of the mechanism. Special attention will be paid to kinematic singularities.

4.1 Workspaces

In industrial practice, the workspace of a robot mechanism is the spatial volume which is reachable by a selected point of the mechanism [77]. From a broader aspect, the workspace is a region of a mechanism's operation where the mechanism possesses certain properties, e.g. the ability to transfer a load or to reach a prescribed velocity. In addition to a mechanism's kinematics, we also consider its static and dynamic properties when discussing its workspace.

The workspace of a mechanism depends on its structure, i.e. on the number of degrees of freedom, their arrangement, the lengths of the segments and constraints in the motion of particular joint coordinates. We will consider only two types of workspaces. The reachable workspace is related to the positions of a control point

Intelligent Systems, Control and Automation: Science and Engineering 60,
DOI 10.1007/978-94-007-4522-3_4,

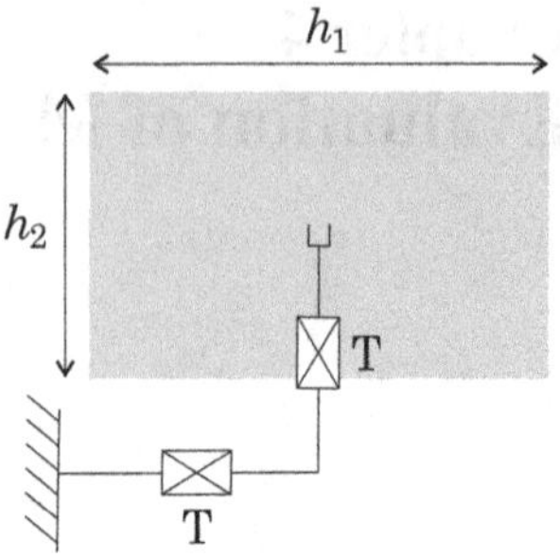

Fig. 4.1 Reachable workspace of 2T mechanism

on the mechanism, while the dexterous workspace is also related to the orientations of the body to which the control point belongs [41]. In kinematics, these are the most important types of workspaces.

4.1.1 Reachable Workspace

The reachable workspace is the region encompassing all positions which can be reached by a selected end-point of a mechanism, regardless of the orientation of the last robot segment [41]. This is a basic kinematic property of a mechanism, which for short is called the reachability. It appears to be reasonable to design mechanisms with large reachable workspaces. However, let us point out early on that reachability is not such an omnipotent criterion, if not related to some other property of a mechanism. Namely, if a mechanism is able to reach a certain point in space, this does not mean that a required task can be accomplished at this point. The mechanism reachability is assessed in positional coordinates. With a planar mechanism, the workspace is represented by an area, while with spatial mechanisms we speak about a volume.

Consider first the reachable workspaces of some planar mechanisms which have two degrees of freedom and where the mechanism's end-point is represented by the center of the gripper. On this basis we later introduce the reachable workspaces of spatial mechanisms with three degrees of freedom. The simplest is the reachable workspace of planar 2T mechanism. The mechanism end-point outlines a plane which is parallel to both translations. When the translations with the coordinates q_1 and q_2 are constrained, the reachable workspace is represented by the rectangle with the sides $h_1 = q_{1\max} - q_{1\min}$ and $h_2 = q_{2\max} - q_{2\min}$, as shown in Fig. 4.1.

The reachable workspace of the planar RT mechanism with rotational coordinate q_1 and translational coordinate q_2 is a sector of a ring with external radius $q_{2\max}$ and internal radius $q_{2\min}$. In this way the thickness of the ring is $h_2 = q_{2\max} - q_{2\min}$, while its angular dimension depends on the constraints (i.e. motion limits) of the rotational coordinate $h_1 = q_{1\max} - q_{1\min}$. Figure 4.2 shows the workspace of the mechanism when $h_1 = 2\pi/3$ (left) and when $h_1 \geq 2\pi$ (right).

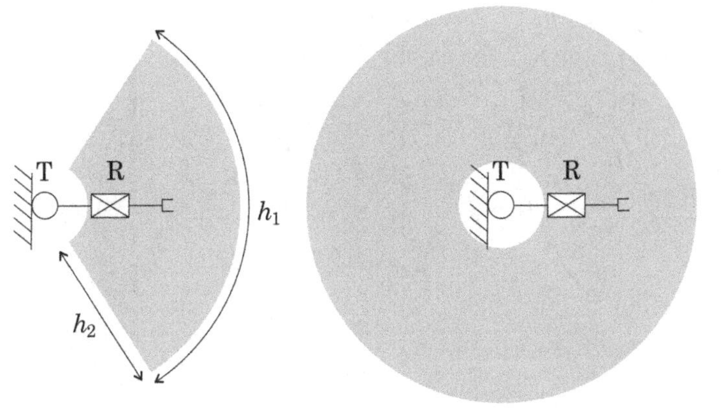

Fig. 4.2 Reachable workspace of the RT mechanism

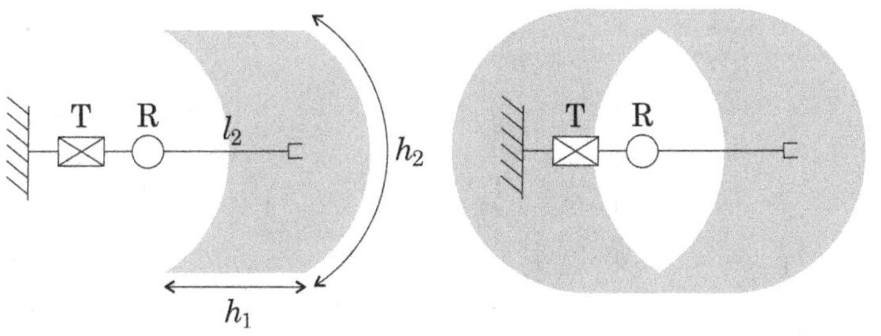

Fig. 4.3 Reachable workspace of the TR mechanism

Quite different is the reachable workspace of the planar TR mechanism with translational coordinate q_1 and rotational coordinate q_2. The shape of this workspace is determined by the arc of the angle $h_2 = q_{2\,\mathrm{max}} - q_{2\,\mathrm{min}}$, while its radius is the length of the segment l_2. The arc is expanded in the direction of the translational coordinate, so that the thickness of the obtained form is $h_1 = q_{1\,\mathrm{max}} - q_{1\,\mathrm{min}}$. Figure 4.3 shows the reachable workspace of the TR mechanism, where $h_2 = \pi/2$ (left) and $h_2 = 2\pi$ (right). Here, the angle h_2 is positioned symmetrically with respect to the translational axis.

The reachable workspace of the planar 2R mechanism is also determined with the angle $h_2 = q_{2\,\mathrm{max}} - q_{2\,\mathrm{min}}$ and the radius represented by the segment length l_2. In this case the arc is expanded in the plane around the first rotational axis at the distance l_1 by the angle $h_1 = q_{1\,\mathrm{max}} - q_{1\,\mathrm{min}}$. The form obtained is shown in Fig. 4.4 for the two cases when $h_1 = 2\pi/3$ and $h_2 = \pi$ (left) and the angle h_2 is placed symmetrically with respect to the first segment axis, and second case when $h_1 \geq 2\pi$ and $h_2 \geq 2\pi$ (right). The inner side of the workspace is determined by the arc with the radius $(l_1^2 + l_2^2 + l_1 l_2 \cos(h_2/2))^{1/2}$, while the external side is the arc with the radius $l_1 + l_2$. In the second case, when the coordinates are not constrained, the

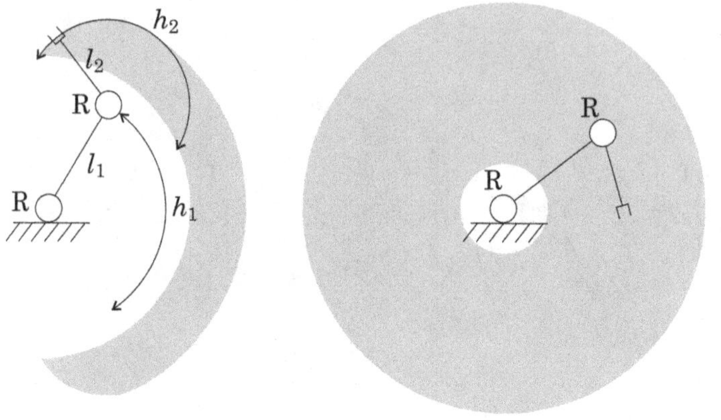

Fig. 4.4 Reachable workspace of the 2R mechanism

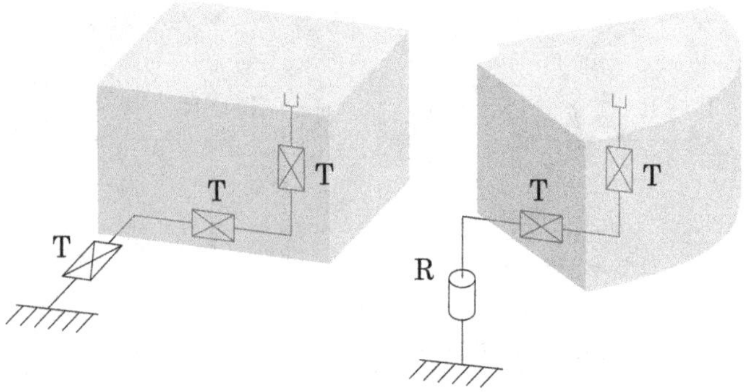

Fig. 4.5 The reachable workspace of the TTT and RTT mechanisms

reachable workspace of the 2R mechanism is a ring with inner radius $|l_1 - l_2|$ and external radius $l_1 + l_2$.

We now add an additional degree of freedom, either a translation T or a rotation R, to the mechanisms 2T, RT, TR and 2R in such a way that the motion of the three degree of freedom mechanism that results will be spatial and the workspace three-dimensional. We consider mechanisms which are most common amongst industrial robot manipulators [77] and were previously shown in Fig. 2.11.

When adding to the 2T mechanism a translation which is perpendicular to the original translation, the Cartesian mechanism TTT is obtained. Its reachable workspace is a rectangular cube as shown in Fig. 4.5. The RTT mechanism is obtained by adding a rotation to the bottom of the 2T mechanism. The reachable workspace results from rotation of a rectangle. As the shape of this workspace is cylindrical (Fig. 4.5), the RTT mechanism is called cylindrical.

The cylindrical mechanism is also obtained when adding a translation to the bottom of the RT mechanism. The reachable workspace of the obtained TRT mecha-

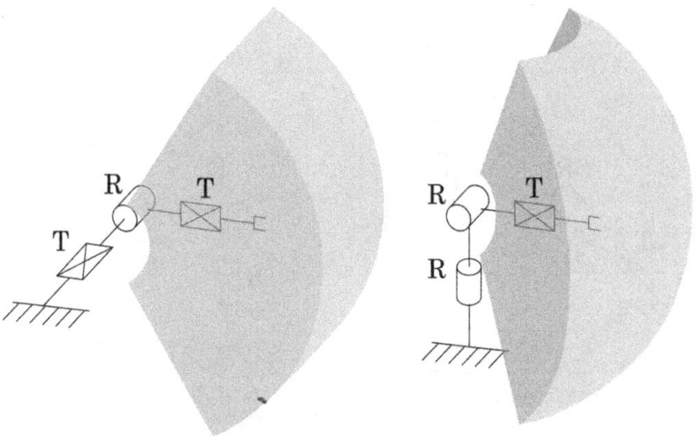

Fig. 4.6 Reachable workspace of the TRT and RTT mechanisms

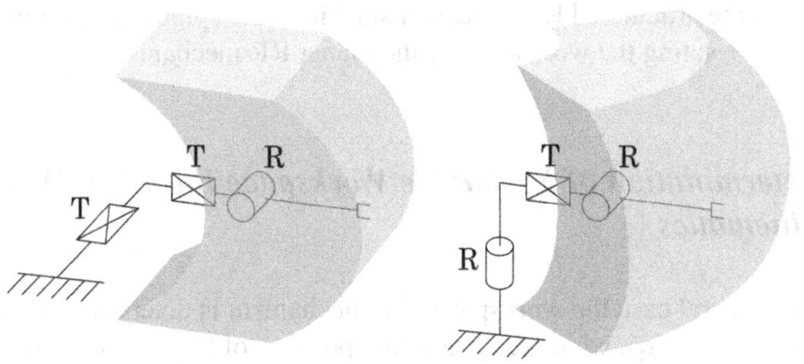

Fig. 4.7 Reachable workspace of the TTR and RTR mechanisms

nism has the cylindrical shape shown in Fig. 4.6. When adding a rotation perpendicularly to the RT mechanism, the RRT mechanism is obtained with the spindle-shaped reachable workspace resulting from rotating the sector of a ring, as shown in Fig. 4.6. The RRT mechanism is therefore called spherical.

The TTR mechanism is obtained when adding a translation to the bottom of the planar TR mechanism. The reachable workspace of the resulting mechanism is shown in Fig. 4.7. However, its shape is not cylindrical. By adding a rotation to the TR mechanism, the reachable workspace of the RTR mechanism is obtained. In special conditions it has the shape of the cut-off ring, as in the case shown in Fig. 4.7.

Among the industrial robot manipulators of today, there prevail those where the positional part of the mechanism has either the TRR or RRR arrangement. The mechanisms are obtained by adding to the original RR planar mechanism either a translation or rotation, as shown in Fig. 4.8. The properties of the reachable workspace of the TRR mechanism, i.e. the so called Scara mechanism, predominantly depend on the properties of the original planar RR mechanism which were described earlier. Similar consideration also holds for the geometry of the reachable

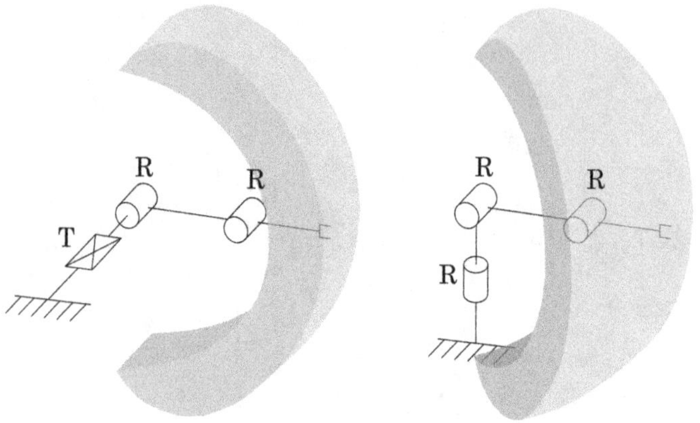

Fig. 4.8 Reachable workspace of the TRR and RRR mechanisms

workspace of the articulated RRR mechanism. Here, the spindle-shaped workspace is obtained by rotating the workspace of the planar RR mechanism.

4.1.2 Determination of Reachable Workspace Based on Direct Kinematics

In the most general case the workspace of a mechanism is determined by calculating the vector $\mathbf{r}_{0,n+1}(\mathbf{q})$, which represents the position of the mechanism end-point, while using (3.5) or (3.7) for all possible values of joint coordinates $\mathbf{q}$. The points described by the vector $\mathbf{r}_{0,n+1}$ determine the reachable workspace of the mechanism. In practical situations such calculation is impossible as it would consist of nested infinite loops in a computer program. The procedure is executed in such a way that the domain of each particular coordinate is divided into a finite number of discrete values. A selected value of joint coordinate q_i is given as a linear combination

$$q_i(k_i) = q_{i\,\min} + k_i \frac{q_{i\,\max} - q_{i\,\min}}{K_i}, \qquad k_i = 0, 1, \ldots, K_i,$$

where $K_i + 1$ different values are assigned to each joint coordinate, while

$$dq_i = \frac{q_{i\,\max} - q_{i\,\min}}{K_i}$$

is the increment between two neighboring values of the coordinate q_i. To determine the reachable workspace, it is necessary to calculate the vector $\mathbf{r}_{0,n+1}$

$$(K_1 + 1)(K_2 + 1) \cdots (K_n + 1)$$

times. The number of necessary calculations steeply increases depending on the number of selected values in the domain and even more on the number of joint coordinates of a mechanism. The calculation of the reachable workspace of a mechanism with multiple degrees of freedom can require a great deal of time.

The computer program, calculating the points of reachable workspace of a serial mechanism based on formula (3.5), encompasses $i = 1, 2, \dots, n$ nested loops and runs as follows:

loop 1 The value of the coordinate q_1 is increased step by step from $q_{1\,\min}$ to $q_{1\,\max}$ in intervals of dq_1. For each new value of the coordinate q_1 we calculate

$$\mathbf{r}_{0,2}^{(0)}(q_1) = \mathbf{r}_{0,1}^{(0)} + \mathbf{r}_{1,2}^{(0)}(q_1)$$

and afterwards we switch into the loop 2. When we sweep through all the values q_1, the procedure can be terminated.

loop 2 At a momentary value of the coordinate q_1, we increment the value of coordinate q_2 from $q_{2\min}$ to $q_{2\max}$ in intervals of dq_2. For each new value of the coordinate q_2, we calculate

$$\mathbf{r}_{0,3}^{(0)}(q_1, q_2) = \mathbf{r}_{0,2}^{(0)}(q_1) + \mathbf{r}_{2,3}^{(0)}(q_1, q_2)$$

and afterwards we switch to the loop 3. After sweeping through all values q_2, we return to the beginning of the loop 1.

$$\vdots$$

loop *n* With the momentary values of the coordinates $q_1, q_2, \dots, q_{n-1}$, we increment the value of the coordinate q_n from $q_{n\,\min}$ to $q_{n\,\max}$ in intervals of dq_n. For each new value of the coordinate q_n, we calculate

$$\mathbf{r}_{0,n+1}^{(0)}(q_1, q_2, \dots, q_n) = \mathbf{r}_{0,n}^{(0)}(q_1, q_2, \dots, q_{n-1}) + \mathbf{r}_{n,n+1}^{(0)}(q_1, q_2, \dots, q_n).$$

After sweeping through all values q_n, we return to the beginning of the loop $n-1$.

Figure 4.9 shows the reachable workspace (area) of the planar 3R mechanism, which was calculated by such computer program. The equations necessary for calculation of vector $\mathbf{r}_{0,4}$ were already derived in Sect. 3.3:

$$\mathbf{r}_{0,2}^{(0)} = \begin{bmatrix} l_1 \cos q_1 \\ l_1 \sin q_1 \\ 0 \end{bmatrix},$$

$$\mathbf{r}_{0,3}^{(0)} = \mathbf{r}_{0,2}^{(0)} + \begin{bmatrix} l_2 \cos(q_1 + q_2) \\ l_2 \sin(q_1 + q_2) \\ 0 \end{bmatrix},$$

$$\mathbf{r}_{0,4}^{(0)} = \mathbf{r}_{0,3}^{(0)} + \begin{bmatrix} l_3 \cos(q_1 + q_2 + q_3) \\ l_3 \sin(q_1 + q_2 + q_3) \\ 0 \end{bmatrix}.$$

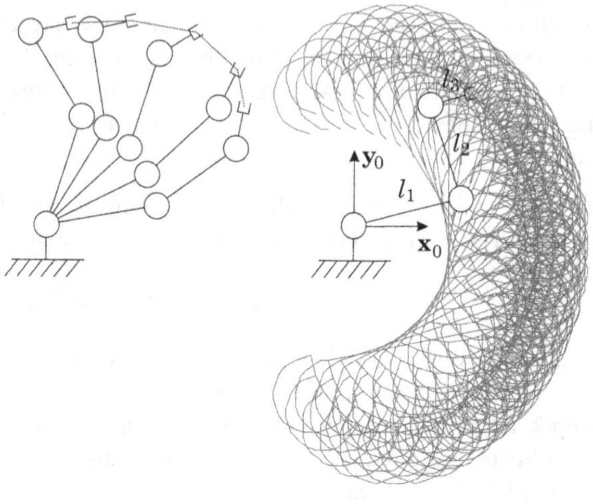

Fig. 4.9 Determining the reachable workspace of planar 3R mechanism by calculating the direct kinematics

While calculating the workspace, the following lengths of the segments were used $l_1 = 0.45$ m, $l_1 = 0.40$ m and $l_3 = 0.15$ m. The range of particular joint rotations was selected in intervals from $q_{1\,\mathrm{min}} = -2\pi/6$ to $q_{1\,\mathrm{max}} = 2\pi/6$, $q_{2\,\mathrm{min}} = -3\pi/6$ and $q_{2\,\mathrm{max}} = 3\pi/6$, and $q_{3\,\mathrm{min}} = -4\pi/6$ and $q_{3\,\mathrm{max}} = 4\pi/6$. The number of values used for particular coordinate was $K_1 = 15$, $K_2 = 15$, and $K_3 = 20$. The workspace in Fig. 4.9 is outlined by the line connecting the tops of the vectors $\mathbf{r}_{0,4}$ between the configurations swept by the program.

4.1.3 Determination of Reachable Workspace Based on Inverse Kinematics

Determining the reachable workspace can also be based on calculation of inverse kinematics. This method is practical only when a closed-form solution of the inverse kinematics problem is available. In this approach we divide the space around the mechanism (Fig. 4.10), where the reachable workspace is expected, into a finite number of positions, defined by vector $\mathbf{r}$ with the components r_x, r_y and r_z, in the following way

$$r_\mathrm{x} = r_\mathrm{xmin} + k_\mathrm{x}\frac{r_\mathrm{xmax} - r_\mathrm{xmin}}{K_\mathrm{x}}, \quad k_\mathrm{x} = 0, 1, \ldots, K_\mathrm{x},$$

$$r_\mathrm{y} = r_\mathrm{ymin} + k_\mathrm{y}\frac{r_\mathrm{ymax} - r_\mathrm{ymin}}{K_\mathrm{y}}, \quad k_\mathrm{y} = 0, 1, \ldots, K_\mathrm{y},$$

$$r_\mathrm{z} = r_\mathrm{zmin} + k_\mathrm{z}\frac{r_\mathrm{zmax} - r_\mathrm{zmin}}{K_\mathrm{z}}, \quad k_\mathrm{z} = 0, 1, \ldots, K_\mathrm{z},$$

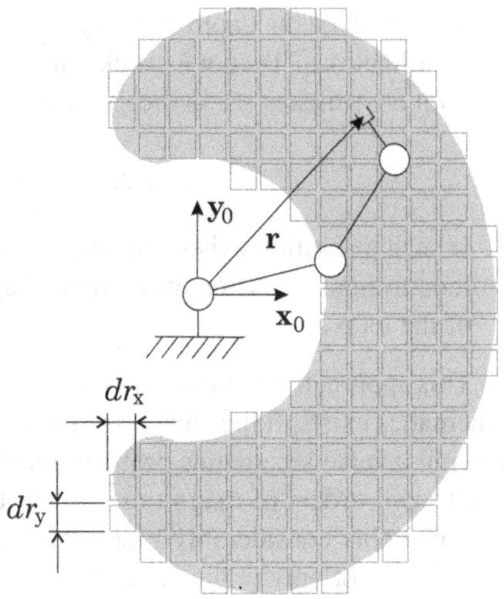

Fig. 4.10 Determining the reachable workspace of planar 3R mechanism by calculating inverse kinematics

where we prescribe $K_x + 1$, $K_y + 1$, and $K_z + 1$ different values to the coordinates r_x, r_y, and r_z respectively. The increments for particular coordinates are as follows

$$dr_x = \frac{r_{xmax} - r_{xmin}}{K_x},$$

$$dr_y = \frac{r_{ymax} - r_{ymin}}{K_y},$$

$$dr_z = \frac{r_{zmax} - r_{zmin}}{K_z}.$$

For each vector $\mathbf{r}$ determined in this way, we solve the inverse kinematics problem. If the solution to the inverse kinematics exists, where the joint coordinates are inside the range of prescribed constraints, then $\mathbf{r}$ is an element of the reachable workspace of the mechanism. The number of necessary calculations is

$$(K_x + 1)(K_y + 1)(K_z + 1).$$

A computer program, which determines the points within the reachable workspace based on the inverse kinematics includes three nested loops:

loop 1 The value of the coordinate r_x is incremented from r_{xmin} to r_{xmax} in intervals of dr_x. With each new value of the coordinate r_x we switch into the loop 2. After sweeping through all values r_x, the procedure is concluded.

loop 2 At a momentary value of r_x, the value of coordinate r_y is incremented from r_{ymin} to r_{ymax} in intervals of dr_y. With each new value of the coordinate r_y we switch into the loop 3. After sweeping through all values r_y, we return to the beginning of the loop 1.

loop 3 At momentary values of r_x and r_y, the value of the coordinate r_z is incremented from r_{zmin} to r_{zmax} in intervals of dr_z. With each new combination of the values of the coordinates r_x, r_y, and r_z we compute all the solutions of joint coordinates $\mathbf{q}$. Then we check whether at least one real solution exists, where all joint coordinates are in the range of constraints

$$q_{i\,\min} \le q_i \le q_{i\,\max}, \quad i = 1, 2, \ldots, n.$$

If such a solution exists, the point $\mathbf{r}$ is an element of the reachable workspace of the mechanism. After sweeping through all values r_z, we return to the beginning of the loop 2.

This approach is valid regardless of whether a closed-form solution of the inverse kinematics exists. If the closed-form solution does not exist, a numerical approach of solving inverse kinematics problem must be used. In this case the procedure will be one to two orders of magnitude slower than when using the closed-form solution, as it contains considerably more calculations. The difficulty of using a numerical solution is not only in the larger number of calculations. It is even more difficult to determine all the possible solutions of the joint coordinates for the given values of the external task coordinates.

Because of the large number of the required repetitions, it is advisable to adapt the approach according to the characteristic properties of the mechanism considered and thus lessen the number of arithmetic operations in particular loops. As an example let us calculate the reachable workspace of the planar 3R mechanism, whose inverse kinematics equations were introduced in Sect. 3.3. In accordance with Fig. 4.10, we write the equations determining the position of the mechanism's end-point

$$r_x = l_1 c_1 + l_2 c_{12} + l_3 c_{123},$$
$$r_y = l_1 s_1 + l_2 s_{12} + l_3 s_{123}.$$

As we are dealing with a planar mechanism, for determining the workspace we do not need the loop for the coordinate r_z. However, we must include an additional loop for the orientation of the mechanism

$$r_\alpha = q_1 + q_2 + q_3,$$

where the coordinate r_α has an arbitrary value.

loop 1 The value of coordinate r_x is incremented from $r_{x\min}$ to $r_{x\max}$ in intervals of dr_x. For each new value of the coordinate r_x we switch into the loop 2. When we have swept through all values r_x, the procedure is concluded.

loop 2 For the momentary value of r_x, the value of coordinate r_y is incremented from $r_{y\min}$ to $r_{y\max}$ in intervals of dr_y. For each new value of the coordinate r_y we switch into the loop 3. After sweeping through all values r_y, we come back to the beginning of the loop 1.

loop 3 For the momentary value of r_x and r_y, the value of the coordinate r_α is increased in steps from 0 to 2π in intervals of dr_α. For each new value of the coordinate r_α, we calculate

$$s_\alpha = \sin(r_\alpha),$$
$$c_\alpha = \cos(r_\alpha),$$

and also

$$p_\mathrm{x} = r_\mathrm{x} - l_3 c_\alpha,$$
$$p_\mathrm{y} = r_\mathrm{y} - l_3 s_\alpha$$

and

$$l^2 = p_\mathrm{x}^2 + p_\mathrm{y}^2.$$

We check if

$$(l_1 - l_2)^2 > l^2 > (l_1 + l_2)^2.$$

When the inequality does not hold, the point with the components r_x and r_y is not an element of the reachable workspace of the mechanism considered (and we return to the beginning of loop 3). Otherwise we calculate the corresponding joint coordinates for the current values r_x, r_y and r_α. As we already know, two solutions are possible, whose components can be written as follows

$$q_2 = \pm\left(\pi - \arccos\left(\frac{l_1^2 + l_2^2 - l^2}{2 l_1 l_2}\right)\right)$$

and

$$q_2 > 0 \quad \Longrightarrow \quad q_1 = \arctan_2 \frac{p_\mathrm{y}}{p_\mathrm{x}} - \arccos\left(\frac{l_1^2 + l^2 - l_2^2}{2 l_1 l}\right),$$

$$q_2 < 0 \quad \Longrightarrow \quad q_1 = \arctan_2 \frac{p_\mathrm{y}}{p_\mathrm{x}} + \arccos\left(\frac{l_1^2 + l^2 - l_2^2}{2 l_1 l}\right),$$

when $l \neq 0$. Finally there is

$$q_3 = r_\alpha - q_1 - q_2.$$

If all coordinates are inside the range of the prescribed constraints

$$q_{1\,\mathrm{min}} \leq q_1 \leq q_{1\,\mathrm{max}},$$
$$q_{2\,\mathrm{min}} \leq q_2 \leq q_{2\,\mathrm{max}},$$
$$q_{3\,\mathrm{min}} \leq q_3 \leq q_{3\,\mathrm{max}},$$

for any of the two combinations of the joint coordinates, then the point with the components r_x and r_y is an element of the reachable workspace (area) of the 3R mechanism considered. After sweeping through all values r_z, we come back to the beginning of the loop 2.

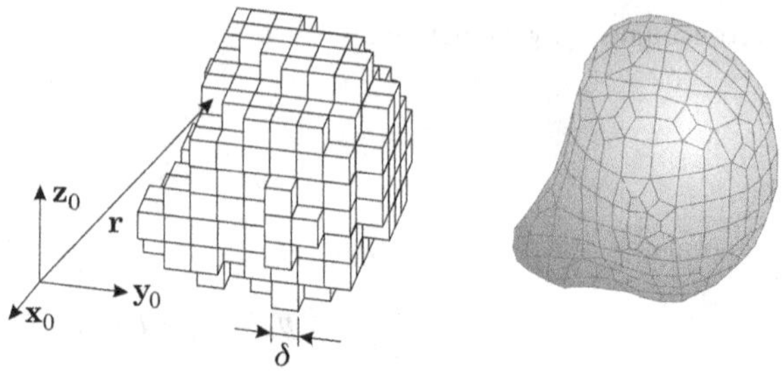

Fig. 4.11 Graphical representation of the workspace with smoothed surface

The reachable workspace of the 3R mechanism, determined in this way, is represented in Fig. 4.10 by a set of squares. The parameters of a mechanism are in accordance with those from Fig. 4.9.

In the literature, different approaches for determining the workspace can be found which provide a considerable reduction in the number of calculations. Usually they are based on the methods of computational geometry and only provide determination of the workspace envelope for a specific group of mechanisms [12]. When determining the robot workspace, also the graphical presentation is of importance. Figure 4.11 shows an example of the workspace given by a set of points represented by the cubes of the side δ, whose surface was additionally numerically smoothed [75].

4.1.4 Dexterous Workspace

The dexterous workspace is a region encompassing all positions which can be reached by a selected end-point of a mechanism with all possible orientations of the last segment [41]. In contrast with the reachable workspace, describing only the reach of a mechanism, the dexterous workspace also reflects the ability of a mechanism to arbitrarily orient the last segment in a selected position. It is evident that the dexterous workspace is a subspace of the reachable space. The point P in Fig. 4.12 is a part of the dexterous workspace when a mechanism can reach it with all orientations of the last segment in the interval from 0 to 2π.

The dexterous workspace can be determined either by solving the direct kinematics or inverse kinematics of a mechanism. For this purpose we can use both afore mentioned approaches. The only difference is that with calculation of the dexterous workspace the orientation of the last segment must also be checked for each position of the end-point. When the orientation is given by the use of Euler or YPR orientational angles ψ, θ and ϕ, which can be divided into the following values

$$\psi = k_\psi \frac{\psi_{\max} - \psi_{\min}}{K_\psi}, \quad k_\psi = 0, 1, \ldots, K_\psi,$$

$$\theta = k_\theta \frac{\theta_{\max} - \theta_{\min}}{K_\theta}, \quad k_\theta = 0, 1, \ldots, K_\theta,$$

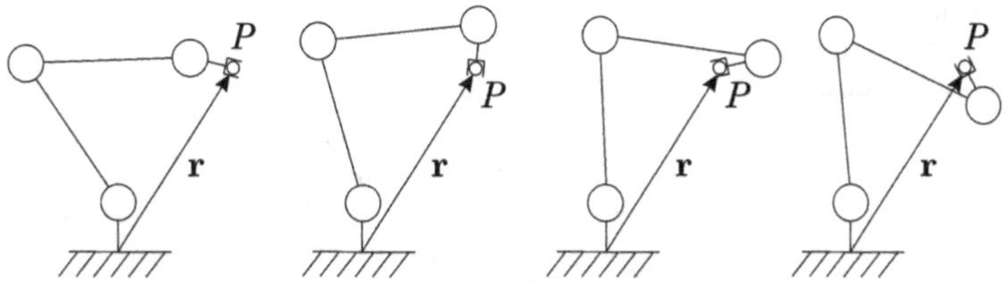

Fig. 4.12 The mechanism can reach point P with all orientations of the gripper

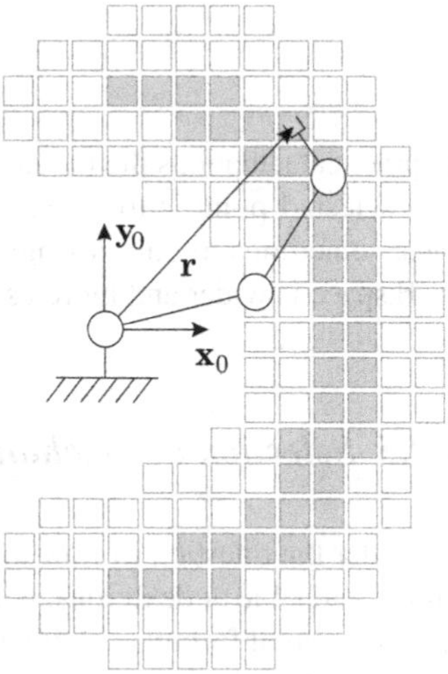

Fig. 4.13 Determining the dexterous workspace of planar 3R mechanism by calculating inverse kinematics

$$\phi = k_\phi \frac{\phi_{\max} - \phi_{\min}}{K_\phi}, \qquad k_\phi = 0, 1, \ldots, K_\phi,$$

we must add for each value of the vector $\mathbf{r}$

$$(K_\psi + 1)(K_\theta + 1)(K_\phi + 1)$$

additional nested loops. The total number of calculations of the inverse kinematics is

$$(K_\mathrm{x} + 1)(K_\mathrm{y} + 1)(K_\mathrm{z} + 1)(K_\psi + 1)(K_\theta + 1)(K_\phi + 1).$$

Figure 4.13 shows the dexterous workspace in comparison to the reachable workspace for a planar 3R mechanism whose parameters were given in the preceding section.

We must know that the last segment attains all the orientations when the orientational angles encompass all the values from the intervals

$$\psi_{\max} - \psi_{\min} = 2\pi,$$

$$\theta_{\max} - \theta_{\min} = \pi,$$

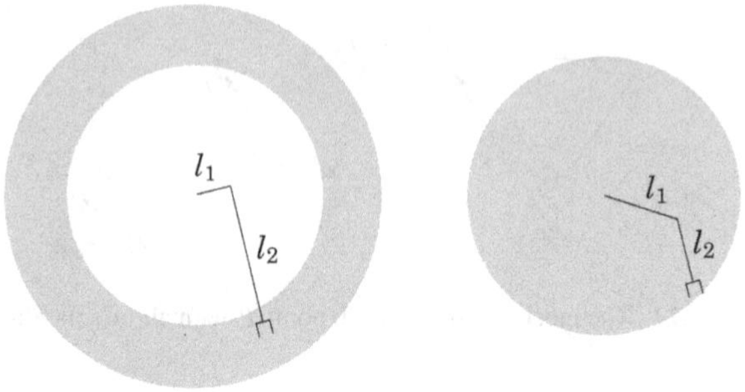

Fig. 4.14 Reachable workspaces of two different 2R mechanisms of equal surface

$$\phi_{\max} - \phi_{\min} = 2\pi.$$

In practical situations the requirement that the mechanism can attain all orientations at a selected point is too strict. We therefore decrease the range of particular angles, sometimes to only a single orientation. Such understanding of the dexterous workspace is wider and more useful than the original definition [41].

4.1.5 Selection of Mechanism with a Desired Workspace

Consider the question of how to select a mechanism which will have a workspace, either reachable or dexterous, whose shape and size will best comply with a selected task. A general formula solving this problem does not exist. In this section we focus on a few of the basic problems and present some of the simplest solutions.

Generally at first consideration, most would agree that a workspace of larger volume is better than a smaller one. Figure 4.14 shows that the truth is not just so. The two planar 2R mechanisms with unconstrained rotations have the same surface of their reachable workspace. The workspace on the left side is expanded far from the center, while the workspace on the right is concentrated around the center. Although the mechanism on the right is even smaller than the one on the left, we can assert that the workspace on the right is more adequate for most tasks. It is evident that the volume is only one of the criteria for evaluation of the workspaces. Another criterion is related to the shape of the workspace.

First we briefly consider the area P of the workspace of a planar 2R mechanism, where both rotations q_1 and q_2 are unconstrained and whose segment lengths are l_1 and l_2 (Fig. 4.4). We already know that the reachable workspace of such a mechanism is a ring with inner radius $R_n = |l_1 - l_2|$ and outer radius $R_z = l_1 + l_2$. We find the ratio of the segment lengths l_1 and l_2 in such a way, that the area of the workspace will be maximal at constant collective length of the mechanism R_z. The problem can be solved in a closed-form as the area of the workspace P can easily be expressed as a function of one of the segment lengths. The working area P is determined as

$$P = \pi R_z^2 - \pi R_n^2$$

after inserting

$$R_n^2 = (l_1 - l_2)^2 = (2l_1 - R_z)^2,$$

we have

$$P = \pi R_z^2 - \pi(2l_1 - R_z)^2.$$

Let us find the derivative of the area P with respect to the segment length l_1 and equate it to zero

$$\frac{\partial P}{\partial l_1} = 2(2l_1 - R_z) = 0.$$

The result is

$$l_1 = \frac{R_z}{2}$$

and

$$l_1 = l_2.$$

The planar 2R mechanism with two segments of equal length has the largest working area. It is shown on the right side of Fig. 4.14. The same answer would be found when calculating the largest reachable volume of the articulated RRR mechanism from Fig. 4.8. When it has unconstrained joint coordinates and segments of equal length, its reachable workspace is a full sphere. When the segments are not of equal length, in the center of the workspace there is a spherical void resulting in a reduced volume for the mechanism workspace.

In most cases it is not possible to calculate the volume of the workspace in a closed-form. When a workspace is represented as a set of points $\mathbf{r}_i$, $i = 1, 2, \ldots, I$, where each point is the center of a cube with the side length δ (Fig. 4.11), we obtain a numerical approximation of the volume by adding the volumes of particular cubes

$$V = \sum_{i=1}^{I} \delta^3 = I\delta^3. \tag{4.1}$$

The second question arises as how to describe the shape of a workspace with a scalar. One of the possibilities is the so called compactness [53]. The compactness of the workspace is a criterion which anticipates that a full workspace of spherical shape is in general preferable to workspaces with more elongated shapes. Let the center of the workspace be expressed by the mathematical expectation

$$\mathbf{s} = \frac{1}{I}\sum_{i=1}^{I} \mathbf{r}_i, \tag{4.2}$$

then the dispersion of the elements of the workspace is equal to

$$D = \frac{1}{I}\sum_{i=1}^{I}(\mathbf{s}-\mathbf{r}_i)^{\mathrm{T}}(\mathbf{s}-\mathbf{r}_i).$$

The equation can be written as follows

$$D = \frac{1}{I}\sum_{i=1}^{I}\mathbf{s}^{\mathrm{T}}\mathbf{s} - \frac{1}{I}\sum_{i=1}^{I}\mathbf{s}^{\mathrm{T}}\mathbf{r}_i - \frac{1}{I}\sum_{i=1}^{I}\mathbf{r}_i^{\mathrm{T}}\mathbf{s} + \frac{1}{I}\sum_{i=1}^{I}\mathbf{r}_i^{\mathrm{T}}\mathbf{r}_i$$

and because there is

$$\frac{1}{I}\sum_{i=1}^{I}\mathbf{s}^{\mathrm{T}}\mathbf{r}_i = \frac{1}{I}\sum_{i=1}^{I}\mathbf{r}_i^{\mathrm{T}}\mathbf{s} = \mathbf{s}^{\mathrm{T}}\mathbf{s},$$

the equation can be simplified as follows

$$D = -\mathbf{s}^{\mathrm{T}}\mathbf{s} + \frac{1}{I}\sum_{i=1}^{I}\mathbf{r}_i^{\mathrm{T}}\mathbf{r}_i. \tag{4.3}$$

The dispersion D is higher when the vectors $\mathbf{r}_i$ are less concentrated near the center. The dispersion of a sphere with radius R can be calculated in a closed-form through integration, the result is

$$D_0 = \frac{3}{5}R^2.$$

When taking

$$V = I\delta^3 = \frac{4}{3}\pi R^3$$

and writing the expression for the radius

$$R = \left(\frac{3I}{4\pi}\right)^{\frac{1}{3}}\delta$$

and inserting it into the expression for the dispersion of the sphere, we obtain

$$D_0 = \frac{3}{5}\left(\frac{3I}{4\pi}\right)^{\frac{2}{3}}\delta^2. \tag{4.4}$$

The compactness of the workspace c is determined as the ratio of the dispersion of the elements of the workspace D and the dispersion of the sphere D_0 of equal volume [53]

$$c = \frac{D_0}{D}, \tag{4.5}$$

where $0 < c < 1$. The workspace is more compact, when its shape is closer to that of the sphere.

The workspace of a planar mechanism is measured by its area, which is

$$P = I\delta^2,$$

the compactness is defined as a ratio of dispersion of the shape, corresponding to this area, and dispersion of a full circle with the same area. Dispersion of the circle of the radius R is calculated as follows

$$D_O = \frac{1}{2}R^2$$

or in discrete form

$$D_O = \frac{I}{2\pi}\delta^2,$$

when taking into account that

$$\pi R^2 = I\delta^2.$$

Go back to the reachable workspace of a 2R mechanism, which has the shape of a ring when the joint coordinates are not constrained. The dispersion of a ring depends on its inner and outer radius

$$D = \frac{R_z^2 + R_n^2}{2},$$

the circle with the same area as this ring has a radius R, which is equal

$$R^2 = R_z^2 - R_n^2.$$

The compactness of the workspace of a 2R planar mechanism is as follows

$$c = \frac{R_z^2 - R_n^2}{R_z^2 + R_n^2}.$$

It has its maximum when $R_n = 0$, which occurs, as we know, when

$$l_1 = l_2.$$

Therefore, the 2R mechanism has a maximum workspace area and a maximum workspace compactness when its segments are of equal length. This is, however, an exceptional example, since usually the volume and the compactness of the workspace are not in accordance with each other. For example, this accordance does not hold in the planar 2R mechanism when the rotations are constrained. Figure 4.15 shows, how the area and the compactness of the reachable workspace of 2R planar

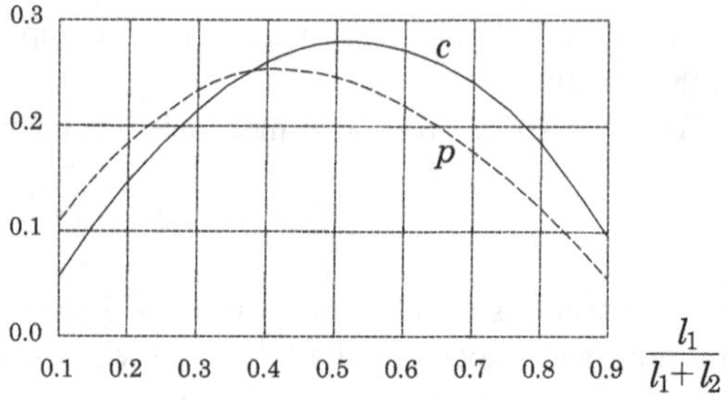

Fig. 4.15 Area and compactness of a reachable workspace of a planar 2R mechanism with constrained rotations presented as functions of the segment lengths ratio

mechanism change with the length of the segments. Here, the collective length of the segments $l_1 + l_2$ is constant, while the rotations assume the following intervals

$$-\frac{\pi}{3} \leq q_1 \leq \frac{\pi}{3},$$
$$-\frac{\pi}{2} \leq q_2 \leq \frac{\pi}{2}.$$

In the figure the relative area is shown, expressed with respect to the largest possible value

$$p = \frac{P}{\pi(l_1 + l_2)^2}.$$

The maximal area was obtained at $l_1/(l_1 + l_2) = 0.415$ while the maximal compactness at $l_1/(l_1 + l_2) = 0.530$. It must be stressed that such calculations cannot be performed in a closed-form. With mechanisms with multiple degrees of freedom the search for the optimum can require a great deal of computation time.

Unfortunately there is no general rule how to select a mechanism for a particular workspace. The problem must be studied for each particular mechanism and different criteria should be used. As we have shown in this short section, the workspace volume and compactness are useful in evaluating robot mechanisms because they both have a clear practical significance. The process of determining the robot workspace is tedious, however, it is very convenient that we can simultaneously compute the workspace volume and compactness in a single process of workspace determination.

4.2 Kinematic Flexibility and Kinematic Singularity

In this section we briefly consider two important kinematic properties, namely kinematic flexibility and kinematic singularity.

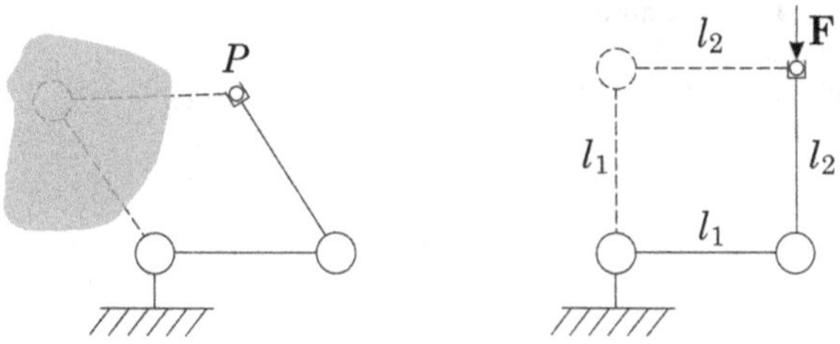

Fig. 4.16 The flexibility of a mechanism with respect to the task

4.2.1 Kinematic Flexibility

The kinematic flexibility of a serial mechanism, from a mathematical point of view, is related to the number of inverse kinematic solutions or in another words, to the number of configurations of a mechanism which correspond to a selected pose of the gripper. The number of inverse kinematic solutions can be seen as something that disturbs computations and introduces additional problems to the control. This was the reason why in early development of robot mechanisms, the preferred mechanisms had minimal numbers of inverse kinematic solutions. Today, in some cases, the view is just the opposite. The aim may be to design mechanisms where the number of the configurations is as large as possible, as such mechanisms more efficiently comply with demands of various tasks.

The task of the mechanism on the left side of Fig. 4.16 is to place the control point at point P. Theoretically this 2R mechanism has two configurations which solve the given task. Because of the presence of an obstacle in the workspace, only one of the configurations can be considered. If the joint rotations of the mechanism are constrained in such a way that only the configuration, where the mechanism bumps into the obstacle, is possible, the task cannot be successfully accomplished. On the right side of Fig. 4.16 the mechanism must act against external force **F**. When ignoring the segments gravity, the mechanism in the upper configuration has the following collective torque of the motors

$$\tau_1 + \tau_2 = l_2 F + l_2 F = 2 l_2 F,$$

while in the lower configuration the torque is considerably less, as the force **F** has no influence on the second motor,

$$\tau_1 + \tau_2 = l_1 F + 0 = l_1 F.$$

When a mechanism is incorporated into a production process, the multitude of configurations yields more options.

Assume that various combinations of joint coordinates $\mathbf{q}, \mathbf{q}', \mathbf{q}'', \ldots$ correspond to the external task coordinates $\mathbf{p}$ and that each combination represents one configuration of a mechanism. Let there be f_{a} different combinations. The magnitude f_{a}, which can assume values $1, 2, 3, \ldots$, is called the absolute kinematic flexibility of the mechanism [54]. A mechanism is more kinematically flexible, when the number of configurations f_{a} is larger.

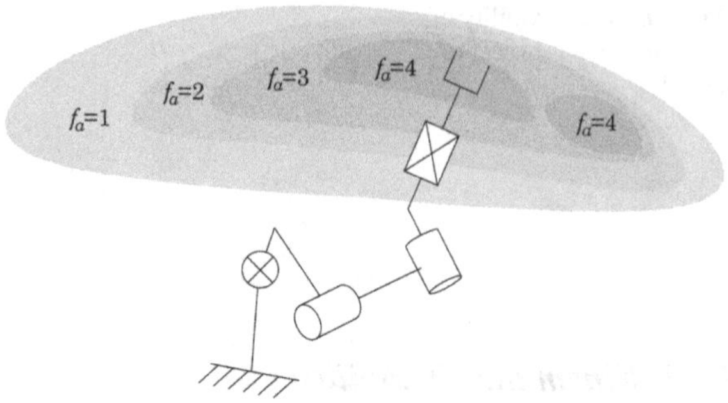

Fig. 4.17 The absolute kinematic flexibility in different regions of the workspace

It is understood that the kinematic flexibility f_a depends on the kinematic structure of a mechanism and constraints on the joint coordinates. Regions with different values of absolute kinematic flexibility f_a can exist within the reachable workspace (Fig. 4.17). The measure of kinematic flexibility, which can be used for evaluation of an entire workspace, is the relative kinematic flexibility, defined as the following sum of the volumes [54]

$$f_r = \frac{1}{V}(V_1 + 2V_2 + 3V_3 + \cdots), \tag{4.6}$$

where V represents the volume of the entire workspace, while V_1 is the volume of the region of the workspace where $f_a = 1$, V_2 the volume of the region of the workspace with $f_a = 2, \ldots$, where $V_1 + V_2 + V_3 + \cdots = V$.

We add the following comment to (4.6). We will determine the workspace of a mechanism with several configurations for each configuration separately. Let the volume of the first configuration be V', the volume of the second configuration V'' etc. The workspace of one configuration may overlap in some part the workspace of another configuration. The union of all those workspaces is the collective workspace of the mechanism with the volume V. The part of the workspace with no overlapping of the workspaces, appertaining to particular configurations, has the volume V_1. The volume of the part of the workspace, where two workspaces of particular configurations overlap, is V_2 etc. With regard to the definition (4.6), it follows

$$f_r = \frac{1}{V}\left(V' + V'' + V''' + \cdots\right).$$

The relative kinematic flexibility is therefore the ratio of the sum of the volumes of the workspaces of particular configurations and the volume of collective workspace of a mechanism. The circumstances for the 2R mechanism are shown in Fig. 4.18. The reachable workspace of the 2R mechanism, which belongs to the configurations where $q_2 > 0$, is shown on the left side of the figure, while the workspace for the configurations $q_2 < 0$ is in the middle. The collective workspace, being the union of the first two workspaces, is shown on the right side of the figure. The absolute kinematic flexibility is doubled in the part of the workspace, where the particular workspaces overlap.

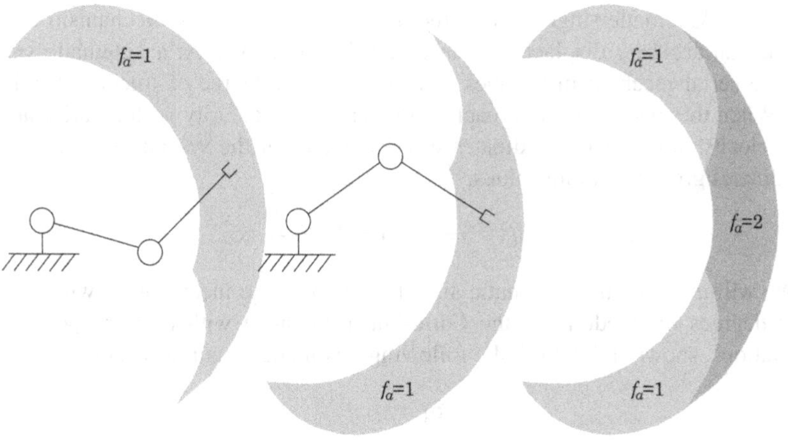

Fig. 4.18 The arrangement of the absolute kinematic flexibility in the reachable workspace of the 2R mechanism

Here we have

$$f_r = \frac{V' + V''}{V} = \frac{V_1 + 2V_2}{V}.$$

By changing the ratio of the segment lengths or the ratio of the ranges of rotations, we can change the relative kinematic flexibility (we increase or decrease V_2 with respect to V). When designing a mechanism and selecting its parameters, we aim for large values of f_r, when it is not in contradiction with other requirements.

4.2.2 *Kinematic Singularity*

Recall that an infinitesimal displacement in the external coordinates $d\mathbf{p}$ is linearly related to an infinitesimal displacement in joint coordinates $d\mathbf{q}$ through Jacobian matrix $\mathbf{J}$ (3.54)

$$d\mathbf{p} = \mathbf{J}d\mathbf{q},$$

while the inverse relation

$$d\mathbf{q} = \mathbf{J}^{-1}d\mathbf{p}$$

only holds when the Jacobian matrix is not degenerate, i.e. not singular. In the other case, at the values of the joint coordinates $\mathbf{q} = \mathbf{q}^*$ for which the Jacobian matrix is singular, this system of the kinematic equations cannot be solved. A singularity of the Jacobian matrix is a mathematical phenomenon, resulting in decreased mobility of a mechanism. The phenomenon is encountered in our daily lives, e.g. when riding a bicycle. When the pedal of a bicycle is in its extreme lower or upper position, the bicycle cannot be propelled by pushing vertically against the pedal with your foot.

This is the kinematic singularity of the mechanism, where the mechanism cannot produce an arbitrary displacement in external coordinates. At a kinematic singularity, a serial mechanism behaves as if at least one degree of freedom has been lost. When the mechanism approaches a kinematic singularity with an arbitrary finite velocity in external coordinates $\dot{\mathbf{p}}$, components of the velocity vector in joint coordinates grow to infinite values.

$$\mathbf{q} \rightarrow \mathbf{q}^* \quad \Longrightarrow \quad \dot{\mathbf{q}} = \mathbf{J}^{-1}\dot{\mathbf{p}} \rightarrow \infty.$$

We will examine the kinematic singularities of some mechanisms with two or three degrees of freedom. In the Cartesian mechanism with three perpendicular translations, shown in Fig. 4.5, the following kinematic equations occur

$$\begin{aligned} p_1 &= q_1, \\ p_2 &= q_2, \\ p_3 &= q_3. \end{aligned}$$

The Jacobian matrix is obtained by partial differentiation with respect to the joint coordinates

$$\mathbf{J} = \begin{bmatrix} 1 & 0 & 0 \\ 0 & 1 & 0 \\ 0 & 0 & 1 \end{bmatrix}.$$

Its determinant equals one. It means that the mechanism has no kinematic singularities.

As a second example, consider the planar RT mechanism (top of Fig. 4.19). Here, the kinematic equations are

$$\begin{aligned} p_1 &= q_2 c_1, \\ p_2 &= q_2 s_1, \end{aligned}$$

so that the Jacobian matrix is

$$\mathbf{J} = \begin{bmatrix} -q_2 s_1 & c_1 \\ q_2 c_1 & s_1 \end{bmatrix}.$$

The mechanism is in a kinematic singularity when

$$\det \mathbf{J} = -q_2 = 0.$$

When the mechanism is in the singular pose (right side of Fig. 4.19), the gripper can only be displaced along the translation q_2, while the rotation q_1 has no influence on the coordinates p_1 and p_2.

With the planar TR mechanism (middle of Fig. 4.19) we have

$$\begin{aligned} p_1 &= q_1 + l_2 c_2, \\ p_2 &= l_2 s_2, \end{aligned}$$

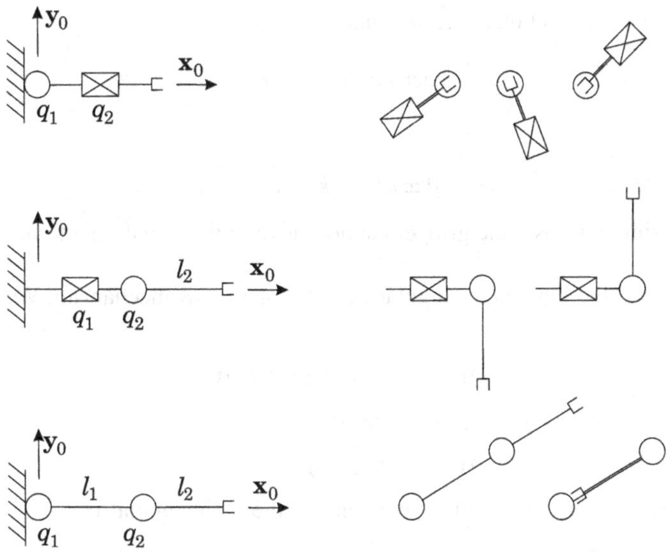

Fig. 4.19 Kinematic singularities of the mechanisms RT, TR and 2R

so that the Jacobian matrix equals

$$\mathbf{J} = \begin{bmatrix} 1 & -l_2 s_2 \\ 0 & l_2 c_2 \end{bmatrix}.$$

This mechanism is in a kinematic singularity when

$$\det \mathbf{J} = l_2 c_2 = 0$$

and

$$q_2 = \pm\left(\frac{\pi}{2} + k\pi\right), \quad k = 1, 2, \ldots.$$

Two singular poses are shown on the right side of Fig. 4.19. The gripper cannot move along the coordinate p_2.

With the planar 2R mechanism (bottom of Fig. 4.19) the kinematic equations are as follows

$$p_1 = l_1 c_1 + l_2 c_{12},$$
$$p_2 = l_1 s_1 + l_2 s_{12}.$$

The Jacobian matrix is

$$\mathbf{J} = \begin{bmatrix} -l_1 s_1 - l_2 s_{12} & -l_2 s_{12} \\ l_1 c_1 + l_2 c_{12} & l_2 c_{12} \end{bmatrix}.$$

The mechanism is in a kinematic singularity when

$$\det \mathbf{J} = l_1 l_2 s_2 = 0$$

and

$$q_2 = 0 \pm k\pi, \quad k = 1, 2, \ldots.$$

When in a singular pose, the gripper cannot move in the radial direction (Fig. 4.19 right).

We will find the kinematic singularities of a planar 3R mechanism. We proceed from (3.74)

$$\begin{aligned}
p_1 &= l_1 c_1 + l_2 c_{12} + l_3 c_{123},\\
p_2 &= l_1 s_1 + l_2 s_{12} + l_3 s_{123},\\
p_3 &= q_1 + q_2 + q_3.
\end{aligned}$$

After differentiation, we obtain the elements of the Jacobian matrix

$$\mathbf{J} = \begin{bmatrix} -l_1 s_1 - l_2 s_{12} - l_3 s_{123} & -l_2 s_{12} - l_3 s_{123} & -l_3 s_{123} \\ l_1 c_1 + l_2 c_{12} + l_3 c_{123} & l_2 c_{12} + l_3 c_{123} & l_3 c_{123} \\ 1 & 1 & 1 \end{bmatrix}.$$

The determinant is

$$\det \mathbf{J} = l_1 l_2 \sin q_2 = 0,$$

and therefore the Jacobian matrix is singular when

$$q_2 = 0 \pm k\pi, \quad k = 0, 1, 2, \ldots$$

as is the case with the 2R mechanism. The singularities of the mechanism were already shown in Fig. 3.10.

Consider now two spatial mechanisms. The kinematic equations for the RRT mechanism (Fig. 4.20 top) are as follows

$$\begin{aligned}
p_1 &= -q_3 s_1 c_2,\\
p_2 &= q_3 c_1 c_2,\\
p_3 &= l_1 + q_3 s_2,
\end{aligned}$$

while the Jacobian matrix is

$$\mathbf{J} = \begin{bmatrix} -q_3 c_1 c_2 & q_3 s_1 s_2 & -s_1 c_2 \\ -q_3 s_1 c_2 & -q_3 c_1 s_2 & c_1 c_2 \\ 0 & q_3 c_2 & s_2 \end{bmatrix}.$$

The determinant of this matrix is

$$\det \mathbf{J} = c_2 q_3^2$$

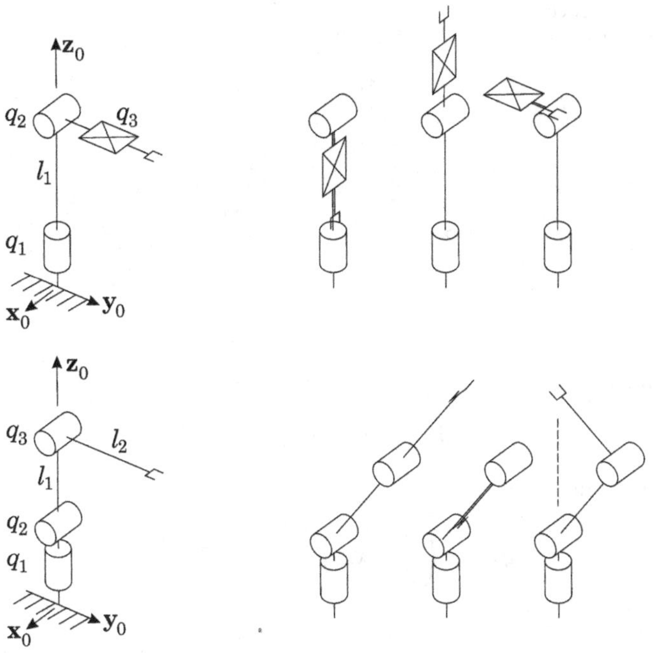

Fig. 4.20 Kinematic singularities of spatial mechanisms RRT and RRR

which equals zero when

$$q_2 = \pm\left(\frac{\pi}{2} + k\pi\right), \quad k = 1, 2, \ldots$$

and when

$$q_3 = 0.$$

In the first case the gripper cannot move along the direction of the coordinate p_1, while in the second case it can only move in the direction defined by the translational joint (Fig. 4.20 right).

The equations for the RRR mechanism can be derived from Fig. 4.20

$$\begin{aligned} p_1 &= -s_1(-l_1 s_2 + l_2 c_{23}), \\ p_2 &= c_1(-l_1 s_2 + l_2 c_{23}), \\ p_3 &= l_1 c_2 + l_2 s_{23}. \end{aligned}$$

The Jacobian matrix is

$$\mathbf{J} = \begin{bmatrix} -c_1(-l_1 s_2 + l_2 c_{23}) & s_1(l_1 c_2 + l_2 s_{23}) & l_2 s_1 s_{23} \\ -s_1(-l_1 s_2 + l_2 c_{23}) & -c_1(l_1 c_2 + l_2 s_{23}) & -l_2 c_1 s_{23} \\ 0 & -l_1 s_2 + l_2 c_{23} & l_2 c_{23} \end{bmatrix}.$$

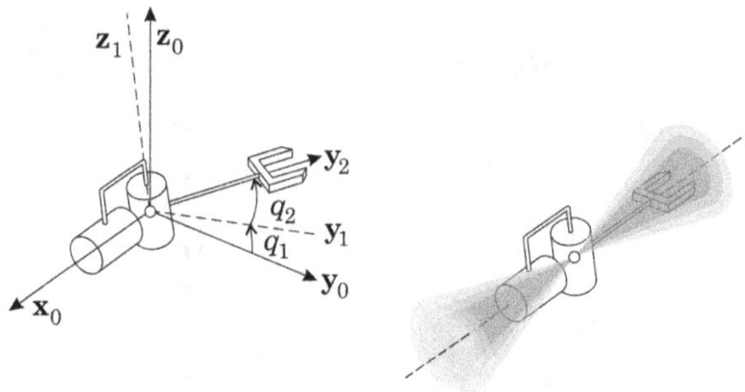

Fig. 4.21 Singularity cones of the universal wrist

The determinant

$$\det \mathbf{J} = l_1 l_2 c_3(-l_1 s_2 + l_2 c_{23})$$

is equal zero when

$$q_3 = \pm\left(\frac{\pi}{2} + k\pi\right), \quad k = 1, 2, \ldots$$

and when

$$l_1 s_2 = l_2 c_{23}.$$

We are already familiar with first condition as it has occurred in the 2R mechanism. According to the second condition the gripper is in contact with the first rotational axis, which annihilates the influence of the first coordinate on the position of the gripper.

Within the mechanisms of industrial robot manipulators, many problems result from the universal joint portion of the wrist (Fig. 4.21). This represents the part of the gripper orientation which is a consequence of the two perpendicular rotations found within a universal joint. According to the figure the first rotation q_1 occurs around the $\mathbf{x}_0$ axis, while the second rotation q_2 around the $\mathbf{z}_1$ axis which is not fixed. The rotation matrix is

$$\begin{aligned}\mathbf{A}_{0,2} &= \mathbf{A}_{0,1}\mathbf{A}_{1,2} \\ &= \begin{bmatrix} 1 & 0 & 0 \\ 0 & c_1 & -s_1 \\ 0 & s_1 & c_1 \end{bmatrix} \begin{bmatrix} c_2 & -s_2 & 0 \\ s_2 & c_2 & 0 \\ 0 & 0 & 1 \end{bmatrix} = \begin{bmatrix} c_2 & -s_2 & 0 \\ c_1 s_2 & c_1 c_2 & -s_1 \\ s_1 s_2 & s_1 c_2 & c_1 \end{bmatrix}.\end{aligned}$$

The purpose of the rotations q_1 and q_2 is to direct (i.e. point or aim) the vector $\mathbf{y}_2$, whose direction is defined by the equation

$$\mathbf{y}_2^{(0)} = \begin{bmatrix} c_2 & -s_2 & 0 \\ c_1 s_2 & c_1 c_2 & -s_1 \\ s_1 s_2 & s_1 c_2 & c_1 \end{bmatrix} \begin{bmatrix} 0 \\ 1 \\ 0 \end{bmatrix} = \begin{bmatrix} -s_2 \\ c_1 c_2 \\ s_1 c_2 \end{bmatrix}.$$

In the equation

$$d\mathbf{y}_2^{(0)} = \mathbf{J}\begin{bmatrix} dq_1 \\ dq_2 \end{bmatrix}$$

the Jacobian matrix is a non-square matrix

$$\begin{bmatrix} 0 & -c_2 \\ -s_1c_2 & -c_1s_2 \\ c_1c_2 & -s_1s_2 \end{bmatrix}$$

and the inverse solution is obtained in the following form

$$\begin{bmatrix} dq_1 \\ dq_2 \end{bmatrix} = (\mathbf{J}^{\mathrm{T}}\mathbf{J})^{-1}\mathbf{J}^{\mathrm{T}}d\mathbf{y}_2^{(0)}.$$

The mechanism is in a kinematic singularity when

$$\det\left(\mathbf{J}^{\mathrm{T}}\mathbf{J}\right) = c_2^2 = 0$$

and

$$q_2 = \pm\left(\frac{\pi}{2} + k\pi\right), \quad k = 1, 2, \ldots.$$

Theoretically this mechanism cannot direct vector $\mathbf{y}_2$, when vector $\mathbf{y}_2$ is collinear with respect to the vector of the first rotation $\mathbf{x}_0$. In practical situations, problems begin to occur in a conical region in the vicinity of the singularity where the Jacobian is very near to being singular. This region around the axis $\mathbf{x}_0$, where the axis $\mathbf{y}_2$ should not enter, is called the singular cone [68]. The universal joint has two diametrically opposed singular cones as shown in Fig. 4.21.

We must be aware of the fact that a mechanism with several degrees of freedom can be singular even when particular parts of the mechanisms do not have singularities. Consider the mechanism of an industrial robot manipulator which usually has three degrees of freedom in its positional part and three degrees of freedom in its orientational part. Even when the positional and orientational parts do not have their own singular configurations, the combination of both systems can come into a singularity. This type of coupled singularity cannot occur if the kinematic equations of the positional part and the kinematic equations of the orientational part are independent. This independence occurs most frequently when the orientational part consists of a spherical wrist.

The regions of kinematic singularity, which can be isolated points, curves or surfaces, can be expressed either in the space of external or joint coordinates. For a long time the opinion prevailed that the kinematic singularity represents the border between two or several configurations of a mechanism. This is evidently valid for the 2R mechanism from Fig. 4.22. The mechanism is displaced from configuration (+) to the configuration (−) through the singular configuration $q_2 = 0 \pm k\pi$, $k = 1, 2, \ldots$. There is no other possibility. More recent investigations have shown

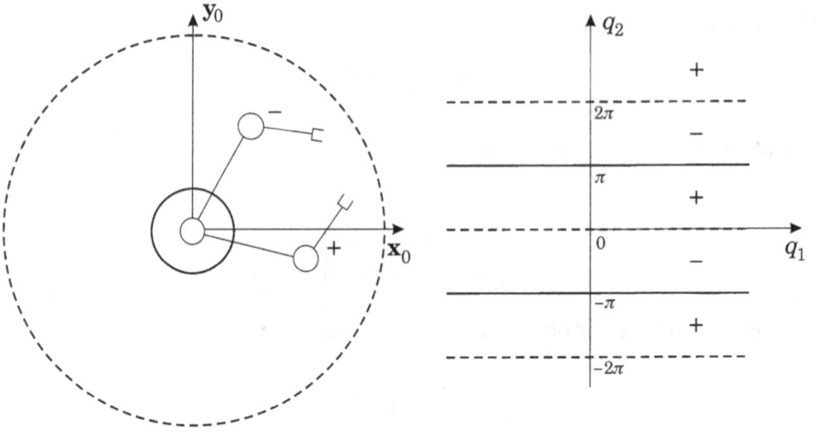

Fig. 4.22 The regions of kinematic singularity for the 2R planar mechanism, expressed in external coordinates, represent the inner side $q_2 = \pm\pi$ and the outer side $q_2 = 0 \pm 2\pi$ of the reachable workspace

that such mechanisms represent exceptions. Usually their degrees of freedom are either successively parallel or perpendicular. A mechanism, where the neighboring axes are inclined one with respect to another, can move from one configuration into another without going through the singular configuration. This property can also be found in mechanisms with successively parallel or perpendicular degrees of freedom [89].

4.3 Manipulability and Kinematic Index

It could be debated whether a singularity is a mathematical or mechanical property of a mechanism. It is a fact, however, that a mechanism behaves differently when in a kinematic singularity. Usually it cannot fully perform the required task and the singularity is viewed negatively. Consequently, we might like to avoid the singularities while designing a mechanism. We might seek to build a mechanism without singular configurations in its workspace [78]. In this section however we will consider exactly the opposite, namely we consider whether the kinematic singularity can also be a useful property of a mechanism.

The task of a mechanism in Fig. 4.23 is to bring an object from a pose P_1 into a pose P_2. As the mechanism has only two degrees of freedom, it is in principle unable to orient the object arbitrarily, as the change in orientation is the consequence of the change in position. But let us execute the task in steps as follows. The object is first transferred to the pose P_0, corresponding to the singular configuration, where the mechanism can rotate around the axis of the coordinate q_1 without changing the position of the gripper (Fig. 3.10). In this configuration we first release the object and after rotating the mechanism by the angle Δq_1, we grasp the object once again

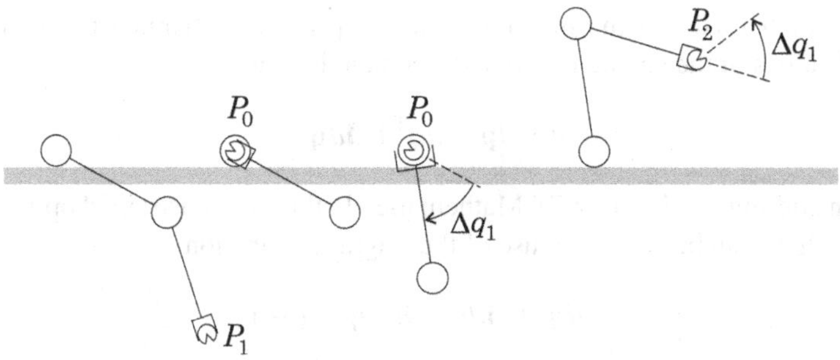

Fig. 4.23 The transfer of an object into required position and orientation by making use of singular configuration of the 2R mechanism

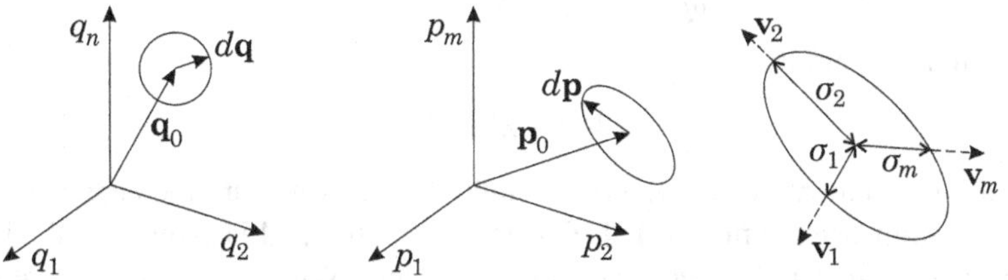

Fig. 4.24 Manipulability ellipsoid

and transfer it to the desired pose P_2. The final orientation of the object is different from the one obtained without regrasping the object in the singular configuration. The orientations differ by the angle Δq_1. In this way we effected both the position and orientation of this object, even when the mechanism has only two degrees of freedom.

In contrast to expectations, we have in some sense gained a degree of freedom by taking advantage of the kinematic singularity. We have made use of a kinematic singularity which has a particular property. We should know that the properties of kinematic singularities are different [34] and there is no general rule on how to handle them.

Before moving to other possibilities of how to take advantage of kinematic singularities, let us prepare some mathematical tools. Let the external coordinates $\mathbf{p}_0$ correspond to the configuration of the mechanism $\mathbf{q}_0$ (Fig. 4.24). A displacement from this configuration in joint coordinates $d\mathbf{q}$ leads to the displacement $d\mathbf{p}$ from the vector $\mathbf{p}_0$ in external coordinates. When the displacements in joint coordinates are of equal length for all directions, i.e. when the vector $d\mathbf{q}$ lies on the surface of an n-dimensional sphere

$$d\mathbf{q}^{\mathrm{T}} d\mathbf{q} = 1, \tag{4.7}$$

then the vector in external coordinates $d\mathbf{p}$ lies on the surface of an m-dimensional ellipsoid, called the manipulability ellipsoid [92].

The principal axes of the manipulability ellipsoid are obtained by finding the maximal values of the square of the displacement length

$$d\mathbf{p}^{\mathrm{T}}d\mathbf{p} = d\mathbf{q}^{\mathrm{T}}\mathbf{J}^{\mathrm{T}}\mathbf{J}d\mathbf{q}, \tag{4.8}$$

while considering condition (4.7). Mathematically this is a constrained optimization problem which can be solved by use of the Lagrange function

$$L = d\mathbf{q}^{\mathrm{T}}\mathbf{J}^{\mathrm{T}}\mathbf{J}d\mathbf{q} - \lambda\left(d\mathbf{q}^{\mathrm{T}}d\mathbf{q} - 1\right),$$

where the scalar λ is a Lagrange multiplier. The derivative of the Lagrange function must equal zero

$$\frac{\partial L}{\partial d\mathbf{q}} = 2\mathbf{J}^{\mathrm{T}}\mathbf{J}d\mathbf{q} - 2\lambda d\mathbf{q} = 0.$$

From here it follows that

$$\mathbf{J}^{\mathrm{T}}\mathbf{J} - \lambda\mathbf{I} = \mathbf{0}. \tag{4.9}$$

There are m solutions for the Lagrange multiplier λ, which are the eigenvalues of the $m \times m$-dimensional matrix $\mathbf{J}^{\mathrm{T}}\mathbf{J}$. Because the matrix $\mathbf{J}^{\mathrm{T}}\mathbf{J}$ is symmetrical, all the eigenvalues are real and greater than or equal to zero. When inserting (4.9) into the expression (4.8), we obtain

$$d\mathbf{p}^{\mathrm{T}}d\mathbf{p} = \lambda d\mathbf{q}^{\mathrm{T}}d\mathbf{q}$$

and after considering (4.7), we have

$$d\mathbf{p}^{\mathrm{T}}d\mathbf{p} = \lambda. \tag{4.10}$$

We know that an m-dimensional eigenvector $\mathbf{v}_i$ corresponds to each eigenvalue λ_i, so that

$$\mathbf{J}^{\mathrm{T}}\mathbf{J}\mathbf{v}_i - \lambda_i\mathbf{v}_i = \mathbf{0},$$

$i = 1, 2, \ldots, m$. The eigenvectors are the principal axes of the manipulability ellipsoid. Considering (4.10), the thickness of the ellipsoid in the direction of the eigenvector $\mathbf{v}_i$, i.e. the length of the displacement $d\mathbf{p}$, is given by

$$\sigma_i = \sqrt{\lambda_i}, \quad i = 1, 2, \ldots, m. \tag{4.11}$$

The σ_i, $i = 1, 2, \ldots, m$, are the singular values of the matrix $\mathbf{J}$ [38]. Singular values of a matrix are especially important in matrix algebra. The number of non-zero singular values represent the rank of the matrix. If at least one of the singular values is zero the matrix is singular.

The manipulability of a mechanism is defined as the product of the singular values of the Jacobian matrix [92]

$$M = \sigma_2\sigma_2 \cdots \sigma_m. \tag{4.12}$$

Manipulability brings information about the volume of the manipulability ellipsoid. Although the manipulability is related to the volumetric size of the ellipsoid, it does not tell much, as the manipulability of a flat ellipsoid equals zero regardless of its largeness. The singular values can be useful in evaluating the largeness of the ellipsoid. Of special interest is the ratio of the maximal and the minimal singular value. Their ratio

$$K = \frac{\sigma_{\min}}{\sigma_{\max}} \tag{4.13}$$

is called the kinematic index. It is a normalized magnitude which describes the sphericality of the manipulability ellipsoid. The ellipsoid is closer to spherical as K approaches one, and flatter as K approaches zero. The kinematic index is reciprocal to the condition number $\sigma_{\max}/\sigma_{\min}$ and is used in mathematics in place of the determinant when in the neighborhood of the singularity of a matrix. When the condition number goes to infinity, the matrix approaches a singularity. In mechanics we prefer a reciprocal magnitude, constrained by zero and one. The Jacobian matrix is singular when $K = 0$, and it is far from singular when $K = 1$.

Unfortunately such a general definition of manipulability ellipsoid is not without its problems. If a mechanism is a mixture of different joint coordinates, i.e. translations and rotations, different external coordinates, i.e. positions and orientations, leads to various combinations of units, making comparisons between mechanisms difficult. The manipulability index has practical meaning when it is related only to a single type of joint coordinates and a single type of external coordinates. For this reason we consider the manipulability ellipsoid only as related to the change in position of the mechanism end-point, i.e. the translational velocity of the selected end-point.

The relation between the translational velocities and the velocities of the joint coordinates is given by the Jacobian matrix $\mathbf{J}_\mathrm{v}$, which was introduced in Sect. 3.2.2. The Jacobian matrix $\mathbf{J}_\mathrm{v}$ also relates the vector of an external force $\mathbf{F}$ to the joint moments

$$\begin{bmatrix} \tau_1 \\ \tau_2 \\ \vdots \\ \tau_n \end{bmatrix} = \boldsymbol{\tau} = \mathbf{J}_\mathrm{v}^\mathrm{T}\mathbf{F}. \tag{4.14}$$

When the joint moments are normalized as follows

$$\boldsymbol{\tau}^\mathrm{T}\boldsymbol{\tau} = 1,$$

the external forces

$$\mathbf{F} = \left(\mathbf{J}_\mathrm{v}\mathbf{J}_\mathrm{v}^\mathrm{T}\right)^{-1}\mathbf{J}_\mathrm{v}\boldsymbol{\tau}$$

are limited by an ellipsoid whose axes coincide with the directions of the axes of the manipulability ellipsoid but their lengths are reciprocal to the singular values of the Jacobian matrix $\mathbf{J}_\mathrm{v}$ (Fig. 4.25). Thus, the manipulability ellipsoid, related to the Jacobian matrix $\mathbf{J}_\mathrm{v}$, represents the translational velocity of the last link $\mathbf{v} = \mathbf{v}_{0,n+1}$

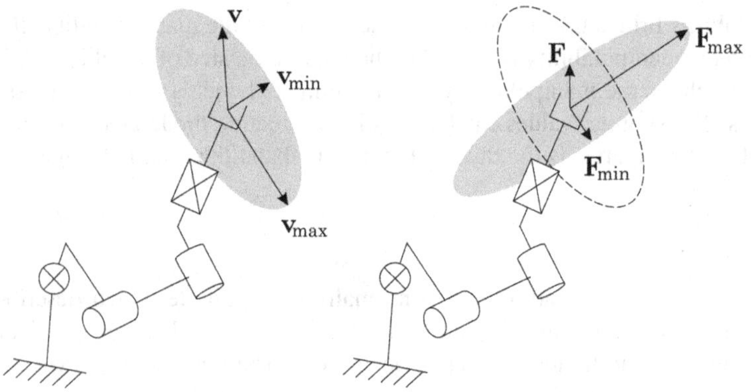

Fig. 4.25 Velocity and force ellipsoid

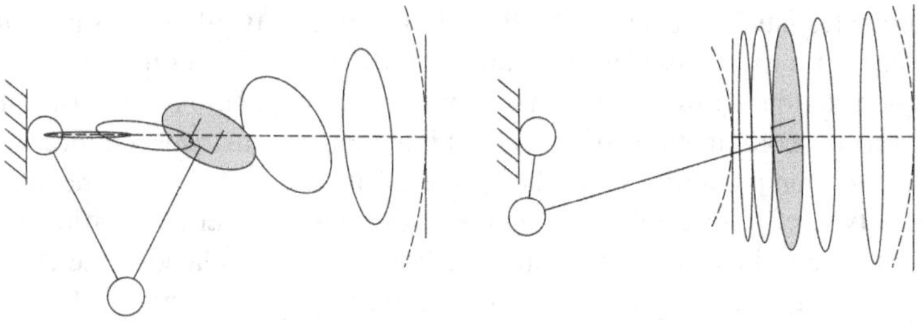

Fig. 4.26 Velocity ellipsoids of 2R mechanism

when the joint coordinates are normalized as $\dot{\mathbf{q}}^{\mathrm{T}}\dot{\mathbf{q}} = 1$. They are therefore called velocity ellipsoids. The reciprocal ellipsoids (ellipsoids which contain reciprocal lengths of the axes) are named force ellipsoids and represent the external force $\mathbf{F}$ corresponding to the joint torques limited by $\boldsymbol{\tau}^{\mathrm{T}}\boldsymbol{\tau} = 1$.

In a kinematic singularity, the manipulability ellipsoid is flat in the direction of the eigenvector $\mathbf{v}_i$, which corresponds to the singular value $\sigma_i = 0$. The mechanism's end point cannot be displaced in this direction. Correspondingly, an infinitely large external force can be exerted on the mechanism's end point in this direction. The more that the movement of a mechanism's end point is diminished in a particular direction, the more mechanical advantage it has against an external force in that direction. From this point of view, the kinematic singularity has both disadvantages and advantages [35].

As a numerical example, we again make use of the planar 2R mechanism. Figure 4.26 shows how the velocity ellipsoid is changing its shape from the bottom of the mechanism towards the end-point [47]. We can notice that the ellipsoid is flat on the border of the workspace, when the mechanism moves into a singularity at $q_2 = 0 \pm \pi$, while in the middle it becomes more and more round.

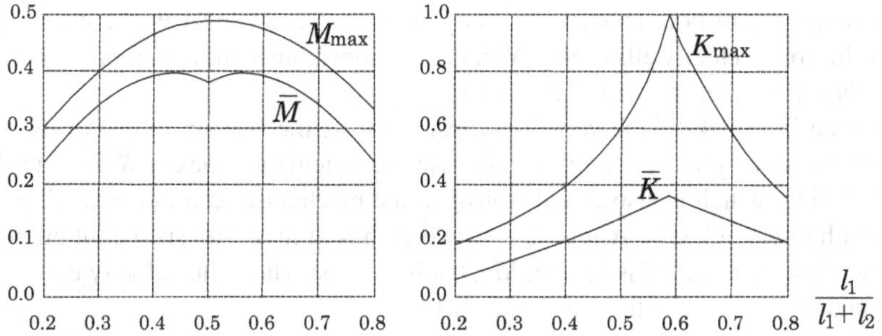

Fig. 4.27 Maximal M_{max} and average manipulability $\overline{M}$ and maximal K_{max} and average kinematic index $\overline{K}$ of 2R mechanism in dependence of the ratio of the segment lengths

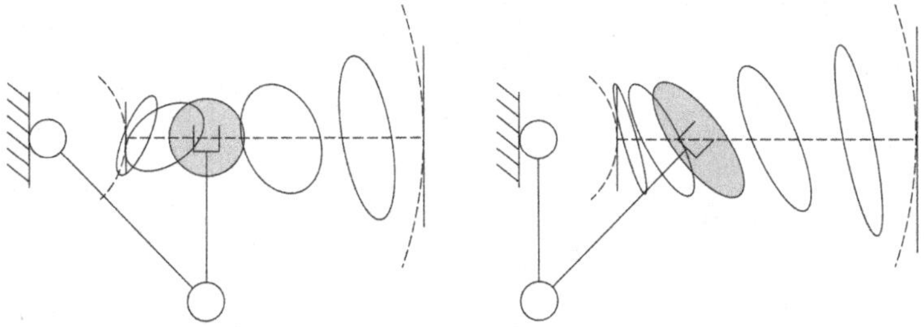

Fig. 4.28 Velocity ellipsoids of 2R mechanism when the segment lengths ratios are $l_1/l_2 = \sqrt{2}$ and $l_1/l_2 = 1/\sqrt{2}$

Manipulability ellipsoids apparently change their size and shape depending on the ratio of the link lengths l_1 and l_2. Figure 4.27 gives the maximum manipulability, M_{max}, in the workspace and the average manipulability, $\overline{M}$, as functions of l_1 and l_2. The highest M_{max} is obtained when $l_1 = l_2$ and the highest $\overline{M}$ is obtained when either $l_1 = \sqrt{2}\,l_2$ or when $l_2 = \sqrt{2}\,l_1$. In Fig. 4.27 we can also see the dependency of the maximum kinematic index, K_{max}, and the average kinematic index, $\overline{K}$, as functions of the ratio of the segment lengths. The ellipsoids are round only when $l_1 = \sqrt{2}\,l_2$.

A mechanism with rounder ellipsoids is desirable when we want its movements to be uniform in all directions. When the ellipsoids are flatter, the mechanism offers more effective resistance against an external force acting in a particular direction. Consider the example in Fig. 4.28. Here, we have two mechanisms. The ratio of the segments lengths of the first mechanism is

$$\frac{l_1}{l_2} = \frac{\sqrt{2}}{1},$$

and that of the second mechanism is

$$\frac{l_1}{l_2} = \frac{1}{\sqrt{2}}.$$

The workspaces of both mechanisms are equally large, however they have significantly different velocity ellipsoids. With the first mechanism they are more roundish, while they are rather flat with the second mechanism.

As seen in Fig. 4.28, the velocity ellipsoid of the mechanism in one of its configurations is completely round. In this case the kinematic index is $K = 1$. In this configuration, which is also called isotropic, the mechanism can move in all directions with equal velocity. A mechanism which has at least one such configuration in its workspace is called isotropic. Mechanisms exist where the velocity ellipsoids are round at all points of the workspace [2].

Chapter 5
Singular Planes and Dexterous Robot Mechanisms

Abstract When the Jacobian matrix connecting the joint displacements with the displacements in the task coordinates degenerates into a singular matrix, the mechanism cannot move in certain directions and the task cannot be performed. These are the kinematic singularities of the mechanism. This chapter deals with describing the kinematic singularities in industrial robots in the form of the singular planes. They represent the loci of singular points on the terminal link of a robotic system. We calculate the singular planes for serial mechanisms that are normally used in industrial robots, such as the articulated arm, the spherical arm, the cylindrical arm and the Scara arm. At the end, the singularities of a spherical wrist are discussed and a singularity-free pointing system is presented.

As seen in Chap. 4, robot singularities are detrimental to dexterity and beneficial to mechanical advantage. A robot singularity can occur either on the robot's workspace boundary or within the interior of the robot's workspace. From the point of view of dexterity, singularities on workspace boundaries are of no concern. They can be avoided by simply bringing the desired operation into the interior of the robot's workspace. However, from the point of view of dexterity, singularities in the interior of the workspace are problematic.

Singularities of serial robots are typically defined in terms of the values of the internal coordinates which correspond to a singular pose. Such a description is dependent on the method of kinematic modeling and the references from which the internal coordinates are measured. This chapter presents a geometric description of singularities, in terms of the external coordinates of a robot. Such a description is generic and independent of the internal coordinates. The description of singularities given in this chapter is in the form of singular planes. The singular planes are the loci of singular points on the terminal link of a robotic mechanism.

Singular planes pertain to robots which have spherical wrists. Such robots may be considered as two distinct mechanisms. They were referred to by [22] as wrist-partitioned and were introduced in Chap. 4. The first mechanism, called the arm, is a spatial three degree of freedom positioning mechanism whose function is to control the spatial motion of the center of the spherical wrist. The second mechanism, called

J. Lenarčič et al., *Robot Mechanisms*,
Intelligent Systems, Control and Automation: Science and Engineering 60,
DOI 10.1007/978-94-007-4522-3_5,

the wrist, is a spherical three degree of freedom mechanism whose function is to control the orientation of the gripper. It is easy to show that the singularities of such robots are also decoupled into distinct singularities of the arm and the wrist.

The singular planes reveal that singularities of industrial manipulators which are interior to their workspace are all pointing singularities. This chapter concludes by describing pointing systems which are singularity-free and showing how these have been implemented in developing singularity-free robots.

5.1 Decoupled Singularities of Robots with Spherical Wrists

This discussion considers six degree of freedom robots whose first three links form a spatial positioning mechanism called the arm and whose last three links form a spherical orienting mechanism called the wrist. A spherical wrist is one where the three axes of the wrist (these would be the fourth, fifth and sixth axes of the robot) intersect each other at a single point, called the wrist center C. The position of the wrist center is unaffected by the three rotations within the wrist since it lies on each of the three axes of rotation. Since the wrist is attached to the third link of the arm (frequently referred to as the forearm), the point C does not move relative to the forearm. The motion of point C on the forearm is affected only by the first three internal coordinates which are found in the arm.

Let $\mathbf{v}_\mathrm{c}$ represent the velocity of point C and let $\boldsymbol{\omega}$ represent the angular velocity of the gripper. Conceptually, one may consider C to be the center of a ball and socket joint between the forearm (link 3) and the gripper (link 6). Let $\dot{\boldsymbol{\theta}}_\mathrm{a}$ be a vector of the three joint velocities of the arm,

$$\dot{\boldsymbol{\theta}}_\mathrm{a} = \begin{bmatrix} \dot{q}_1 \\ \dot{q}_2 \\ \dot{q}_3 \end{bmatrix}$$

and $\dot{\boldsymbol{\theta}}_\mathrm{w}$ be a vector of the three joint velocities of the wrist,

$$\dot{\boldsymbol{\theta}}_\mathrm{w} = \begin{bmatrix} \dot{q}_4 \\ \dot{q}_5 \\ \dot{q}_6 \end{bmatrix}.$$

Since point C lies on the axes of the three wrist rotations, $\mathbf{v}_\mathrm{c}$ is affected only by $\dot{\boldsymbol{\theta}}_\mathrm{a}$ and the forward velocity equations of a wrist-partitioned robot have the form,

$$\begin{bmatrix} \mathbf{v}_\mathrm{c} \\ \boldsymbol{\omega} \end{bmatrix} = \mathbf{J} \begin{bmatrix} \dot{\boldsymbol{\theta}}_\mathrm{a} \\ \dot{\boldsymbol{\theta}}_\mathrm{w} \end{bmatrix} \tag{5.1}$$

where the 6×6 matrix $\mathbf{J}$ is partitioned into four 3×3 submatrices

$$\mathbf{J} = \begin{bmatrix} \mathbf{J}_a & \mathbf{0} \\ \mathbf{J}_{21} & \mathbf{J}_w \end{bmatrix} \quad \text{and,} \tag{5.2}$$

- $\mathbf{J}_a$ maps $\dot{\boldsymbol{\theta}}_a$ into $\mathbf{v}_c$,
- $\mathbf{0}$ is the 3×3 null matrix (consequence of spherical wrist condition),
- $\mathbf{J}_{21}$ maps $\dot{\boldsymbol{\theta}}_a$ into the angular velocity of the forearm, and
- $\mathbf{J}_w$ maps $\dot{\boldsymbol{\theta}}_w$ into the angular velocity of the gripper relative to the forearm.

From (5.1) and (5.2) we can solve the inverse velocity problem,

$$\dot{\boldsymbol{\theta}}_a = \mathbf{J}_a^{-1}\mathbf{v}_c \quad \text{and}$$
$$\dot{\boldsymbol{\theta}}_w = \mathbf{J}_w^{-1}(\boldsymbol{\omega} - \mathbf{J}_{21}\dot{\boldsymbol{\theta}}_a).$$

From which we see that a wrist-partitioned robot is singular when either

$$\det \mathbf{J}_a = 0$$

which corresponds to singularities of the arm, or

$$\det \mathbf{J}_w = 0$$

which corresponds to singularities of the wrist. Thus the singularities of wrist-partitioned robots are decoupled into distinct singularities of the arm and singularities of the wrist. The next sections consider these singularities and describe them in terms of singular planes.

5.2 Singularities of Serial Robot Arms

We will consider individually the five types of serial robot arms we described earlier, namely the articulated, spherical, cylindrical, scara and Cartesian.

5.2.1 Singular Planes of the Articulated Arm

The joints of the articulated arm are all rotational and the axis vectors $\mathbf{e}_i$ $(i = 1, 2, 3)$ are assigned as shown. The first two axes are orthogonally intersecting and the second and third axes are parallel. The mechanism's internal coordinates are the three rotational variables θ_1, θ_2 and θ_3. Frames are attached to the links in accordance to Sect. 2.2.3. The vector parameters of the arm are summarized in Table 5.1.

Due to our choice of reference frames and reference position, the simplified expressions for the rotation matrices (1.51), (1.52), (1.53), can be used. This gives the

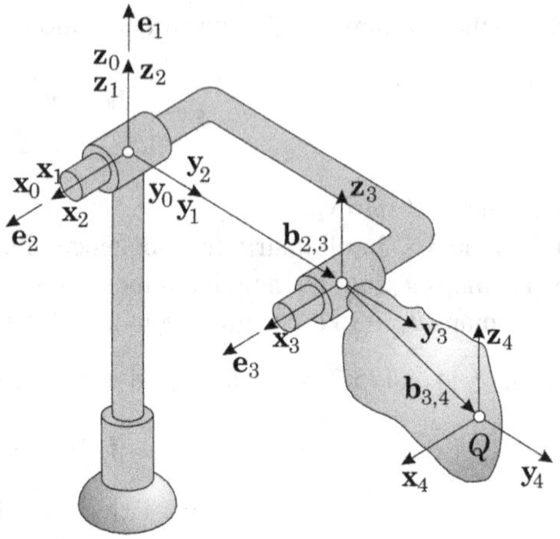

Fig. 5.1 The articulated arm, frame assignments and definition of vector parameters

Table 5.1 Vector parameters and variables of the articulated arm in Fig. 5.1

i	1	2	3
θ_i	θ_1	θ_2	θ_3
d_i	0	0	0

i	1	2	3
$\mathbf{e}_i^{(i-1)}$	0	1	1
	0	0	0
	1	0	0

i	1	2	3
$\mathbf{b}_{i-1,i}^{(i-1)}$	0	0	0
	0	0	$b_{2,3}$
	0	0	0

homogeneous transformation matrices $\mathbf{H}_{i-1,i}$ from (2.12) as,

$$\mathbf{H}_{0,1} = \begin{bmatrix} c_1 & -s_1 & 0 & 0 \\ s_1 & c_1 & 0 & 0 \\ 0 & 0 & 1 & 0 \\ 0 & 0 & 0 & 1 \end{bmatrix},$$

$$\mathbf{H}_{1,2} = \begin{bmatrix} 1 & 0 & 0 & 0 \\ 0 & c_2 & -s_2 & 0 \\ 0 & s_2 & c_2 & 0 \\ 0 & 0 & 0 & 1 \end{bmatrix},$$

$$\mathbf{H}_{2,3} = \begin{bmatrix} 1 & 0 & 0 & 0 \\ 0 & c_3 & -s_3 & b_{2,3} \\ 0 & s_3 & c_3 & 0 \\ 0 & 0 & 0 & 1 \end{bmatrix}.$$

Now consider an arbitrary point of reference on the terminal link, point Q in Fig. 5.1. To locate Q we will use a fourth frame which is always parallel to the third frame but is translated by a vector $\mathbf{b}_{3,4}$ where

$$\mathbf{b}_{3,4}^{(3)} = \begin{bmatrix} x \\ y \\ z \end{bmatrix},$$

and x, y and z are the coordinates of Q relative to coordinate system $\mathbf{x}_3$, $\mathbf{y}_3$, $\mathbf{z}_3$. In the case of an industrial manipulator Q would be the wrist center. Thus we have the fourth transformation

$$\mathbf{H}_{3,4} = \begin{bmatrix} 1 & 0 & 0 & x \\ 0 & 1 & 0 & y \\ 0 & 0 & 1 & z \\ 0 & 0 & 0 & 1 \end{bmatrix},$$

which is a constant. Following the procedure outlined in Sect. 3.1.1, the coordinates of Q relative to coordinate system $\mathbf{x}_3$, $\mathbf{y}_3$, $\mathbf{z}_3$ are given by,

$$\mathbf{r}_{3,4}^{(3)} = \begin{bmatrix} x \\ y \\ z \end{bmatrix},$$

while the coordinates of Q relative to coordinate system $\mathbf{x}_2$, $\mathbf{y}_2$, $\mathbf{z}_2$ are given by,

$$\mathbf{r}_{2,4}^{(2)} = \begin{bmatrix} x \\ yc_3 - zs_3 + b_{2,3} \\ ys_3 + zc_3 \end{bmatrix}, \tag{5.3}$$

and the coordinates of Q relative to coordinate system $\mathbf{x}_1$, $\mathbf{y}_1$, $\mathbf{z}_1$ are given by,

$$\mathbf{r}_{1,4}^{(1)} = \begin{bmatrix} x \\ yc_{23} - zs_{23} + b_{2,3}c_2 \\ ys_{23} + zc_{23}b_{2,3}s_2 \end{bmatrix}, \tag{5.4}$$

where $c_{23} = \cos(\theta_2 + \theta_3)$ and $s_{23} = \sin(\theta_2 + \theta_3)$. Finally, the coordinates of Q relative to coordinate system $\mathbf{x}_0$, $\mathbf{y}_0$, $\mathbf{z}_0$ are given by,

$$\mathbf{r}_{0,4}^{(0)} = \begin{bmatrix} xc_1 - (yc_{23} - zs_{23} + b_{2,3}c_2)s_1 \\ xs_1 + (yc_{23} - zs_{23} + b_{2,3}c_2)c_1 \\ ys_{23} + zc_{23}b_{2,3}s_2 \end{bmatrix}. \tag{5.5}$$

For the upcoming discussion, it is worthwhile to now make the following two observations,

- from (5.3) note the $\mathbf{z}_2$ coordinate of Q is given by $(ys_3 + zc_3)$ and
- from (5.4) note the $\mathbf{y}_1$ coordinate of Q is given by $(yc_{23} - zs_{23} + b_{2,3}c_2)$.

You will soon see these two terms again.

Taking a time derivative of (5.5) and following the procedure in Sect. 3.2.2 yields a Jacobian matrix of the arm, $\mathbf{J}_\mathrm{a}$, which maps the velocities of the internal coordinates into the velocity of point Q, $\dot{\mathbf{r}}_{0,4}$.

$$\dot{\mathbf{r}}_{0,4} = \mathbf{J}_\mathrm{a} \begin{bmatrix} \dot{\theta}_1 \\ \dot{\theta}_2 \\ (\dot{\theta}_2 + \dot{\theta}_3) \end{bmatrix}, \tag{5.6}$$

where

$$\mathbf{J}_\mathrm{a} = \begin{bmatrix} \dfrac{\partial \mathbf{r}_{0,4}}{\partial \theta_1} & \dfrac{\partial \mathbf{r}_{0,4}}{\partial \theta_2} & \dfrac{\partial \mathbf{r}_{0,4}}{\partial (\theta_2 + \theta_3)} \end{bmatrix}. \tag{5.7}$$

In formulating the Jacobian for the articulated arm, a common "trick" which simplifies the kinematic equations is being used. In the articulated arm, internal coordinate θ_3 always appears as a sum with θ_2. So, we replace θ_3 with a new internal coordinate which is $(\theta_2 + \theta_3)$, i.e. we make this change of variables. So, in formulating the Jacobian we treat the articulated arm as having three independent internal coordinates θ_1, θ_2 and $(\theta_2 + \theta_3)$ and specifically, when we take the partial derivative with respect to θ_2, we ignore the terms that include $(\theta_2 + \theta_3)$. Following this change of variables we find $\mathbf{J}_\mathrm{a}$ as,

$$\mathbf{J}_\mathrm{a} = \begin{bmatrix} -xs_1 - (yc_{23} - zs_{23} + b_{2,3}c_2)c_1 & b_{2,3}s_2s_1 & (ys_{23} + zc_{23})s_1 \\ xc_1 - (yc_{23} - zs_{23} + b_{2,3}c_2)s_1 & -b_{2,3}s_2c_1 & -(ys_{23} + zc_{23})c_1 \\ 0 & b_{2,3}c_2 & yc_{23} - zs_{23} \end{bmatrix}. \tag{5.8}$$

We can now take the determinant of $\mathbf{J}_\mathrm{a}$ and enumerate the kinematic singularities of the articulated arm. It is easiest to expand the determinant down the first column.

$$\begin{aligned} \det \mathbf{J}_\mathrm{a} = &\big(-xs_1 - (yc_{23} - zs_{23} + b_{2,3}c_2)c_1\big)D_1 \\ &- \big(xc_1 - (yc_{23} - zs_{23} + b_{2,3}c_2)s_1\big)D_2 \end{aligned} \tag{5.9}$$

where

$$D_1 = \det \begin{bmatrix} -b_{2,3}s_2c_1 & -(ys_{23} + zc_{23})c_1 \\ b_{2,3}c_2 & yc_{23} - zs_{23} \end{bmatrix} = b_{2,3}c_1(ys_3 + zc_3) \tag{5.10}$$

and

$$D_2 = \det \begin{bmatrix} b_{2,3}s_2s_1 & (ys_{23} + zc_{23})s_1 \\ b_{2,3}c_2 & yc_{23} - zs_{23} \end{bmatrix} = -b_{2,3}s_1(ys_3 + zc_3). \tag{5.11}$$

Substituting D_1 and D_2 into the expression for $\mathbf{J}_\mathrm{a}$ gives our result,

$$\det \mathbf{J}_\mathrm{a} = -b_{2,3}(ys_3 + zc_3)(yc_{23} - zs_{23} + b_{2,3}c_2). \tag{5.12}$$

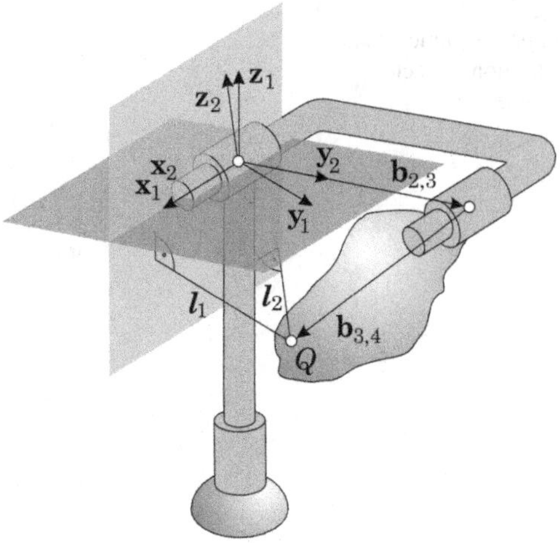

Fig. 5.2 Singular planes of the articulated arm

From this result we see there are three conditions under which the articulated arm is singular.

$$
\begin{aligned}
&\text{Condition 1:} \quad b_{2,3} = 0, \\
&\text{Condition 2:} \quad (ys_3 + zc_3) = 0, \quad \text{or} \\
&\text{Condition 3:} \quad (yc_{23} - zs_{23} + b_{2,3}c_2) = 0.
\end{aligned}
$$

Condition 1 is a trivial result. In this case joints 2 and 3 are coincident and are effectively a single joint. In this condition the articulated arm has only two degrees of freedom and is incapable of producing a three dimensional motion of point Q.

Condition 2 is in terms of the internal coordinate θ_3, however, recall our earlier observation that the term $(ys_3 + zc_3)$ is the external coordinate which is the $\mathbf{z}_2$ coordinate of Q. This coordinate being equal to zero corresponds to points Q which lie on the $\mathbf{x}_2$–$\mathbf{y}_2$ plane. This plane is a locus of points on body 3 that are instantaneously singular. This is a singular plane of the articulated arm and whenever the control point on the terminal lies on this plane, the articulated arm is singular. We refer to this as "the plane of the upper arm". This is the darker shaded plane seen in Fig. 5.2.

Condition 3 is also in terms of the internal coordinates θ_2 and $(\theta_2 + \theta_3)$, however, recall our earlier observation that the term $(yc_{23} - zs_{23} + b_{2,3}c_2)$ is the external coordinate which is the $\mathbf{y}_1$ coordinate of Q. This coordinate being equal to zero correspond to points Q which lie on the $\mathbf{x}_1$–$\mathbf{z}_1$ plane. This plane is also a locus of points on body 3 that are instantaneously singular. This is a singular plane of the articulated arm and whenever the control point on the terminal lies on this plane, the articulated arm is singular. We refer to this as "the plane of the shoulder". This is the vertical and lighter shaded plane seen in Fig. 5.2.

Figure 5.2 shows two lengths, l_1 and l_2. These are the perpendicular distances from point Q to the plane of the shoulder and to the plane of the upper arm, respectively, i.e. the $\mathbf{y}_1$ and the $\mathbf{z}_2$ coordinates of Q, respectively. We have shown that the

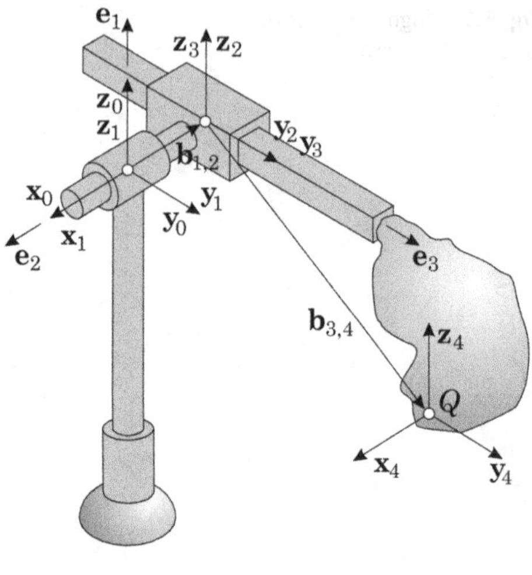

Fig. 5.3 The spherical arm, frame assignments and definition of vector parameters

$\det \mathbf{J}_a$ is simply the product of these two distances, each of which is a metric that measures the proximity of the articulated arm to its kinematic singularities. If either distance l_1 or l_2 is zero, the articulated arm is singular.

5.2.2 Singular Planes of the Spherical Arm

Figure 5.3 shows the spherical arm, sometimes called the Stanford arm, since an early version was developed at Stanford University for research purposes. The first two joints are rotational and their axes are orthogonally intersecting. The third axis is translational and is orthogonal to the second. The axis vectors $\mathbf{e}_i$ $(i = 1, 2, 3)$ are assigned as shown. The mechanism's internal coordinates are the two rotational variables θ_1, θ_2 and the translational variable d_3. Frames are attached to the links in accordance to Sect. 2.2.3. The vector parameters of the spherical arm are summarized in Table 5.2.

Note that in Table 5.2, $b_{1,2}$ has a negative value because it is in the negative $\mathbf{x}_2$ direction.

Due to our choice of reference frames and reference position, the simplified expressions for the rotation matrices (1.51), (1.52), (1.53), can be used. This gives the homogeneous transformation matrices $\mathbf{H}_{i-1,i}$ from (2.12) as,

$$\mathbf{H}_{0,1} = \begin{bmatrix} c_1 & -s_1 & 0 & 0 \\ s_1 & c_1 & 0 & 0 \\ 0 & 0 & 1 & 0 \\ 0 & 0 & 0 & 1 \end{bmatrix},$$

Table 5.2 Vector parameters and variables of the spherical arm in Fig. 5.3

i	1	2	3
θ_i	θ_1	θ_2	0
d_i	0	0	d_3

i	1	2	3
$\mathbf{e}_i^{(i-1)}$	0	1	0
	0	0	1
	1	0	0

i	1	2	3
$\mathbf{b}_{i-1,i}^{(i-1)}$	0	$-b_{1,2}$	0
	0	0	0
	0	0	0

$$\mathbf{H}_{1,2} = \begin{bmatrix} 1 & 0 & 0 & -b_{1,2} \\ 0 & c_2 & -s_2 & 0 \\ 0 & s_2 & c_2 & 0 \\ 0 & 0 & 0 & 1 \end{bmatrix},$$

$$\mathbf{H}_{2,3} = \begin{bmatrix} 1 & 0 & 0 & 0 \\ 0 & 1 & 0 & d_3 \\ 0 & 0 & 1 & 0 \\ 0 & 0 & 0 & 1 \end{bmatrix}.$$

Now consider an arbitrary point of reference on the terminal link, point Q in Fig. 5.1. To locate Q we use a fourth frame which is always parallel to the third frame but is translated by a vector $\mathbf{b}_{3,4}$ where

$$\mathbf{b}_{3,4}^{(3)} = \begin{bmatrix} x \\ y \\ z \end{bmatrix},$$

and x, y and z are the coordinates of Q relative to coordinate system $\mathbf{x}_3$, $\mathbf{y}_3$, $\mathbf{z}_3$. In the case of an industrial manipulator Q would be the wrist center. Thus we have the fourth transformation

$$\mathbf{H}_{3,4} = \begin{bmatrix} 1 & 0 & 0 & x \\ 0 & 1 & 0 & y \\ 0 & 0 & 1 & z \\ 0 & 0 & 0 & 1 \end{bmatrix},$$

which is a constant. Following the procedure outlined in Sect. 3.1.1, the coordinates of Q relative to coordinate system $\mathbf{x}_3, \mathbf{y}_3, \mathbf{z}_3$ are given by,

$$\mathbf{r}_{3,4}^{(3)} = \begin{bmatrix} x \\ y \\ z \end{bmatrix},$$

while the coordinates of Q relative to coordinate system $\mathbf{x}_2, \mathbf{y}_2, \mathbf{z}_2$ are given by,

$$\mathbf{r}_{2,4}^{(2)} = \begin{bmatrix} x \\ y + d_3 \\ z \end{bmatrix}, \tag{5.13}$$

and the coordinates of Q relative to coordinate system $\mathbf{x}_1, \mathbf{y}_1, \mathbf{z}_1$ are given by,

$$\mathbf{r}_{1,4}^{(1)} = \begin{bmatrix} x - b_{1,2} \\ (y + d_3)c_2 - zs_2 \\ (y + d_3)s_2 + zc_2 \end{bmatrix}. \tag{5.14}$$

Finally, the coordinates of Q relative to coordinate system $\mathbf{x}_0, \mathbf{y}_0, \mathbf{z}_0$ are given by,

$$\mathbf{r}_{0,4}^{(0)} = \begin{bmatrix} (x - b_{1,2})c_1 - \big((y + d_3)c_2 - zs_2\big)s_1 \\ (x - b_{1,2})s_1 + \big((y + d_3)c_2 - zs_2\big)c_1 \\ (y + d_3)s_2 + zc_2 \end{bmatrix}, \tag{5.15}$$

For the upcoming discussion, it is worthwhile to now make the following two observations,

- from (5.3) note the $\mathbf{y}_2$ coordinate of Q is given by $(y + d_3)$ and
- from (5.4) note the $\mathbf{y}_1$ coordinate of Q is given by $(y + d_3)c_2 - zs_2$.

As in the articulated arm, these terms will appear in the determinant of the Jacobian for the spherical arm.

Taking a time derivative of (5.15) and following the procedure in Sect. 3.2.2 yields a Jacobian matrix which maps the velocities of the internal coordinates into $\dot{\mathbf{r}}_{0,4}$, which is the velocity of point Q.

$$\mathbf{J}_{\mathrm{a}} = \begin{bmatrix} \frac{\partial \mathbf{r}_{0,4}}{\partial \theta_1} & \frac{\partial \mathbf{r}_{0,4}}{\partial \theta_2} & \frac{\partial \mathbf{r}_{0,4}}{\partial d_3} \end{bmatrix}, \tag{5.16}$$

i.e.

$$\mathbf{J}_{\mathrm{a}} = \begin{bmatrix} -(x - b_{1,2})s_1 - \big((y + d_3)c_2 - zs_2\big)c_1 & \big((y + d_3)s_2 + zc_2\big)s_1 & -c_2s_1 \\ (x - b_{1,2})c_1 - \big((y + d_3)c_2 - zs_2\big)s_1 & -\big((y + d_3)s_2 + zc_2\big)c_1 & c_2c_1 \\ 0 & (y + d_3)c_2 - zs_2 & s_2 \end{bmatrix}. \tag{5.17}$$

Taking the determinant of $\mathbf{J}$ and expanding the determinant down the first column gives,

$$\det \mathbf{J}_a = \big\{-(x-b_{1,2})s_1 - \big((y+d_3)c_2 - zs_2\big)c_1\big\}D_1 - \big\{(x-b_{1,2})c_1 - \big((y+d_3)c_2 - zs_2\big)s_1\big\}D_2, \tag{5.18}$$

where

$$D_1 = \det\begin{bmatrix} \big(-(y+d_3)s_2 - zc_2\big)c_1 & c_2c_1 \\ (y+d_3)c_2 - zs_2 & s_2 \end{bmatrix} = -(y+d_3)c_1 \tag{5.19}$$

and

$$D_2 = \det\begin{bmatrix} \big((y+d_3)s_2 + zc_2\big)s_1 & -c_2s_1 \\ (y+d_3)c_2 - zs_2 & s_2 \end{bmatrix} = (y+d_3)s_1. \tag{5.20}$$

Substituting D_1 and D_2 into the expression for $\det \mathbf{J}_a$ gives,

$$\det \mathbf{J}_a = \big((y+d_3)c_2 - zs_2\big)(y+d_3). \tag{5.21}$$

So there are two conditions under which the spherical arm is singular,

$$\text{Condition 1:}\quad (y+d_3)c_2 - zs_2 = 0 \quad \text{and}$$
$$\text{Condition 2:}\quad (y+d_3) = 0.$$

Condition 1 corresponds to when the $\mathbf{y}_1$ coordinate of point Q is zero, which means that Q lies in the $\mathbf{x}_1$–$\mathbf{z}_1$ plane. This singular plane is again the plane of the shoulder, just as in the articulated arm. It is the vertical lighter shaded plane shown in Fig. 5.4. The locus of points on body 3 which lie on this plane are instantaneously singular.

Condition 2 corresponds to when the $\mathbf{y}_2$ coordinate of point Q is zero, which means that Q lies in the $\mathbf{x}_2$–$\mathbf{z}_2$ plane. The locus of points on body 3 which lie on this plane are instantaneously singular. This singular plane is the plane that contains the axis of the second rotational joint and is perpendicular to the translational axis of the third joint.

Figure 5.2 shows two lengths, l_1 and l_2. These are the perpendicular distances from point Q the singular planes of conditions 1 and 2, respectively, i.e. the $\mathbf{y}_1$ and the $\mathbf{y}_2$ coordinates of Q, respectively. We have shown that $\det \mathbf{J}_a$ is the product of these two distances, each of which is a metric that measures the proximity of the articulated arm to its kinematic singularities.

5.2.3 Singular Plane of the Cylindrical Arm

Figure 5.5 shows the cylindrical arm. The first joint is rotational and the last two joints are translational. The axes of the first and second joints are parallel and the axes of the second and third joints are perpendicular. The axis vectors $\mathbf{e}_i$ $(i = 1,2,3)$

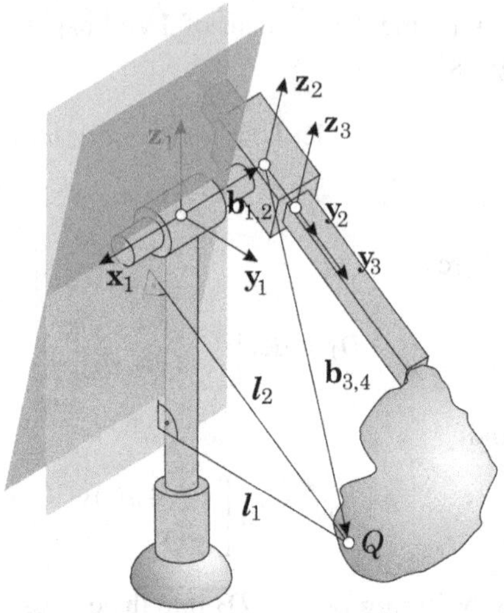

Fig. 5.4 Singular planes of the spherical arm

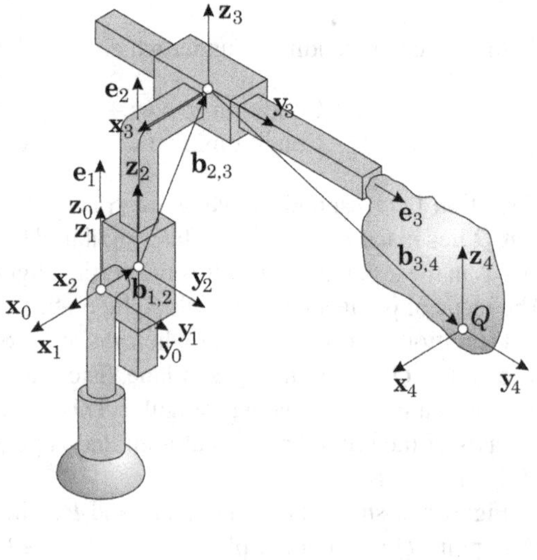

Fig. 5.5 The cylindrical arm, frame assignments and definition of vector parameters

are assigned as shown. The mechanism's internal coordinates are the rotational variable θ_1 and the translational variables d_2 and d_3. Frames are attached to the links in accordance to Sect. 2.2.3. The vector parameters of the articulated arm are summarized in Table 5.1.

Here it is understood that

$$\mathbf{b}_{i-1,i} = (b_{i-1,ix}, b_{i-1,iy}, b_{i-1,iz})^{\mathrm{T}}.$$

Note that in Table 5.3, $b_{2,3x}$ acts in the negative $\mathbf{x}_2$ direction and is a negative number whereas $b_{2,3z}$ acts in the positive $\mathbf{z}_2$ direction and is a positive number. This gives

Table 5.3 Vector parameters and variables of the cylindrical arm in Fig. 5.4

i	1	2	3
θ_i	θ_1	0	0
d_i	0	d_2	d_3

i	1	2	3
$\mathbf{e}_i^{(i-1)}$	0	0	0
	0	0	1
	1	1	0

i	1	2	3
$\mathbf{b}_{i-1,i}^{(i-1)}$	0	$b_{1,2x}$	$b_{2,3x}$
	0	0	0
	0	0	$b_{2,3z}$

the homogeneous transformation matrices $\mathbf{H}_{i-1,i}$ as,

$$\mathbf{H}_{0,1} = \begin{bmatrix} c_1 & -s_1 & 0 & 0 \\ s_1 & c_1 & 0 & 0 \\ 0 & 0 & 1 & 0 \\ 0 & 0 & 0 & 1 \end{bmatrix}, \tag{5.22}$$

$$\mathbf{H}_{1,2} = \begin{bmatrix} 1 & 0 & 0 & b_{1,2x} \\ 0 & 1 & 0 & 0 \\ 0 & 0 & 1 & d_2 \\ 0 & 0 & 0 & 1 \end{bmatrix}, \tag{5.23}$$

$$\mathbf{H}_{2,3} = \begin{bmatrix} 1 & 0 & 0 & b_{2,3x} \\ 0 & 1 & 0 & d_3 \\ 0 & 0 & 1 & b_{2,3z} \\ 0 & 0 & 0 & 1 \end{bmatrix}, \tag{5.24}$$

and x, y and z are the coordinates of Q relative to coordinate system $\mathbf{x}_3$, $\mathbf{y}_3$, $\mathbf{z}_3$. In the case of an industrial manipulator Q would be the wrist center. Thus we have the fourth transformation

$$\mathbf{H}_{3,4} = \begin{bmatrix} 1 & 0 & 0 & x \\ 0 & 1 & 0 & y \\ 0 & 0 & 1 & z \\ 0 & 0 & 0 & 1 \end{bmatrix},$$

which is a constant. Following the procedure outlined in Sect. 3.1.1, the coordinates of Q relative to coordinate system $\mathbf{x}_3$, $\mathbf{y}_3$, $\mathbf{z}_3$ are given by,

$$\mathbf{r}_{3,4}^{(3)} = \begin{bmatrix} x \\ y \\ z \end{bmatrix},$$

while the coordinates of Q relative to coordinate system $\mathbf{x}_2$, $\mathbf{y}_2$, $\mathbf{z}_2$ are given by,

$$\mathbf{r}_{2,4}^{(2)} = \begin{bmatrix} x + b_{2,3x} \\ y + d_3 \\ z + b_{2,3z} \end{bmatrix}, \tag{5.25}$$

and the coordinates of Q relative to coordinate system $\mathbf{x}_1$, $\mathbf{y}_1$, $\mathbf{z}_1$ are given by,

$$\mathbf{r}_{1,4}^{(1)} = \begin{bmatrix} x + b_{2,3x} + b_{1,2x} \\ (y + d_3) \\ z + b_{2,3z} + d_2 \end{bmatrix}. \tag{5.26}$$

Finally, the coordinates of Q relative to coordinate system $\mathbf{x}_0$, $\mathbf{y}_0$, $\mathbf{z}_0$ are given by,

$$\mathbf{r}_{0,4}^{(0)} = \begin{bmatrix} (x + b_{2,3x} + b_{1,2x})c_1 - (y + d_3)s_1 \\ (x + b_{2,3x} + b_{1,2x})s_1 + (y + d_3)c_1 \\ (z + b_{2,3z} + d_2) \end{bmatrix}. \tag{5.27}$$

Taking a time derivative of (5.27) and following the procedure in Sect. 3.2.2 yields a Jacobian matrix which maps the velocities of the internal coordinates into $\dot{\mathbf{r}}_{0,4}$, which is the velocity of point Q.

$$\mathbf{J}_\mathrm{a} = \begin{bmatrix} \frac{\partial \mathbf{r}_{0,4}}{\partial \theta_1} & \frac{\partial \mathbf{r}_{0,4}}{\partial d_2} & \frac{\partial \mathbf{r}_{0,4}}{\partial d_3} \end{bmatrix}, \tag{5.28}$$

i.e.

$$\mathbf{J}_\mathrm{a} = \begin{bmatrix} (x + b_{2,3x} + b_{1,2x})s_1 - (y + d_3)c_1 & 0 & -s_1 \\ (x + b_{2,3x} + b_{1,2x})c_1 - (y + d_3)s_1 & 0 & c_1 \\ 0 & 1 & 0 \end{bmatrix}. \tag{5.29}$$

Taking the determinant of $\mathbf{J}_\mathrm{a}$ and expanding the determinant across the bottom row gives,

$$\det \mathbf{J}_\mathrm{a} = (y + d_3), \tag{5.30}$$

and only one condition under which the cylindrical arm is singular,

$$(y + d_3) = 0.$$

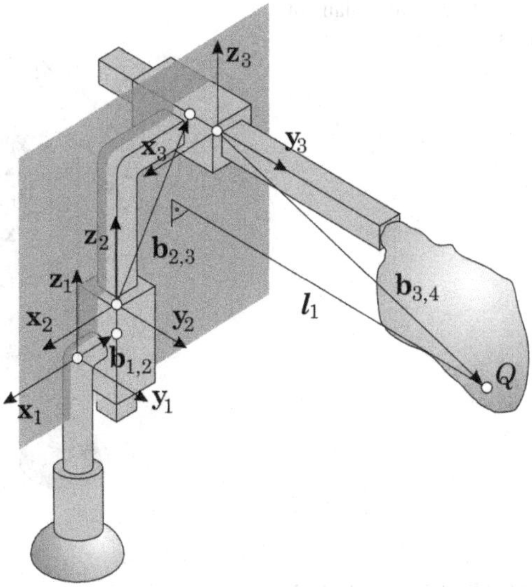

Fig. 5.6 Singular plane of the cylindrical arm

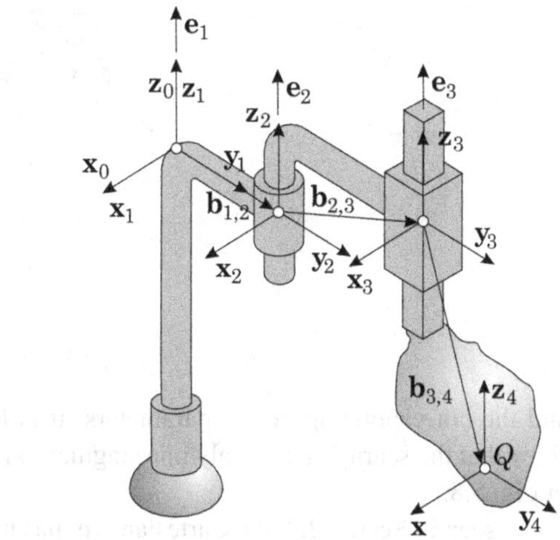

Fig. 5.7 The scara arm, frame assignments and definition of vector parameters

From (5.25), we see that this corresponds to when the $\mathbf{y}_1$ coordinate of Q is zero, thus Q lies in the $\mathbf{x}_1$–$\mathbf{z}_1$ plane. This is the only singular plane of the cylindrical arm and it is shown in Fig. 5.6. The locus of points on body 3 which lie on this plane are instantaneously singular.

5.2.4 Singular Plane of the Scara Arm

Figure 5.7 shows the scara arm. All three joint axes are parallel, with the first two being rotational and the third translational. The figure shows the frames assignments

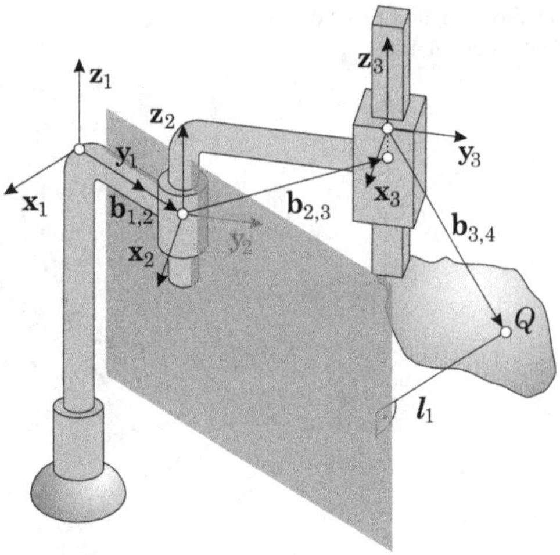

Fig. 5.8 Singular plane of the scara arm

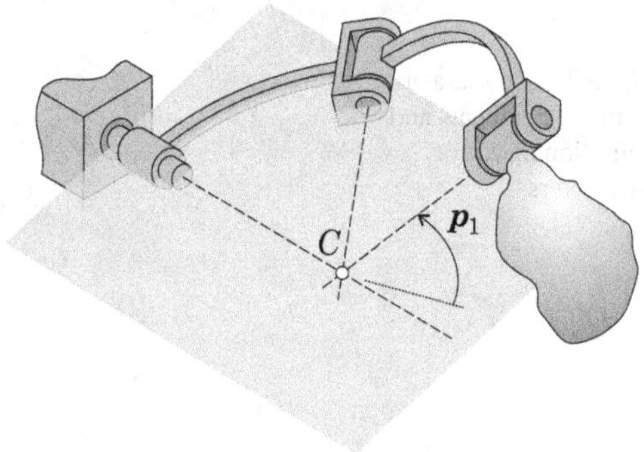

Fig. 5.9 The spherical wrist and its singular plane

and the corresponding vector parameters. It is left as an exercise to the reader to show that the scara arm has only one singular plane, which is the $\mathbf{y}_1$–$\mathbf{z}_1$ plane shown in Fig. 5.8.

As seen in Sect. 4.2.2, the Cartesian arm has no singularities and thus no singular planes.

5.3 Singularities of Spherical Wrists

Figure 5.9 shows a spherical wrist. It consists of three rotational joints whose axes intersect at the point C, which is the center of the wrist. As shown by [68], when the third axis of rotation lies in the plane formed by the first two axes of rotation $\det \mathbf{J}_\mathrm{w} = 0$ and the wrist is singular. Being a spherical mechanism, the metric for measuring proximity to this singular plane is no longer a distance, as it was with the various arms, but is now the angle p_1 formed between the third axis and the singular

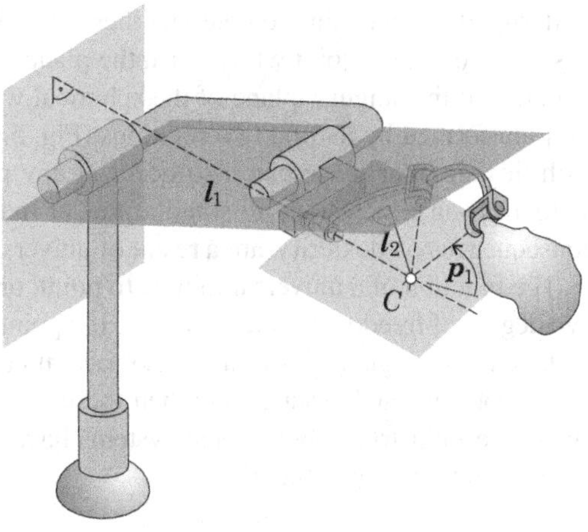

Fig. 5.10 Singularity metrics of a common six degree of freedom industrial robot

plane, as shown in Fig. 5.9. This angle measures the distance from the singular plane of the wrist.

5.4 Singularity Metrics of Industrial Robots

Figure 5.10 shows a six degree of freedom industrial robot which is very common. Versions of it are manufactured by Staubli such as the model seen in Fig. 2.10. It is also the mechanism found in the well known Puma 560. There are three metrics which measure proximity of the mechanism to its kinematic singularities. They are the two distances from the wrist center to the plane of the shoulder and the plane of the forearm, l_1 and l_2 respectively, and the angle p_1 from the roll axis to the singular plane of the wrist. If any of these metrics go to zero, the mechanism is singular. The concept of singular planes can be applied to describing the singularities of six degree of freedom industrial robots in terms of metrics which can be measured and are not purely mathematical.

5.4.1 Pointing Singularities

With the exception of the singular plane which is the plane of the shoulder found in the articulated arm (Fig. 5.2), spherical arm (Fig. 5.4) all the remaining singular planes of robot arms correspond to either the innermost or outermost workspace boundaries of the arms. Workspace boundary singularities are inconsequential to dexterity. They are avoided by bringing the operation into the interior of the workspace.

However, the singular planes which are the plane of the shoulder result in singularities which are interior to the workspace and these are a significant problem to dexterity. It is interesting to note that these singularities correspond to the singularities of the universal joint which form the plane of the shoulder. It is also interesting to note that the singular plane of the spherical wrist also corresponds to the universal joint formed by its first two rotations (Fig. 5.9) and that singularities associated with this singular plane can also occur at any point in the workspace. This leads us to a strong conclusion; all singularities of industrial robots which have negative consequences on dexterity, are a result of universal joint based shoulders and wrists.

The function of a universal joint is to point, or aim a line. The universal joint is a two degree of freedom pointing system. If a pointing system can be found which has no kinematic singularities in its workspace, it can serve as the basis for singularity free robot arms and wrists, and when these are combined the result will be singularity free industrial robots. Such systems have been developed and the upcoming section we will look at them.

5.5 Singularity Free Pointing Systems

Several individuals have developed singularity free pointing systems, [61, 72, 78, 82]. They all operate on a principle of symmetric actuation. In this section we will show how symmetric actuation applied to a redundant system results in singularity free pointing.

Figure 5.11 shows a spherical three degree of freedom mechanism which consists of one fixed and three moving bodies. The bodies are connected by rotational joints which are successively orthogonal and mutually intersecting at the point C, the center of the spherical mechanism.

The mechanism will be used to point a line which is the central axis of the cylindrical shape at the end of the system, which is seen in the figure. The external coordinates which define a pointing direction can be two angles, or a unit vector. Either representation indicates that a pointing direction is described by two independent external coordinates.

The mechanism in Fig. 5.11 has three independent internal coordinates. Since the number of independent internal coordinates exceeds the number of independent external coordinates by one, there are an infinite number of solutions to the inverse pointing (kinematics) problem and the mechanism is kinematically redundant. Chapter 6 studies redundant mechanisms in great detail. Assign coordinate systems to the four bodies as shown, where system $\mathbf{x}_i$, $\mathbf{y}_i$, $\mathbf{z}_i$ is attached to body i $(i = 0, 1, 2, 3)$. All frames have their origins at C. The pointing direction the mechanism controls is the direction of the $\mathbf{z}_3$ axis, which is permanently aligned with the central axis of the cylindrical shape at the end of the system. The rotation matrices which describe the relative orientations of the frames are given by,

$$\mathbf{A}_{0,1} = \begin{bmatrix} 1 & 0 & 0 \\ 0 & c_1 & -s_1 \\ 0 & s_1 & c_1 \end{bmatrix},$$

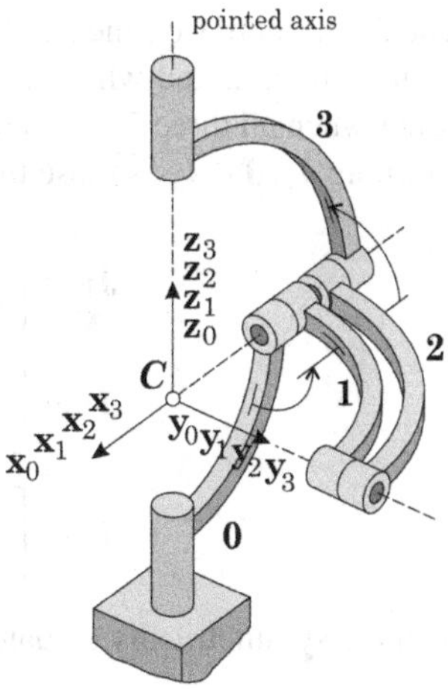

Fig. 5.11 A symmetrically actuated double pointing system

$$\mathbf{A}_{1,2} = \begin{bmatrix} c_2 & 0 & s_2 \\ 0 & 1 & 0 \\ -s_2 & 0 & c_2 \end{bmatrix},$$

$$\mathbf{A}_{2,3} = \begin{bmatrix} 1 & 0 & 0 \\ 0 & c_3 & -s_3 \\ 0 & s_3 & c_3 \end{bmatrix}.$$

Symmetric actuation requires the internal angle between bodies 0 and 1 and that between bodies 2 and 3, indicated in the figure, be equal. This translates into the rotations about the first and third axes being equal, i.e. $\theta_1 = \theta_3$. This constraint reduces the system to two independent internal coordinates, which resolves the kinematic redundancy and now $\mathbf{A}_{2,3} = \mathbf{A}_{0,1}$. Multiplying the rotation matrices gives,

$$\mathbf{A}_{0,3} = \mathbf{A}_{0,1}\mathbf{A}_{1,2}\mathbf{A}_{2,3} = \begin{bmatrix} c_2 & s_1 s_2 & c_1 s_2 \\ s_1 s_2 & c_1^2 - s_1^2 c_2 & -s_1 c_1 (1 + c_2) \\ -c_1 s_2 & s_1 c_1 (1 + c_2) & -s_1^2 + c_1^2 c_2 \end{bmatrix}.$$

The third column of $\mathbf{A}_{0,3}$ is the pointing direction $\mathbf{z}_3^{(0)}$,

$$\mathbf{z}_3^{(0)} = \begin{bmatrix} c_1 s_2 \\ -s_1 c_1 (1 + c_2) \\ -s_1^2 + c_1^2 c_2 \end{bmatrix}. \tag{5.31}$$

Given internal coordinates θ_1 and θ_2 the above equation computes the external coordinates $\mathbf{z}_3$. Let us investigate the singularities of this symmetrically actuated pointing system. Differentiating $\mathbf{z}_3^{(0)}$ gives,

$$\dot{\mathbf{z}}_3^{(0)} = \begin{bmatrix} \frac{\partial \mathbf{z}_3^{(0)}}{\partial \theta_1} & \frac{\partial \mathbf{z}_3^{(0)}}{\partial \theta_2} \end{bmatrix} \begin{bmatrix} \dot{\theta}_1 \\ \dot{\theta}_2 \end{bmatrix} = \begin{bmatrix} -s_1 s_2 & c_1 c_2 \\ (-c_1^2 + s_1^2)(1 + c_2) & s_1 c_1 s_2 \\ -2 s_1 c_1 (1 + c_2) & -c_1^2 s_2 \end{bmatrix} \begin{bmatrix} \dot{\theta}_1 \\ \dot{\theta}_2 \end{bmatrix}.$$

The 3×2 matrix on the right side of this equation is the Jacobian matrix. Its maximum rank is 2 and when singular it becomes rank deficient. Rank deficiency will occur when all three 2×2 matrices formed by the rows become zero simultaneously. Defining $\mathbf{J}_1, \mathbf{J}_2, \mathbf{J}_3$ as these three matrices,

$$\mathbf{J}_1 = \begin{bmatrix} -s_1 s_2 & c_1 c_2 \\ (-c_1^2 + s_1^2)(1 + c_2) & s_1 c_1 s_2 \end{bmatrix},$$

$$\mathbf{J}_2 = \begin{bmatrix} -s_1 s_2 & c_1 c_2 \\ -2 s_1 c_1 (1 + c_2) & -c_1^2 s_2 \end{bmatrix},$$

$$\mathbf{J}_3 = \begin{bmatrix} (-c_1^2 + s_1^2)(1 + c_2) & s_1 c_1 s_2 \\ -2 s_1 c_1 (1 + c_2) & -c_1^2 s_2 \end{bmatrix},$$

then the symmetrically actuated pointing system is singular when

$$\det \mathbf{J}_1 = \det \mathbf{J}_2 = \det \mathbf{J}_3 = 0.$$

Evaluating the determinants gives,

$$\det \mathbf{J}_1 = \det \begin{bmatrix} -s_1 s_2 & c_1 c_2 \\ (-c_1^2 + s_1^2)(1 + c_2) & s_1 c_1 s_2 \end{bmatrix} = c_1 \left(-s_1^2 + c_1^2 c_2\right)(1 + c_2),$$

$$\det \mathbf{J}_2 = \det \begin{bmatrix} -s_1 s_2 & c_1 c_2 \\ -2 s_1 c_1 (1 + c_2) & -c_1^2 s_2 \end{bmatrix} = s_1 c_1^2 (1 + c_2)^2,$$

$$\det \mathbf{J}_3 = \det \begin{bmatrix} (-c_1^2 + s_1^2)(1 + c_2) & s_1 c_1 s_2 \\ -2 s_1 c_1 (1 + c_2) & -c_1^2 s_2 \end{bmatrix} = c_1^2 s_2 (1 + c_2).$$

There are three conditions under which all three determinants vanish simultaneously The first two of these are,

$$\begin{array}{lll} \text{Condition 1:} & (1 + c_2) = 0 & \longrightarrow \quad \theta_2 = \pi \quad \text{and} \\ \text{Condition 2:} & c_1 = 0 & \longrightarrow \quad \theta_1 = \pm(\pi/2). \end{array}$$

Observe that the internal coordinates corresponding to these two conditions result in external coordinates given by (5.31) as,

$$\mathbf{z}_3^{(0)} = \begin{bmatrix} 0 \\ 0 \\ -1 \end{bmatrix},$$

which means that when the symmetrically actuated pointing system is singular, the $\mathbf{z}_3$ axis is pointing in the opposite direction of the $\mathbf{z}_0$ axis. Observe from Fig. 5.11 that this corresponds to when then pointed axis (the cylindrical shape at the end of the system) is occupying the space of the cylindrical shape which is the supporting base. This supporting structure is not considered a usable part of the pointing system's workspace. The conclusion is that the maximum possible usable workspace of the symmetrically actuated pointing system is singularity-free.

What makes this symmetric pointing system different from a universal joint pointing system is that it has only this one singular pointing direction. The universal joint pointing system has two diametrically opposed singular pointing directions, as seen in Fig. 4.21. In Fig. 4.21 we see that one of the two singular pointing directions occupies the space of the supporting structure, however the other singular pointing direction is in the workspace of the system. It has been proven mathematically any pointing system has a minimum of one pointing direction [87] and the symmetrically actuated pointing system has realized this minimum. Since the symmetrically actuated pointing system has only this one singular pointing direction, it can be made to coincide with the supporting base, leading to no singular directions existing in the workspace.

There is a third condition under which all three determinants vanish,

$$\begin{aligned} \text{Condition 3:} \quad -s_1^2 + c_1^2 c_2 &= 0 \quad \text{while simultaneously} \\ s_1 &= 0 \quad \text{while simultaneously} \\ s_2 &= 0. \end{aligned}$$

However we see from (5.31) that this corresponds to $\mathbf{z}_3^{(0)} = [0\,0\,0]^{\mathrm{T}}$, which does not exist. This third condition is meaningless.

5.5.1 Singularity Free Robot Wrists

A singularity free spherical robot wrist results when the symmetrically actuated pointing system is used as its basis. The wrist would come from adding a fourth rotation about the pointed axis in Fig. 5.11 [61, 72, 78, 82].

5.5.2 Singularity Free Robot Arms

Adding a translational joint in the direction of the pointed axis of the symmetrically actuated pointing system in Fig. 5.11, results in a singularity free spherical arm [19]. Adding an RR dyad to span across the translational joint of the spherical arm results in a singularity free articulated arm [70]. As we saw earlier, the remaining robot arms have no workspace singularities. Their singularities occur on their workspace boundaries.

A closing note about singularity free robots. As we have seen in Chap. 4, dexterity and mechanical advantage are opposing objectives. Thus, although singularity free robots have enhanced dexterity, their mechanical advantage is reduced. Furthermore, in order to achieve the symmetry constraint they require either an extra actuator or a complicated linkage. These translate into higher costs and a reduced reliability.

In most robotic tasks the process can be preplanned to avoid the singular poses of the robot. The preplanning is typically done using a robot simulation package.

Singularity avoidance is only an issue in impromptu tasks, tasks which cannot be preplanned, tasks such as robotic surgery. It is only in such impromptu tasks that singularity free robots may be required.

Chapter 6
Redundant Mechanisms

Abstract A redundant mechanism is one that contains more degrees of freedom than are needed to perform a given task. Redundant mechanisms can solve a given primary task in an infinite number of ways. This feature allows the robot to simultaneously solve additional secondary tasks. The system of differential equations defining the kinematics of a redundant mechanism is underdetermined. This requires special mathematical approaches to solve the inverse kinematics problem. One of these is the so-called task-priority approach, in which the secondary task is subordinated to the primary task. We show in this chapter that humans and animals also take advantage of kinematic redundancy to optimize their motion.

A mechanism with n degrees of freedom, whose joint coordinates are $q_1, q_2, \ldots, q_n$, can in general perform a task, which is described by m external coordinates p_1, $p_2, \ldots, p_m$, when $n \geq m$. Up to now we only considered mechanisms with $n = m$ degrees of freedom. In this chapter we will study the case where $n \neq m$. We say that a mechanism does not have enough degrees of freedom with respect to the task when

$$n < m \tag{6.1}$$

and that it has too many degrees of freedom with respect to the task when

$$n > m. \tag{6.2}$$

Such a mechanism is referred to as kinematically redundant and the difference

$$D = n - m \tag{6.3}$$

is the degree of kinematic redundancy [7, 65].

As an example let us consider a mechanism with $n = 5$ degrees of freedom, which is supposed to perform the task of positioning the gripper, which only requires $m = 3$ degrees of freedom. As there is $n > m$, the mechanism is kinematically

J. Lenarčič et al., *Robot Mechanisms*,
Intelligent Systems, Control and Automation: Science and Engineering 60,
DOI 10.1007/978-94-007-4522-3_6, © Springer Science+Business Media Dordrecht 2013

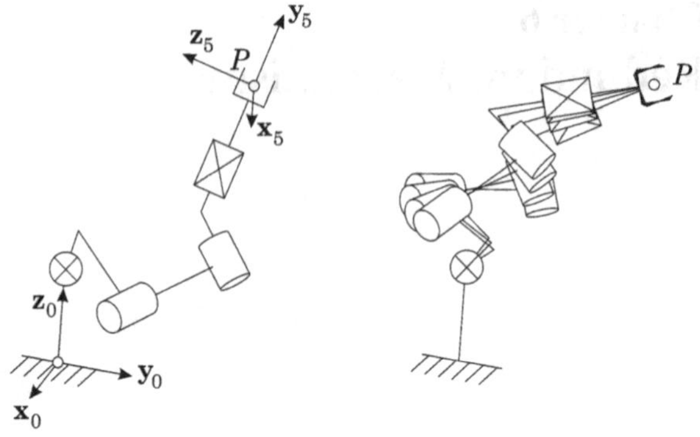

Fig. 6.1 Mechanism with kinematic redundancy ($n = 5$, $m = 3$, $D = 2$)

redundant. The system of kinematic equations

$$\mathbf{p} = \mathbf{p}_q(\mathbf{q}) \tag{6.4}$$

relates three external coordinates p_1, p_2, and p_3 to the five joint coordinates $q_1, q_2, \ldots, q_5$. There is no difficulty in solving the direct kinematics problem, where the values of joint coordinates are given and the values of the external coordinates are to be calculated. The situation becomes more complicated when solving the inverse kinematics, where the five joint coordinates are to be calculated from three equations describing the external coordinates.

The system of equations is underconstrained and there exists an infinite number of solutions, i.e. an infinite number of configurations which belong to a selected position of the gripper. As a result, a redundant mechanism can move from one configuration into another without influencing the values of the given external coordinates (Fig. 6.1). Such movements are called the self motion of the redundant mechanism. This self motion is a characteristic property of redundant mechanisms which enables them to solve the required task in an infinite number of ways. Remember that for a non-redundant mechanism there are a finite number of inverse kinematic solutions that correspond to a pose of the gripper and that a non-redundant mechanism cannot pass from one configuration into another configuration without changing values of the external coordinates.

One possible method of solving the inverse kinematics problem of a redundant mechanism is to select the values of D joint coordinates, while the rest of the coordinates are then calculated from the kinematic equations, in the same manner as in a non-redundant mechanism. In this way we have mathematical control over the redundant variables, however we are not taking advantage of them. Using this method of solution there is no advantage to the redundancy. As we have described it, a redundant mechanism possesses too many degrees of freedom for the execution of a primary task. So, alternately, we can prescribe additional, secondary tasks to a redundant mechanism in order to exploit its redundancy, such as performing a movement with minimal energy consumption, avoiding obstacles, optimizing dexterity, or optimizing mechanical advantage.

Consider our example in Fig. 6.1, which has $D = 2$ for the prescribed task. Let the secondary task of our mechanism be orienting its gripper. This task is defined by three additional external coordinates denoted as s_1, s_2, and s_3. They are related to the joint coordinates as follows

$$\mathbf{s} = \mathbf{s}_{\mathrm{q}}(\mathbf{q}). \tag{6.5}$$

In general we assume that a mechanism has m primary external coordinates, l secondary external coordinates and n joint coordinates.

How to deal with a system of equations, which describe the primary and the secondary task, depends on the relationship between the parameters n and $m + l$. In our case we are dealing with three primary external coordinates and three secondary external coordinates which are included in the vectors $\mathbf{p}$ and $\mathbf{s}$, and five joint coordinates $\mathbf{q}$. As there is $n < m + l$, the collective system of equations is overconstrained. The exact solution only exists in special circumstances. Usually, only an approximate solution can be determined.

Our discussion is founded on the system of equations relating the infinitesimal displacements of the primary and secondary external coordinates to the infinitesimal displacements of the joint coordinates

$$d\mathbf{p} = \mathbf{J}_{\mathrm{p}} d\mathbf{q} \tag{6.6}$$

and

$$d\mathbf{s} = \mathbf{J}_{\mathrm{s}} d\mathbf{q}, \tag{6.7}$$

where the Jacobian matrix $\mathbf{J}_{\mathrm{p}}$ is of dimension $m \times n$

$$\mathbf{J}_{\mathrm{p}} = \begin{bmatrix} \frac{\partial p_1}{\partial q_1} & \frac{\partial p_1}{\partial q_2} & \cdots & \frac{\partial p_1}{\partial q_n} \\ \frac{\partial p_2}{\partial q_1} & \frac{\partial p_2}{\partial q_2} & \cdots & \frac{\partial p_2}{\partial q_n} \\ \vdots & \vdots & & \vdots \\ \frac{\partial p_m}{\partial q_1} & \frac{\partial p_m}{\partial q_2} & \cdots & \frac{\partial p_m}{\partial q_n} \end{bmatrix} \tag{6.8}$$

and the Jacobian matrix $\mathbf{J}_{\mathrm{s}}$ is of dimension $l \times n$

$$\mathbf{J}_{\mathrm{s}} = \begin{bmatrix} \frac{\partial s_1}{\partial q_1} & \frac{\partial s_1}{\partial q_2} & \cdots & \frac{\partial s_1}{\partial q_n} \\ \frac{\partial s_2}{\partial q_1} & \frac{\partial s_2}{\partial q_2} & \cdots & \frac{\partial s_2}{\partial q_n} \\ \vdots & \vdots & & \vdots \\ \frac{\partial s_l}{\partial q_1} & \frac{\partial s_l}{\partial q_2} & \cdots & \frac{\partial s_l}{\partial q_n} \end{bmatrix}. \tag{6.9}$$

The solution of the inverse kinematics problem will be found numerically by the use of the gradient iterative method.

6.1 Independent Solution of Primary and Secondary Tasks

In this section, we search for the independent solutions of the primary and secondary task. In a mathematical sense we are interested in independent solutions to the systems of equations (6.6) and (6.7). The system (6.6) contains m equations, while we are searching for n components of the vector $d\mathbf{q}$. Because we have $n > m$, the Jacobian matrix $\mathbf{J}_\mathrm{p}$ is rectangular. We describe this matrix as horizontal, as the horizontal side is longer. We can determine its generalized inverse in the following form [36]

$$\mathbf{J}_\mathrm{p}^{+} = \mathbf{J}_\mathrm{p}^\mathrm{T}\left(\mathbf{J}_\mathrm{p}\mathbf{J}_\mathrm{p}^\mathrm{T}\right)^{-1}. \tag{6.10}$$

The dimension of the generalized inverse $\mathbf{J}_\mathrm{p}^{+}$ is $n \times m$. It is valid when the matrix $\mathbf{J}_\mathrm{p}$ has a rank m, so that the matrix $\mathbf{J}_\mathrm{p}\mathbf{J}_\mathrm{p}^\mathrm{T}$ of size $m \times m$, which is to be inverted, is of full rank. It is not difficult to realize that

$$\mathbf{J}_\mathrm{p}\mathbf{J}_\mathrm{p}^{+} = \mathbf{J}_\mathrm{p}\mathbf{J}_\mathrm{p}^\mathrm{T}\left(\mathbf{J}_\mathrm{p}\mathbf{J}_\mathrm{p}^\mathrm{T}\right)^{-1} = \mathbf{I},$$

while in the case of a product of the reversed order, the equality does not hold

$$\mathbf{J}_\mathrm{p}^{+}\mathbf{J}_\mathrm{p} = \mathbf{J}_\mathrm{p}^\mathrm{T}\left(\mathbf{J}_\mathrm{p}\mathbf{J}_\mathrm{p}^\mathrm{T}\right)^{-1}\mathbf{J}_\mathrm{p} \neq \mathbf{I}.$$

The matrix $\mathbf{J}_\mathrm{p}^{+}$ is the so called non-weighted generalized inverse. The weighted generalized inverse is

$$\mathbf{J}_\mathrm{pA}^{+} = \mathbf{A}^{-1}\mathbf{J}_\mathrm{p}^\mathrm{T}\left(\mathbf{J}_\mathrm{p}\mathbf{A}^{-1}\mathbf{J}_\mathrm{p}^\mathrm{T}\right)^{-1}. \tag{6.11}$$

The matrix $\mathbf{A}$ is the matrix of weights with dimension $n \times n$. It should be nonsingular and positive definite. The dimension of the matrix $\mathbf{J}_\mathrm{pA}^{+}$ is $n \times m$ and likewise

$$\mathbf{J}_\mathrm{p}\mathbf{J}_\mathrm{pA}^{+} = \mathbf{J}_\mathrm{p}\mathbf{A}^{-1}\mathbf{J}_\mathrm{p}^\mathrm{T}\left(\mathbf{J}_\mathrm{p}\mathbf{A}^{-1}\mathbf{J}_\mathrm{p}^\mathrm{T}\right)^{-1} = \mathbf{I}$$

and

$$\mathbf{J}_\mathrm{pA}^{+}\mathbf{J}_\mathrm{p} = \mathbf{A}^{-1}\mathbf{J}_\mathrm{p}^\mathrm{T}\left(\mathbf{J}_\mathrm{p}\mathbf{A}^{-1}\mathbf{J}_\mathrm{p}^\mathrm{T}\right)^{-1}\mathbf{J}_\mathrm{p} \neq \mathbf{I}.$$

The non-weighted generalized inverse is a special case of the weighted generalized inverse where $\mathbf{A} = \mathbf{I}$. Note that we cannot find a generalized inverse of a horizontal rectangular matrix $\mathbf{J}_\mathrm{p}$

$$\mathbf{J}_\mathrm{p}^{*} = \left(\mathbf{J}_\mathrm{p}^\mathrm{T}\mathbf{J}_\mathrm{p}\right)^{-1}\mathbf{J}_\mathrm{p}^\mathrm{T},$$

as the matrix $\mathbf{J}_\mathrm{p}^\mathrm{T}\mathbf{J}_\mathrm{p}$ is of dimension $n \times n$, while its highest rank is m.

The solution of the primary task, which does not pay regard to the requirements of the secondary task, can be based on the generalized inverse and formulated as follows

$$d\mathbf{q}_{\mathrm{p}} = \mathbf{J}_{\mathrm{pA}}^{+} d\mathbf{p}. \tag{6.12}$$

There exists an infinite number of such solutions since we can arbitrarily change the weighting matrix $\mathbf{A}$. We only have to be careful that the weighting matrix is nonsingular and positive definite.

We can show that the generalized inverse $\mathbf{J}_{\mathrm{pA}}^{+}$ corresponds to the least squares of the joint increments which are weighted with the matrix $\mathbf{A}$. To do this we will minimize the following quadratic form of the vector of joint increments

$$\frac{1}{2} d\mathbf{q}^{\mathrm{T}} \mathbf{A} d\mathbf{q}$$

with the condition

$$d\mathbf{p} - \mathbf{J}_{\mathrm{p}} d\mathbf{q} = \mathbf{0}.$$

The problem is solved by use of the Lagrange function

$$L = \frac{1}{2} d\mathbf{q}^{\mathrm{T}} \mathbf{A} d\mathbf{q} + \boldsymbol{\lambda}^{\mathrm{T}} (d\mathbf{p} - \mathbf{J}_{\mathrm{p}} d\mathbf{q}),$$

where $\boldsymbol{\lambda} = (\lambda_1, \lambda_2, \ldots, \lambda_m)^{\mathrm{T}}$ is vector of Lagrange multipliers. The condition

$$\frac{\partial L}{\partial d\mathbf{q}} = \mathbf{A} d\mathbf{q} - \mathbf{J}_{\mathrm{p}}^{\mathrm{T}} \boldsymbol{\lambda} = \mathbf{0},$$

should be satisfied. From which

$$d\mathbf{q} = \mathbf{A}^{-1} \mathbf{J}_{\mathrm{p}}^{\mathrm{T}} \boldsymbol{\lambda},$$

when the matrix $\mathbf{A}$ is of full rank. After multiplying with the matrix $\mathbf{J}_{\mathrm{p}}$ and after considering the condition $d\mathbf{p} - \mathbf{J}_{\mathrm{p}} d\mathbf{q} = \mathbf{0}$ it follows that

$$\mathbf{J}_{\mathrm{p}} d\mathbf{q} = d\mathbf{p} = \mathbf{J}_{\mathrm{p}} \mathbf{A}^{-1} \mathbf{J}_{\mathrm{p}}^{\mathrm{T}} \boldsymbol{\lambda}$$

and

$$\boldsymbol{\lambda} = \left(\mathbf{J}_{\mathrm{p}} \mathbf{A}^{-1} \mathbf{J}_{\mathrm{p}}^{\mathrm{T}}\right)^{-1} d\mathbf{p}.$$

Here, the rank of the matrix $\mathbf{J}_{\mathrm{p}}$ must equal m. After inserting this expression into $d\mathbf{q}$, we have

$$d\mathbf{q} = \mathbf{A}^{-1} \mathbf{J}_{\mathrm{p}}^{\mathrm{T}} \left(\mathbf{J}_{\mathrm{p}} \mathbf{A}^{-1} \mathbf{J}_{\mathrm{p}}^{\mathrm{T}}\right)^{-1} d\mathbf{p},$$

where we recognize the weighted generalized inverse since $d\mathbf{q} = \mathbf{J}_{\mathrm{pA}}^{+} d\mathbf{p}$.

We can see that the joint coordinates are displaced by the generalized inverse $\mathbf{J}_{\mathrm{pA}}^{+}$ in such a way, that the least squares

$$\dot{\mathbf{q}}^{\mathrm{T}}\mathbf{A}\dot{\mathbf{q}}$$

are minimal. The products $\dot{q}_i\dot{q}_j$, $i, j = 1, 2, \ldots, n$ are in this quadratic form weighted by the elements a_{ij} of the matrix $\mathbf{A}$. We often use the weighting matrix $\mathbf{A}$, where only the diagonal elements are non-zero. Then the role of the weighting factors is clearer since

$$\dot{\mathbf{q}}^{\mathrm{T}}\mathbf{A}\dot{\mathbf{q}} = a_{11}\dot{q}_1^2 + a_{22}\dot{q}_2^2 + \cdots + a_{nn}\dot{q}_n^2,$$

and the matrix $\mathbf{A}$ is of full rank and positive definite when all its diagonal elements are positive and non-zero.

The secondary task remains to be satisfied, which requires satisfying conditions (6.7). When $n > l$, the system of equations is underconstrained, the matrix $\mathbf{J}_{\mathrm{s}}$ is rectangular and horizontal, so the following generalized inverse is used

$$\mathbf{J}_{\mathrm{s}}^{+} = \mathbf{J}_{\mathrm{s}}^{\mathrm{T}}\left(\mathbf{J}_{\mathrm{s}}\mathbf{J}_{\mathrm{s}}^{\mathrm{T}}\right)^{-1} \tag{6.13}$$

with dimension $n \times l$. When $n < l$, the system of equations (6.7) is overconstrained, the matrix $\mathbf{J}_{\mathrm{s}}$ is rectangular and vertical, in which case the following generalized inverse is to be used

$$\mathbf{J}_{\mathrm{s}}^{*} = \left(\mathbf{J}_{\mathrm{s}}^{\mathrm{T}}\mathbf{J}_{\mathrm{s}}\right)^{-1}\mathbf{J}_{\mathrm{s}}^{\mathrm{T}}. \tag{6.14}$$

It is also possible that $n = l$, when the ordinary inverse $\mathbf{J}_{\mathrm{s}}^{-1}$ is calculated.

In our case we have $n > l$, so that the solution of the secondary task, which also takes into account the requirements of the primary task, is the following

$$d\mathbf{q}_{\mathrm{s}} = \mathbf{J}_{\mathrm{s}}^{+}d\mathbf{s}. \tag{6.15}$$

Suppose that the mechanism is in its initial configuration $\mathbf{q}(0)$, while it is our aim to displace it into the configuration which corresponds to the desired position, defined by primary external coordinates $\mathbf{p}$, and the desired orientation, defined by secondary external coordinates $\mathbf{s}$. To obtain the corresponding joint coordinates we make use of the Newton-Raphson iterative method with the following steps

$$\mathbf{q}(\kappa + 1) = \mathbf{q}(\kappa) + d\mathbf{q}(\kappa), \quad \kappa = 0, 1, 2, \ldots$$

We observe the course of iterations by solving each task separately. In the first case, let us first assume $d\mathbf{q} = d\mathbf{q}_{\mathrm{p}}$ in accordance with (6.12). Here, we have $d\mathbf{p}(\kappa) = \mathbf{p} - \mathbf{p}_{\mathrm{q}}(\kappa)$ and the error is

$$\varepsilon_{\mathrm{p}}(\kappa) = \left\|\mathbf{p} - \mathbf{p}_{\mathrm{q}}(\kappa)\right\|.$$

In the second case we impose $d\mathbf{q} = d\mathbf{q}_{\mathrm{s}}$ as defined in (6.15). Now, $d\mathbf{s}(\kappa) = \mathbf{s} - \mathbf{s}_{\mathrm{q}}(\kappa)$ and the error is computed by

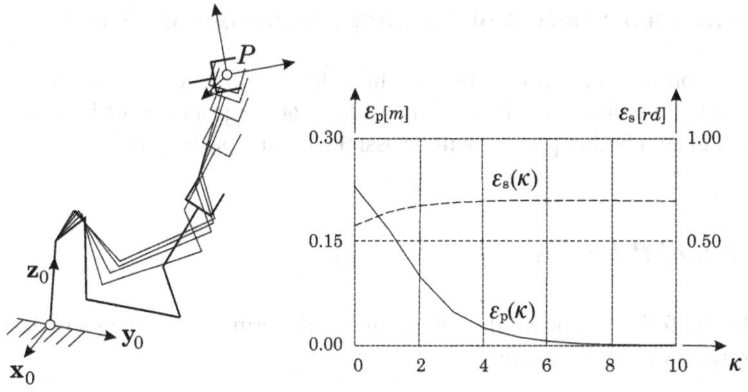

Fig. 6.2 Changing the mechanism configuration and the errors ε_p and ε_s as functions of number of iterations, when $d\mathbf{q} = d\mathbf{q}_p$

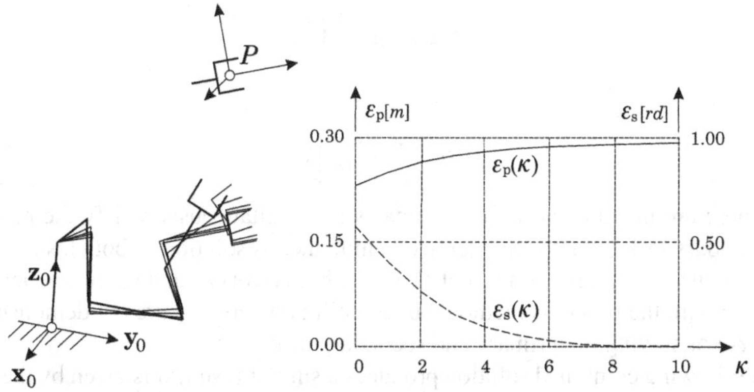

Fig. 6.3 Changing the mechanism configuration and the errors ε_p and ε_s as functions of number of iterations, when $d\mathbf{q} = d\mathbf{q}_s$

$$\varepsilon_s(\kappa) = \left\| \mathbf{s} - \mathbf{s}_q(\kappa) \right\|.$$

The configurations calculated in the described way are shown in Figs. 6.2 and 6.3. In Fig. 6.2, when $d\mathbf{q} = d\mathbf{q}_p$, the difference of initial and desired position ε_p is gradually decreased, while we have no influence on orientation so that the error ε_s even increases with iterations. In Fig. 6.3, when $d\mathbf{q} = d\mathbf{q}_s$, the difference between the initial and desired orientation ε_s is gradually decreased, while we have no influence on gripper position and the error ε_p is increasing. Clearly, such a result is not satisfactory. It appears that the solution must be found by considering the requirements of both the primary and the secondary task simultaneously.

6.2 Combined Solution of Primary and Secondary Tasks

In this section we study different possibilities for solving the primary and the secondary task simultaneously. First we consider that the priorities of both tasks are equal and later, a higher priority will be assigned to the primary task.

6.2.1 Equal Priorities

Consider the following possibility. We combine the primary and the secondary task into a single system of equations

$$\begin{bmatrix} d\mathbf{p} \\ d\mathbf{s} \end{bmatrix} = \begin{bmatrix} \mathbf{J}_{\mathrm{p}} \\ \mathbf{J}_{\mathrm{s}} \end{bmatrix} = \mathbf{J}d\mathbf{q}.$$

Now we are dealing with an overconstrained system of equations, as we have $m + l = 6$ and $n = 5$. The inverse solution is based on the following generalized inverse

$$\mathbf{J}^* = \left(\mathbf{J}^{\mathrm{T}}\mathbf{J}\right)^{-1}\mathbf{J}^{\mathrm{T}},$$

so that

$$d\mathbf{q} = \mathbf{J}^* \begin{bmatrix} d\mathbf{p} \\ d\mathbf{s} \end{bmatrix}.$$

Since there are insufficient joint coordinates to simultaneously satisfy the primary and secondary tasks exactly, the iterative simultaneous solution of both tasks brings the mechanism into a final configuration where both errors ε_{p} and ε_{s} are smaller than they were with the previous method, but are still non zero. Figure 6.4 demonstrates the course of solving the primary and secondary task.

The following combined solution produces a similar result. It is given by the sum of the independent solutions of the primary and secondary task

$$d\mathbf{q} = d\mathbf{q}_{\mathrm{p}} + d\mathbf{q}_{\mathrm{s}},$$

which equals

$$d\mathbf{q} = \mathbf{J}_{\mathrm{pA}}^{+} d\mathbf{p} + \mathbf{J}_{\mathrm{s}}^{+} d\mathbf{s}.$$

The solutions $d\mathbf{q}_{\mathrm{p}}$ and $d\mathbf{q}_{\mathrm{s}}$ are hindering each other. In the best case, the result is similar to that in Fig. 6.4. In order to observe the displacement $d\mathbf{p}_{\mathrm{q}}$ in primary external coordinates produced by the displacement in the joint coordinates $d\mathbf{q}$, we multiply the above equation by the matrix $\mathbf{J}_{\mathrm{p}}$

$$\mathbf{J}_{\mathrm{p}} d\mathbf{q} = \mathbf{J}_{\mathrm{p}} \mathbf{J}_{\mathrm{pA}}^{+} d\mathbf{p} + \mathbf{J}_{\mathrm{p}} \mathbf{J}_{\mathrm{s}}^{+} d\mathbf{s}.$$

Since $\mathbf{J}_{\mathrm{p}}\mathbf{J}_{\mathrm{pA}}^{+} = \mathbf{I}$, we have

$$d\mathbf{p}_{\mathrm{q}} = d\mathbf{p} + \mathbf{J}_{\mathrm{p}} \mathbf{J}_{\mathrm{s}}^{+} d\mathbf{s}.$$

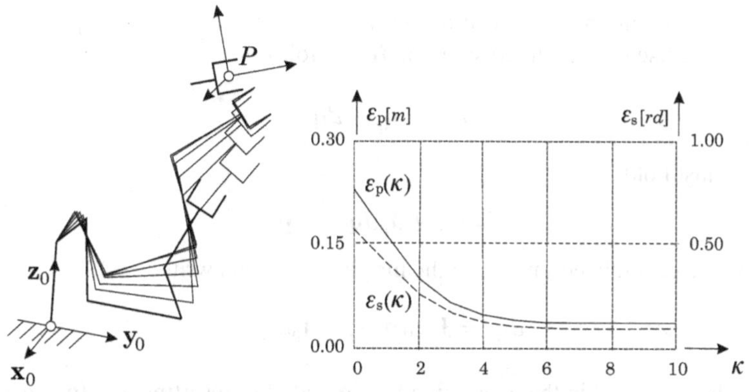

Fig. 6.4 Changing the mechanism configuration and errors ε_p and ε_s as functions of the number of iterations, when the primary and secondary tasks are of equal priority

This means that the actual displacement in the primary external coordinates does not equal the desired displacement

$$d\mathbf{p}_q \neq d\mathbf{p}.$$

In order to determine the actual displacement $d\mathbf{s}_q$ in secondary external coordinates which is produced by the displacement in joint coordinates $d\mathbf{q}$, the equation must be multiplied by the matrix $\mathbf{J}_s$

$$\mathbf{J}_s d\mathbf{q} = \mathbf{J}_s \mathbf{J}_{pA}^{+} d\mathbf{p} + \mathbf{J}_s \mathbf{J}_s^{+} d\mathbf{s}.$$

Since $\mathbf{J}_s \mathbf{J}_s^{+} = \mathbf{I}$, it follows that

$$d\mathbf{s}_q = \mathbf{J}_s \mathbf{J}_{pA}^{+} d\mathbf{p} + d\mathbf{s} \neq d\mathbf{s}.$$

Such a combined solution is neither advantageous for solving the primary nor for solving the secondary task.

6.2.2 Primary Task with Higher Priority

In this case as well, the displacement in joint coordinates is expressed as the sum of the solutions of the primary and secondary task. However, now we seek the vector $d\mathbf{q}$ that produces a displacement in the primary external coordinates which is exact, i.e.

$$d\mathbf{p}_q = d\mathbf{p}.$$

To meet this requirement, instead of the independent solution of the primary task $d\mathbf{q}_\mathrm{p}$,we make use of a modified solution $d\mathbf{q}'_\mathrm{p}$ as follows

$$d\mathbf{q} = d\mathbf{q}'_\mathrm{p} + d\mathbf{q}_\mathrm{s},$$

where it must hold

$$\mathbf{J}_\mathrm{p} d\mathbf{q}'_\mathrm{p} + \mathbf{J}_\mathrm{p} d\mathbf{q}_\mathrm{s} = d\mathbf{p}.$$

By using the generalized inverse of the matrix $\mathbf{J}_\mathrm{p}$, we can write

$$d\mathbf{q}'_\mathrm{p} = \mathbf{J}^+_\mathrm{pA} d\mathbf{p} - \mathbf{J}^+_\mathrm{pA}\mathbf{J}_\mathrm{p} d\mathbf{q}_\mathrm{s}.$$

This can be now used in the expression for $d\mathbf{q}$ and after inserting $\mathbf{J}^+_\mathrm{pA} d\mathbf{p} = d\mathbf{q}_\mathrm{p}$, we have

$$d\mathbf{q} = d\mathbf{q}_\mathrm{p} + \left(\mathbf{I} - \mathbf{J}^+_\mathrm{pA}\mathbf{J}_\mathrm{p}\right) d\mathbf{q}_\mathrm{s}.$$

The matrix

$$\mathbf{N}_\mathrm{pA} = \mathbf{I} - \mathbf{J}^+_\mathrm{pA}\mathbf{J}_\mathrm{p}$$

has dimension $n \times n$ and linearly maps the solution of the secondary task $d\mathbf{q}_\mathrm{s}$ into a new solution $d\mathbf{q}_\mathrm{N}$

$$d\mathbf{q}_\mathrm{N} = \mathbf{N}_\mathrm{pA} d\mathbf{q}_\mathrm{s},$$

so that we have

$$d\mathbf{q} = d\mathbf{q}_\mathrm{p} + d\mathbf{q}_\mathrm{N}.$$

It is convenient that $d\mathbf{q}_\mathrm{N}$ does not produce a displacement in primary external coordinates, as there is

$$\mathbf{J}_\mathrm{p} d\mathbf{q}_\mathrm{N} = \mathbf{J}_\mathrm{p}\mathbf{N}_\mathrm{pA} d\mathbf{q}_\mathrm{s} = \mathbf{0}.$$

This can be examined by multiplying the matrix $\mathbf{N}_\mathrm{pA}$ by the matrix $\mathbf{J}_\mathrm{p}$, while using $\mathbf{J}_\mathrm{p}\mathbf{J}^+_\mathrm{pA} = \mathbf{I}$. It follows

$$\mathbf{J}_\mathrm{p}\mathbf{N}_\mathrm{pA} = \mathbf{J}_\mathrm{p} - \mathbf{J}_\mathrm{p}\mathbf{J}^+_\mathrm{pA}\mathbf{J}_\mathrm{p} = \mathbf{J}_\mathrm{p} - \mathbf{J}_\mathrm{p} = \mathbf{0}.$$

The product of these two nonzero matrices is zero if all the rows of matrix $\mathbf{J}_\mathrm{p}$ are orthogonal to all the columns of matrix $\mathbf{N}_\mathrm{pA}$. We say that $\mathbf{N}_\mathrm{pA}$ is the orthogonal complement of matrix $\mathbf{J}_\mathrm{p}$.

This leads to a joint solution of the primary and the secondary task of the form

$$d\mathbf{q} = \mathbf{J}^+_\mathrm{pA} d\mathbf{p} + \mathbf{N}_\mathrm{pA}\mathbf{J}^+_\mathrm{s} d\mathbf{s}. \tag{6.16}$$

This is the called the task priority method [62], where $\mathbf{J}^+_\mathrm{pA} d\mathbf{p}$ corresponds to the primary task and represents the particular solution, while $\mathbf{N}_\mathrm{pA}\mathbf{J}^+_\mathrm{s} d\mathbf{s}$ belongs to the secondary task and represents the homogeneous solution of the inverse kinematics of a redundant mechanism.

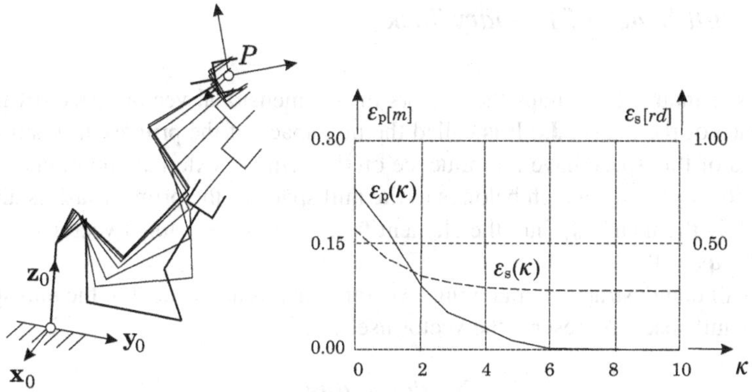

Fig. 6.5 Changing of mechanism configurations and errors ε_p and ε_s as functions of the number of iterations, when the primary task has higher priority

The matrix $\mathbf{N}_{pA}$ provides that the secondary task is subordinated to the primary task. This becomes evident when the expression (6.16) is multiplied by the matrix $\mathbf{J}_p$

$$\mathbf{J}_p d\mathbf{q} = d\mathbf{p}_q = \mathbf{J}_p \mathbf{J}_{pA}^+ d\mathbf{p} + \mathbf{J}_p \mathbf{N}_{pA} \mathbf{J}_s^+ d\mathbf{s}.$$

Taking into account that $\mathbf{J}_p \mathbf{J}_{pA}^+ = \mathbf{I}$ and $\mathbf{J}_p \mathbf{N}_{pA} = \mathbf{0}$ gives

$$d\mathbf{p}_q = d\mathbf{p}.$$

Thus $d\mathbf{q}$ produces a displacement in the primary external coordinates which is exactly as desired, as nothing is contributed by the secondary part of the solution. The situation is different for the displacement in secondary external coordinates, as it is influenced by the primary and secondary part of the solution

$$\mathbf{J}_s d\mathbf{q} = d\mathbf{s}_q = \mathbf{J}_s \mathbf{J}_{pA}^+ d\mathbf{p} + \mathbf{J}_s \mathbf{N}_{pA} \mathbf{J}_s^+ d\mathbf{s},$$

therefore

$$d\mathbf{s}_q \neq d\mathbf{s}.$$

Figure 6.5 illustrates the course of solving the primary and secondary task with this method. The advantage of the task priority method is that it solves the primary task problem unconditionally, while the secondary task is solved to the extent that it does not disturb or interfere with the execution of the primary task. Through the iteration, error ε_p approaches zero while error ε_s is decreased to a value which cannot be determined in advance. We shall now learn how to influence the solution of the secondary task and how to minimize ε_s.

6.2.3 Null Space of Primary Task

The $n \times n$ matrix $\mathbf{N}_{\mathrm{pA}}$ maps the vectors of n-dimensional vector space $d\mathbf{q}$ into a null space of the matrix $\mathbf{J}_{\mathrm{p}}$. It is called the null space of the primary task since the elements of this space have no influence on the primary external coordinates. Any vector $d\mathbf{q}_{\mathrm{N}} = \mathbf{N}_{\mathrm{pA}} d\mathbf{q}$ which belongs to the null space of the primary task is always mapped by the matrix $\mathbf{J}_{\mathrm{p}}$ into the element $\mathbf{0}$ of the m-dimensional vector space $d\mathbf{p}$ since $\mathbf{J}_{\mathrm{p}} d\mathbf{q}_{\mathrm{N}} = \mathbf{0}$.

We will demonstrate that mapping a vector which is an element of the null space, into the null space, represents the vector itself, i.e.

$$\mathbf{N}_{\mathrm{pA}} d\mathbf{q}_{\mathrm{N}} = d\mathbf{q}_{\mathrm{N}}.$$

We can write

$$\mathbf{N}_{\mathrm{pA}} \mathbf{N}_{\mathrm{pA}} d\mathbf{q}_{\mathrm{N}} = \mathbf{N}_{\mathrm{pA}} d\mathbf{q}_{\mathrm{N}}$$

and

$$\mathbf{N}_{\mathrm{pA}} \mathbf{N}_{\mathrm{pA}} = \left(\mathbf{I} - \mathbf{J}_{\mathrm{pA}}^{+} \mathbf{J}_{\mathrm{p}}\right)\left(\mathbf{I} - \mathbf{J}_{\mathrm{pA}}^{+} \mathbf{J}_{\mathrm{p}}\right) = \mathbf{N}_{\mathrm{pA}}.$$

The statement is valid as the matrix $\mathbf{N}_{\mathrm{pA}}$ is idempotent. It is not difficult to also observe the property of symmetry

$$\mathbf{N}_{\mathrm{p}} = \mathbf{N}_{\mathrm{p}}^{\mathrm{T}},$$

when the weighting matrix $\mathbf{A} = \mathbf{I}$.

It is important to know that the matrix $\mathbf{N}_{\mathrm{pA}}$ is not of full rank. Its rank equals the degree of redundancy $D = n - m$. This can be expected as the rank of the matrix $\mathbf{J}_{\mathrm{p}}$ equals m. We know from linear algebra, that the basis vectors of the null space of matrix $\mathbf{J}_{\mathrm{p}}$ are the eigenvectors of the matrix $\mathbf{N}_{\mathrm{pA}} \mathbf{N}_{\mathrm{pA}}^{\mathrm{T}}$ which correspond to the non-zero eigenvalues; these are identical to the eigenvectors of the matrix $\mathbf{J}_{\mathrm{p}}^{\mathrm{T}} \mathbf{J}_{\mathrm{p}}$ which correspond to zero eigenvalues. Suppose that these are n-dimensional orthonormal vectors $\mathbf{v}_1, \mathbf{v}_2, \ldots, \mathbf{v}_{n-m}$. Each element of the null space can be determined as a linear combination of basis vectors

$$d\mathbf{q}_{\mathrm{N}} = \gamma_1 \mathbf{v}_1 + \gamma_2 \mathbf{v}_2 + \cdots + \gamma_{n-m} \mathbf{v}_{n-m},$$

where $\gamma_1, \gamma_2, \ldots, \gamma_{n-m}$ are arbitrary quantities. This can be written in vector matrix form

$$d\mathbf{q}_{\mathrm{N}} = \mathbf{V}_{\mathrm{p}} \boldsymbol{\gamma}, \tag{6.17}$$

where $\boldsymbol{\gamma} = (\gamma_1, \gamma_2, \ldots, \gamma_{n-m})^{\mathrm{T}}$ and

$$\mathbf{V}_{\mathrm{p}} = \begin{bmatrix} \mathbf{v}_1 & \mathbf{v}_2 & \cdots & \mathbf{v}_{n-m} \end{bmatrix}$$

is a matrix of dimension $n \times (n - m)$.

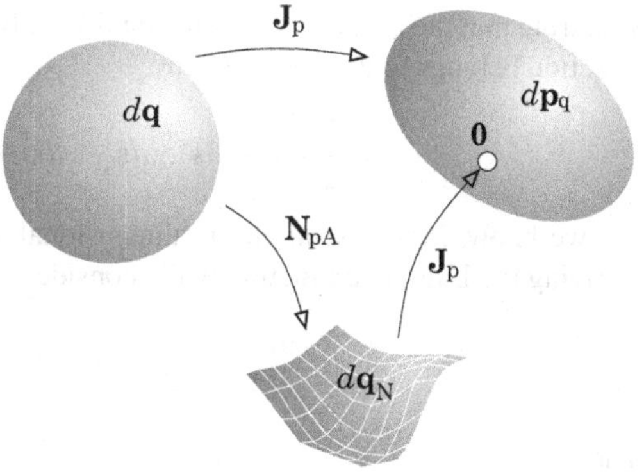

Fig. 6.6 The matrix $\mathbf{N}_{\mathrm{pA}}$ maps the vectors $d\mathbf{q}$ into the null space of the primary task represented by the vectors $d\mathbf{q}_{\mathrm{N}}$, whose map in the space of primary external coordinates is $\mathbf{0}$

The null space of the primary task is therefore an $n-m$ parametric space of n-dimensional vectors (Fig. 6.6). No vector $d\mathbf{q}_{\mathrm{N}}$, selected in the described way, can produce a displacement in primary external coordinates. The null space of the primary task can be therefore considered as a domain of displacements in the joint coordinates, which can be used to solve the secondary task without hindering the execution of the primary task.

There are an infinite number of matrices $\mathbf{N}_{\mathrm{pA}}$. They are selected by changing the weighting matrix $\mathbf{A}$. It is sufficient that the matrix $\mathbf{A}$ is of full rank and positive definite. We must be aware that by changing weighting matrix $\mathbf{A}$, we do not change the null space of the primary task, which only depends on the Jacobian matrix $\mathbf{J}_{\mathrm{p}}$. By the weights of the matrix $\mathbf{A}$ we only change the projection of a vector into the null space of the matrix $\mathbf{J}_{\mathrm{p}}$.

6.2.4 Best Weighting Matrix

The method of the task priorities requires that the secondary task is unconditionally subordinated to the primary task. Therefore, only the null space of the matrix $\mathbf{J}_{\mathrm{p}}$ is available for the execution of the secondary task. We are interested in finding such a weighting matrix $\mathbf{A}$ which will bring the actual displacement in the secondary external coordinates $d\mathbf{s}_{\mathrm{q}} = \mathbf{J}_{\mathrm{s}} d\mathbf{q}$ as near as possible to the desired displacement $d\mathbf{s}$.

This is done by finding the minimum of the quadratic function

$$\frac{1}{2}(d\mathbf{s}_{\mathrm{q}} - d\mathbf{s})^{\mathrm{T}}(d\mathbf{s}_{\mathrm{q}} - d\mathbf{s}),$$

where the following condition must be valid

$$d\mathbf{p} - \mathbf{J}_{\mathrm{p}} d\mathbf{q} = \mathbf{0},$$

requiring that the primary task is unconditionally executed. The following Lagrange function belongs to this requirement

$$L = \frac{1}{2}(d\mathbf{s}_\mathrm{q} - d\mathbf{s})^\mathrm{T}(d\mathbf{s}_\mathrm{q} - d\mathbf{s}) + \boldsymbol{\lambda}^\mathrm{T}(d\mathbf{p} - \mathbf{J}_\mathrm{p} d\mathbf{q}).$$

As we know, $\boldsymbol{\lambda}$ represents the m-dimensional vector of Lagrange multipliers. By deriving the Lagrange function while considering $d\mathbf{s}_\mathrm{q} = \mathbf{J}_\mathrm{s} d\mathbf{q}$, we obtain

$$\frac{\partial L}{\partial d\mathbf{q}} = \mathbf{J}_\mathrm{s}^\mathrm{T}(\mathbf{J}_\mathrm{s} d\mathbf{q} - d\mathbf{s}) - \mathbf{J}_\mathrm{p}^\mathrm{T}\boldsymbol{\lambda} = \mathbf{0}$$

and

$$\mathbf{J}_\mathrm{s}^\mathrm{T}\mathbf{J}_\mathrm{s} d\mathbf{q} = \mathbf{J}_\mathrm{s}^\mathrm{T} d\mathbf{s} + \mathbf{J}_\mathrm{p}^\mathrm{T}\boldsymbol{\lambda}.$$

From the last expression we calculate $d\mathbf{q}$ as follows

$$d\mathbf{q} = (\mathbf{J}_\mathrm{s}^\mathrm{T}\mathbf{J}_\mathrm{s})^{-1}\mathbf{J}_\mathrm{s}^\mathrm{T} d\mathbf{s} + (\mathbf{J}_\mathrm{s}^\mathrm{T}\mathbf{J}_\mathrm{s})^{-1}\mathbf{J}_\mathrm{p}^\mathrm{T}\boldsymbol{\lambda}. \tag{6.18}$$

At this point we assume that the matrix $\mathbf{J}_\mathrm{s}^\mathrm{T}\mathbf{J}_\mathrm{s}$ is not singular. Its singularity will be discussed later in this section.

The imposed condition of the primary task is satisfied by

$$\frac{\partial L}{\partial \boldsymbol{\lambda}} = d\mathbf{p} - \mathbf{J}_\mathrm{p} d\mathbf{q} = \mathbf{0}.$$

Now we multiply $d\mathbf{q}$ in (6.18) by $\mathbf{J}_\mathrm{p}$ and then use $d\mathbf{p} = \mathbf{J}_\mathrm{p} d\mathbf{q}$ to obtain

$$d\mathbf{p} = \mathbf{J}_\mathrm{p}(\mathbf{J}_\mathrm{s}^\mathrm{T}\mathbf{J}_\mathrm{s})^{-1}\mathbf{J}_\mathrm{s}^\mathrm{T} d\mathbf{s} + \mathbf{J}_\mathrm{p}(\mathbf{J}_\mathrm{s}^\mathrm{T}\mathbf{J}_\mathrm{s})^{-1}\mathbf{J}_\mathrm{p}^\mathrm{T}\boldsymbol{\lambda}.$$

From the last expression we can calculate the Lagrange multipliers as follows

$$\boldsymbol{\lambda} = \left(\mathbf{J}_\mathrm{p}(\mathbf{J}_\mathrm{s}^\mathrm{T}\mathbf{J}_\mathrm{s})^{-1}\mathbf{J}_\mathrm{p}^\mathrm{T}\right)^{-1} d\mathbf{p} - \left(\mathbf{J}_\mathrm{p}(\mathbf{J}_\mathrm{s}^\mathrm{T}\mathbf{J}_\mathrm{s})^{-1}\mathbf{J}_\mathrm{p}^\mathrm{T}\right)^{-1}\mathbf{J}_\mathrm{p}(\mathbf{J}_\mathrm{s}^\mathrm{T}\mathbf{J}_\mathrm{s})^{-1}\mathbf{J}_\mathrm{s}^\mathrm{T} d\mathbf{s}.$$

Now we substitute $\boldsymbol{\lambda}$ in (6.18) and obtain

$$\begin{aligned} d\mathbf{q} = {} & (\mathbf{J}_\mathrm{s}^\mathrm{T}\mathbf{J}_\mathrm{s})^{-1}\mathbf{J}_\mathrm{s}^\mathrm{T} d\mathbf{s} + (\mathbf{J}_\mathrm{s}^\mathrm{T}\mathbf{J}_\mathrm{s})^{-1}\mathbf{J}_\mathrm{p}^\mathrm{T}\left(\mathbf{J}_\mathrm{p}(\mathbf{J}_\mathrm{s}^\mathrm{T}\mathbf{J}_\mathrm{s})^{-1}\mathbf{J}_\mathrm{p}^\mathrm{T}\right)^{-1} d\mathbf{p} \\ & - (\mathbf{J}_\mathrm{s}^\mathrm{T}\mathbf{J}_\mathrm{s})^{-1}\mathbf{J}_\mathrm{p}^\mathrm{T}\left(\mathbf{J}_\mathrm{p}(\mathbf{J}_\mathrm{s}^\mathrm{T}\mathbf{J}_\mathrm{s})^{-1}\mathbf{J}_\mathrm{p}^\mathrm{T}\right)^{-1}\mathbf{J}_\mathrm{p}(\mathbf{J}_\mathrm{s}^\mathrm{T}\mathbf{J}_\mathrm{s})^{-1}\mathbf{J}_\mathrm{s}^\mathrm{T} d\mathbf{s}. \end{aligned}$$

Taking into account $\mathbf{J}_\mathrm{s}\mathbf{J}_\mathrm{s}^+ = \mathbf{I}$ we can write

$$(\mathbf{J}_\mathrm{s}^\mathrm{T}\mathbf{J}_\mathrm{s})^{-1}\mathbf{J}_\mathrm{s}^\mathrm{T} = (\mathbf{J}_\mathrm{s}^\mathrm{T}\mathbf{J}_\mathrm{s})^{-1}\mathbf{J}_\mathrm{s}^\mathrm{T}\mathbf{J}_\mathrm{s}\mathbf{J}_\mathrm{s}^+ = \mathbf{J}_\mathrm{s}^+$$

and then

$$\begin{aligned} d\mathbf{q} = {} & \mathbf{J}_\mathrm{s}^+ d\mathbf{s} + (\mathbf{J}_\mathrm{s}^\mathrm{T}\mathbf{J}_\mathrm{s})^{-1}\mathbf{J}_\mathrm{p}^\mathrm{T}\left(\mathbf{J}_\mathrm{p}(\mathbf{J}_\mathrm{s}^\mathrm{T}\mathbf{J}_\mathrm{s})^{-1}\mathbf{J}_\mathrm{p}^\mathrm{T}\right)^{-1} d\mathbf{p} \\ & - (\mathbf{J}_\mathrm{s}^\mathrm{T}\mathbf{J}_\mathrm{s})^{-1}\mathbf{J}_\mathrm{p}^\mathrm{T}\left(\mathbf{J}_\mathrm{p}(\mathbf{J}_\mathrm{s}^\mathrm{T}\mathbf{J}_\mathrm{s})^{-1}\mathbf{J}_\mathrm{p}^\mathrm{T}\right)^{-1}\mathbf{J}_\mathrm{p}\mathbf{J}_\mathrm{s}^+ d\mathbf{s}. \end{aligned}$$

The expression can be rearranged as follows

$$d\mathbf{q} = (\mathbf{J}_\mathrm{s}^\mathrm{T}\mathbf{J}_\mathrm{s})^{-1}\mathbf{J}_\mathrm{p}^\mathrm{T}(\mathbf{J}_\mathrm{p}(\mathbf{J}_\mathrm{s}^\mathrm{T}\mathbf{J}_\mathrm{s})^{-1}\mathbf{J}_\mathrm{p}^\mathrm{T})^{-1}d\mathbf{p}$$
$$+ (\mathbf{I} - (\mathbf{J}_\mathrm{s}^\mathrm{T}\mathbf{J}_\mathrm{s})^{-1}\mathbf{J}_\mathrm{p}^\mathrm{T}(\mathbf{J}_\mathrm{p}(\mathbf{J}_\mathrm{s}^\mathrm{T}\mathbf{J}_\mathrm{s})^{-1}\mathbf{J}_\mathrm{p}^\mathrm{T})^{-1}\mathbf{J}_\mathrm{p})\mathbf{J}_\mathrm{s}^{+}d\mathbf{s}.$$

If we denote

$$\mathbf{A} = \mathbf{J}_\mathrm{s}^\mathrm{T}\mathbf{J}_\mathrm{s}$$

we get

$$d\mathbf{q} = \mathbf{A}^{-1}\mathbf{J}_\mathrm{p}^\mathrm{T}(\mathbf{J}_\mathrm{p}\mathbf{A}^{-1}\mathbf{J}_\mathrm{p}^\mathrm{T})^{-1}d\mathbf{p} + (\mathbf{I} - \mathbf{A}^{-1}\mathbf{J}_\mathrm{p}^\mathrm{T}(\mathbf{J}_\mathrm{p}\mathbf{A}^{-1}\mathbf{J}_\mathrm{p}^\mathrm{T})^{-1}\mathbf{J}_\mathrm{p})\mathbf{J}_\mathrm{s}^{+}d\mathbf{s}.$$

In this equation we recognize the weighted general inverse

$$\mathbf{J}_\mathrm{pA}^{+} = \mathbf{A}^{-1}\mathbf{J}_\mathrm{p}^\mathrm{T}(\mathbf{J}_\mathrm{p}\mathbf{A}^{-1}\mathbf{J}_\mathrm{p}^\mathrm{T})^{-1},$$

so that

$$d\mathbf{q} = \mathbf{J}_\mathrm{pA}^{+}d\mathbf{p} + (\mathbf{I} - \mathbf{J}_\mathrm{pA}^{+}\mathbf{J}_\mathrm{p})\mathbf{J}_\mathrm{s}^{+}d\mathbf{s},$$

and the orthogonal complement of the Jacobian matrix $\mathbf{J}_\mathrm{p}$

$$\mathbf{N}_\mathrm{pA} = \mathbf{I} - \mathbf{J}_\mathrm{pA}^{+}\mathbf{J}_\mathrm{p}.$$

As a result we obtain

$$d\mathbf{q} = \mathbf{J}_\mathrm{pA}^{+}d\mathbf{p} + \mathbf{N}_\mathrm{pA}\mathbf{J}_\mathrm{s}^{+}d\mathbf{s}$$

which is the well known task priority method [62]. We can conclude that the weighting matrix $\mathbf{A} = \mathbf{J}_\mathrm{s}^\mathrm{T}\mathbf{J}_\mathrm{s}$ results in the least square difference of the vectors of actual displacement $d\mathbf{s}_\mathrm{q}$ and desired displacement $d\mathbf{s}$ in secondary external coordinates [50]. Therefore, the solution of the secondary task leads to a lower final error ε_s, while the error ε_p is not compromised (Fig. 6.7). The method is most efficient when the vectors $d\mathbf{s}_\mathrm{q}$ and $d\mathbf{s}$ are similar at the initial iteration.

Unfortunately, the matrix $\mathbf{J}_\mathrm{s}^\mathrm{T}\mathbf{J}_\mathrm{s}$ is not singular only if $n \leq l$. To overcome this difficulty the following approximation for the weighting matrix is used

$$\mathbf{A} \approx \mathbf{J}_\mathrm{s}^\mathrm{T}\mathbf{J}_\mathrm{s} + \varepsilon\mathbf{I}, \tag{6.19}$$

where ε is a small positive value. As the matrix $\mathbf{J}_\mathrm{s}^\mathrm{T}\mathbf{J}_\mathrm{s}$ is symmetric, all its eigenvalues are positive or equal zero. By adding the matrix $\varepsilon\mathbf{I}$, the eigenvalues are increased by ε, so that regardless of the properties of matrix $\mathbf{J}_\mathrm{s}$, the matrix $\mathbf{A}$ is of full rank and positive definite.

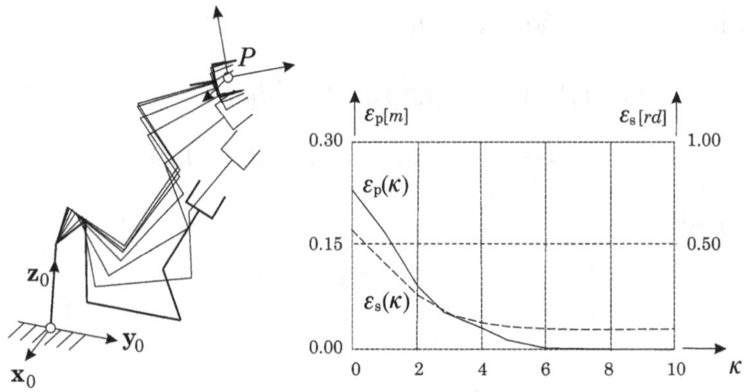

Fig. 6.7 Changing of the mechanism configuration and of the errors ε_p and ε_s as functions of the number of iterations, when the primary task is of higher priority and there is $\mathbf{A} = \mathbf{J}_s^T\mathbf{J}_s + \varepsilon\mathbf{I}$

6.2.5 Best Solution of Secondary Task

In this section we examine an approach which speeds up the process of solving the secondary task. This is accomplished by replacing the independent solution of the secondary task $d\mathbf{q}_s$ by a modified $d\mathbf{q}_s'$ which minimizes the effect of the solution of the primary task $d\mathbf{q}_p$ on the solution of the secondary task. We start with the known expression

$$d\mathbf{q} = \mathbf{J}_{pA}^{+}d\mathbf{p} + \mathbf{N}_{pA}d\mathbf{q}_s',$$

where we suppose that beside $\mathbf{J}_p d\mathbf{q} = d\mathbf{p}$, as well, $\mathbf{J}_s d\mathbf{q} = d\mathbf{s}$ is valid. It is not difficult to realize that the first requirement is valid. This is assured by the method of task priorities. The second requirement will be satisfied to the largest extent possible.

We multiply the initial expression by the Jacobian matrix $\mathbf{J}_s$ and equate it to the desired displacement in the secondary coordinates

$$\mathbf{J}_s d\mathbf{q} = \mathbf{J}_s\mathbf{J}_{pA}^{+}d\mathbf{p} + \mathbf{J}_s\mathbf{N}_{pA}d\mathbf{q}_s' = d\mathbf{s}.$$

From here we calculate the required $d\mathbf{q}_s'$. Unfortunately, the matrix $\mathbf{J}_s\mathbf{N}_{pA}$, which is to be inverted, is in general rectangular, while its dimension is $l \times n$. We therefore determine its generalized inverse as follows

$$(\mathbf{J}_s\mathbf{N}_{pA})^{+} = (\mathbf{J}_s\mathbf{N}_{pA})^{T}\left(\mathbf{J}_s\mathbf{N}_{pA}\mathbf{N}_{pA}^{T}\mathbf{J}_s^{T}\right)^{-1}.$$

Note that the matrix $\mathbf{J}_s\mathbf{N}_{pA}\mathbf{N}_{pA}^{T}\mathbf{J}_s^{T}$ is singular since the rank of the matrix $\mathbf{N}_{pA}$ is $D = n - m < l$. As we are dealing with a symmetric matrix, we can avoid the singularity by using the approximation of the generalized inverse

$$(\mathbf{J}_s\mathbf{N}_{pA})^{+} \approx (\mathbf{J}_s\mathbf{N}_{pA})^{T}\left(\mathbf{J}_s\mathbf{N}_{pA}\mathbf{N}_{pA}^{T}\mathbf{J}_s^{T} + \varepsilon\mathbf{I}\right)^{-1},$$

where ε is a small positive value. We calculate a new solution of the secondary task in the following form

$$d\mathbf{q}_s' = (\mathbf{J}_s\mathbf{N}_{pA})^+ d\mathbf{s} - (\mathbf{J}_s\mathbf{N}_{pA})^+ \mathbf{J}_s\mathbf{J}_{pA}^+ d\mathbf{p}$$

and insert it into the initial expression

$$d\mathbf{q} = \mathbf{J}_{pA}^+ d\mathbf{p} + \mathbf{N}_{pA}(\mathbf{J}_s\mathbf{N}_{pA})^+ d\mathbf{s} - (\mathbf{J}_s\mathbf{N}_{pA})^+ \mathbf{J}_s\mathbf{J}_{pA}^+ d\mathbf{p}.$$

This expression is rewritten into the following well known equation [27]

$$d\mathbf{q} = \mathbf{J}_{pA}^+ d\mathbf{p} + \mathbf{N}_{pA}(\mathbf{J}_s\mathbf{N}_{pA})^+ \left(d\mathbf{s} - \mathbf{J}_s\mathbf{J}_{pA}^+ d\mathbf{p}\right). \tag{6.20}$$

We almost could not hope for more. We can easily verify that $\mathbf{J}_p d\mathbf{q} = d\mathbf{p}$. At first glance it even appears that $\mathbf{J}_s d\mathbf{q} = d\mathbf{s}$. All this enthusiasm however is spoiled by the fact that because of the singularity in the generalized inverse of the matrix $\mathbf{J}_s\mathbf{N}_{pA}$ we must use its approximation. Therefore, the actual displacement in the secondary external coordinates is only an approximation of the desired

$$\mathbf{J}_s d\mathbf{q} = d\mathbf{s}_q \approx d\mathbf{s}.$$

We cannot expect that, as in our example, a mechanism which has $n = 5$ degrees of freedom where $m = 3$ degrees of freedom are needed to solve the primary task, to simultaneously execute the secondary task having $l = 3$ coordinates. The method (6.20) works in a similar way as the ordinary method of the task priorities (6.16) with the weighting matrix (6.19), as derived in the previous section. The course of the errors as functions of the number of iterations is illustrated in Fig. 6.7. Both methods produce better results when a sufficiently large null space of the primary task is available.

The expression (6.20) can be rearranged by the use of (6.17). In the same way as we calculated the displacement $d\mathbf{q}_s'$ we now also find the vector $\boldsymbol{\gamma}$, which will be most efficient in solving the secondary task. The result is

$$\boldsymbol{\gamma} = (\mathbf{J}_s\mathbf{V}_p)^+ d\mathbf{s} - (\mathbf{J}_s\mathbf{V}_p)^+ \mathbf{J}_s\mathbf{J}_{pA}^+ d\mathbf{p},$$

which gives

$$d\mathbf{q} = \mathbf{J}_{pA}^+ d\mathbf{p} + \mathbf{V}_p(\mathbf{J}_s\mathbf{V}_p)^+ \left(d\mathbf{s} - \mathbf{J}_s\mathbf{J}_{pA}^+ d\mathbf{p}\right). \tag{6.21}$$

Also, problems arise because of the singularities in the generalized inverse of the matrix $\mathbf{J}_s\mathbf{V}_p$. We must make use of its approximation

$$(\mathbf{J}_s\mathbf{V}_p)^+ \approx (\mathbf{J}_s\mathbf{V}_p)^T \left(\mathbf{J}_s\mathbf{V}_p\mathbf{V}_p^T\mathbf{J}_s^T + \varepsilon\mathbf{I}\right)^{-1}.$$

The expression (6.21) differs from the expression (6.20) in the size of matrices. The matrix $\mathbf{V}_p$ has lesser dimension, i.e. $n \times (n - m)$, than the matrix $\mathbf{N}_{pA}$ with dimension $n \times n$. When using expression (6.21) we therefore need fewer arithmetic operations. It is, however, necessary to determine the eigenvectors of the matrix $\mathbf{N}_{pA}\mathbf{N}_{pA}^T$, i.e. the vectors with corresponding nonzero eigenvalues, in order to be in a position to produce the matrix $\mathbf{V}_p$.

6.3 Use of Kinematic Redundancy

Kinematic redundancy enables a mechanism to adapt to the requirements of the secondary task while executing the primary task. The higher the redundancy, the more adaptive is the mechanism. In the following section we determine if we can assess the kinematic redundancy and where it can be usefully exploited.

6.3.1 Kinematic Flexibility and Self-motion Curves of a 3R Mechanism

Self motion is defined as the displacement of a mechanism which corresponds to unchanging values of the primary external coordinates. The domain of vectors of joint increments which correspond to the self-motion are all vectors $d\mathbf{q}_{\mathrm{N}}$ which form the null space of the primary Jacobian matrix $\mathbf{J}_{\mathrm{p}}$. Suppose that the mechanism is in a selected configuration $\mathbf{q}(\kappa)$ and is displaced into an another configuration as follows

$$\mathbf{q}(\kappa+1) = \mathbf{q}(\kappa) + d\mathbf{q}_{\mathrm{N}}(\kappa).$$

Since $d\mathbf{p} = 0$, the values of the primary external coordinates remain fixed

$$\mathbf{p}_{\mathrm{q}}(\kappa+1) = \mathbf{p}_{\mathrm{q}}(\kappa).$$

When the excess of the degrees of freedom equals $D = n - m$, the self motion is D-parametric.

Self motion of a mechanism is a tool which enables the redundant mechanisms to execute the secondary task along with the primary task. The range of self motion is assessed in the space of joint coordinates, unfortunately it can only be visualized when $n \leq 3$. As an example we consider a planar 3R mechanism, whose primary task is positioning of the gripper. In this way we have $n = 3$, $m = 2$, and $D = 1$. The self motion of a mechanism is 1-parametric, i.e. it can be visualized by a curve in the space of the joint coordinates.

The configurations of a mechanism $\mathbf{q}(\kappa)$, $\kappa = 0, 1, 2, \ldots$, which correspond to a selected vector of the primary external coordinates $\mathbf{p}$, represent in the coordinate frame q_1, q_2, and q_3 the points on the curve of the self motion (Fig. 6.8). With the 3R mechanism two different areas of self motion occur [51]. First, the closed loop curve of self motion (α) belongs to the vector of the external coordinates $\mathbf{p}$. In the second case two separated open loop curves (β) of half length belong to the $\mathbf{p}$ coordinates. The curve α is at the transition of the mechanism through the kinematic singularity split into a pair of curves β.

Observe the self motion curves of a 3R mechanism for different positions of the gripper $\mathbf{p}$ along the abscissa from the center of the workspace towards the external border. Let the height be zero

$$p_2 = 0,$$

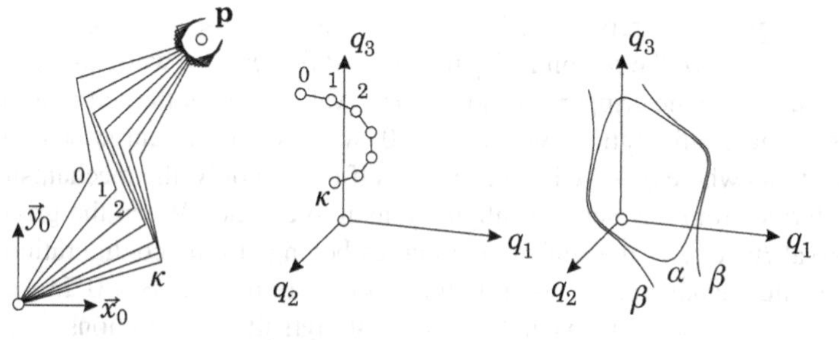

Fig. 6.8 The curves of self motion

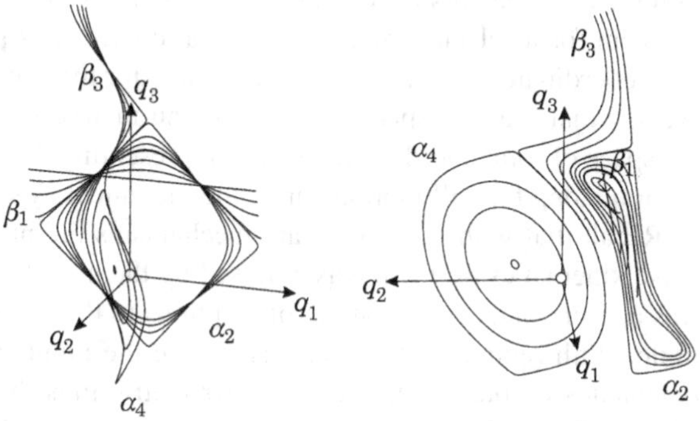

Fig. 6.9 The self motion curves of a 3R mechanism for different positions of the gripper along the horizontal axis

while the horizontal positional component should vary in the interval

$$0 \leq p_1 \leq l_1 + l_2 + l_3.$$

When the lengths of the mechanism segments are different, there occur four areas of the self motion curves along the abscissa. In two areas two open loop curves belong to each position of the gripper. They are denoted as β_1 and β_3. In the other two areas a single closed loop curve belongs to each position of the gripper. The areas are denoted as α_2 and α_4 (Fig. 6.9).

The curves from Fig. 6.9 belong to a 3R mechanism where the segment lengths are

$$l_1 > l_2 > l_3.$$

The regions of the curves α and β are divided by the following kinematic singularities

$$p_1 = -l_1 + l_2 + l_3 = a,$$
$$p_1 = l_1 - l_2 + l_3 = b,$$
$$p_1 = l_1 + l_2 - l_3 = c,$$
$$p_1 = l_1 + l_2 + l_3 = d,$$

where the region β_1 is represented by the interval $0 < p_1 < a$, the region α_2 by the interval $a < p_1 < b$, the region β_3 by the interval $b < p_1 < c$ while the region α_4 is $c < p_1 < d$. When the gripper is inside the interval β_1, one of the self motion curves belongs to the configurations where $q_3 > 0$, while the other curve belongs to the configurations where $q_3 < 0$. By the use of self motion only, the mechanism cannot be transferred from one set of configurations into another. When the gripper is in the interval β_3, one of the self motion curves belongs to the configurations where $q_2 > 0$, while the other corresponds to the configurations with $q_2 < 0$. In the regions α_2 and α_4 the 3R mechanism can be rotated through all configurations.

One curve or a pair of self motion curves, which are plotted by an infinite number of joint coordinates vectors $\mathbf{q}$, belong to a single external coordinates vector $\mathbf{p}$. This infinite set of vectors encompasses all possible solutions of the inverse kinematics problem for a selected vector of external coordinates $\mathbf{p}$. This is the domain of the joint coordinates vectors, which is available for the execution of the secondary task. We say that the null space of the Jacobian matrix $\mathbf{J}_\mathrm{p}$ is tangential with respect to the space of joint coordinates vectors, representing the self motion curve, as at each point $\mathbf{q}$ along the self motion curve, the increment $d\mathbf{q}_\mathrm{N}$ is tangential to the curve.

Recall that with non-redundant mechanisms the number of solutions of the inverse kinematics problem was assessed by the use of kinematic flexibility f_a. This was appropriate, as the number of solutions to the inverse kinematics problem was finite. With redundant mechanisms, where the number of solutions to the inverse kinematics problem is infinite, another metric must be introduced, as it is impossible to distinguish between two infinite magnitudes. With the mechanisms where $D = 1$, we define kinematic flexibility as the length ϱ of the self motion curve belonging to the vector of the primary external coordinates $\mathbf{p}$,

$$f_\mathrm{a} = \varrho(\mathbf{p}). \tag{6.22}$$

Figure 6.10 shows the kinematic flexibility of a 3R mechanism, whose primary task is positioning of a gripper, as a function of the gripper position along the abscissa. When the mechanism is fully extended $p_1 = d$, the self motion curve is represented by a point, its length is therefore zero. The length of the curve increases towards the center of the workspace until we reach the singularity $p_1 = c$, where the curve is split into two curves of half length. The situation remains unchanged until we reach the singularity $p_1 = b$, where the curves flow into a loop, which is again split in the singular point $p_1 = a$. In the region between $p_1 = a$ and $p_1 = 0$ there exist two half curves of the self motion.

The property of a redundant mechanism, that in some parts of the workspace there exists one longer self motion curve and in other parts two shorter curves occur, is much more important than it appears at first glance. It can occur that a favorable solution of the secondary task is possible, however only along a region of the curve which is momentarily not reachable by the mechanism. Therefore in the regions with two curves the kinematic flexibility is considerably smaller. It is more advantageous to plan the primary task in such a way that it is executed in the α regions, where there are more possibilities for adapting to the requirements of the secondary task.

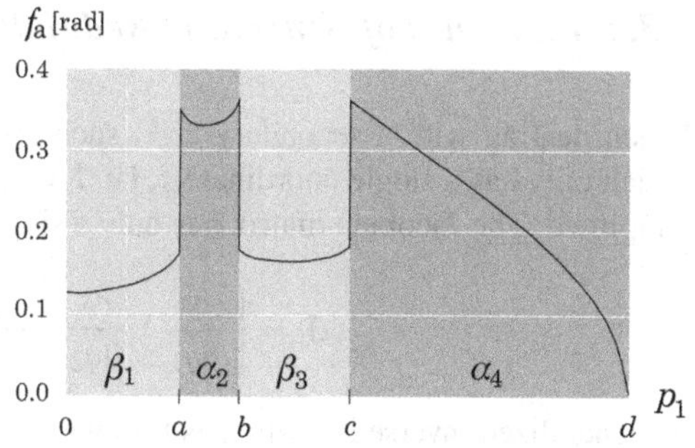

Fig. 6.10 Kinematic flexibility of a 3R mechanism as function of the coordinate p_1 in the interval $0 \leq p_1 \leq l_1 + l_2 + l_3$ ($p_2 = 0$)

Calculating the self motion curve is rather simple. Let $\mathbf{q}(\kappa)$ be the vector of the joint coordinates corresponding to the vector of the primary external coordinates $\mathbf{p}$. We calculate the corresponding Jacobian matrix $\mathbf{J}_\mathrm{p}(\kappa)$ and its orthogonal complement $\mathbf{N}_\mathrm{pA}(\kappa)$ with an arbitrary weighting matrix $\mathbf{A}$. If $D = 1$, there exists only one eigenvector of the matrix $\mathbf{N}_\mathrm{pA}(\kappa)\mathbf{N}_\mathrm{pA}(\kappa)^\mathrm{T}$, which defines the null space of the matrix $\mathbf{J}_\mathrm{p}(\kappa)$. It thus follows that

$$d\mathbf{q}_\mathrm{N}(\kappa) = \gamma \mathbf{v}(\kappa).$$

From where it follows that

$$\mathbf{q}(\kappa + 1) = \mathbf{q}(\kappa) + \gamma \mathbf{v}(\kappa),$$

where $\kappa = 0, 1, \ldots$, and γ is a small positive scalar. The calculation is continued as long as the curve returns to the initial point and the error drops below the predefined value ε

$$\left\| \mathbf{q}(\kappa) - \mathbf{q}(0) \right\| < \varepsilon.$$

By decreasing γ, we increase the accuracy of the calculation and at the same time the number of steps $\kappa = 0, 1, \ldots$. The kinematic flexibility of a mechanism, corresponding to the vector of the primary external coordinates $\mathbf{p}$, is

$$f_\mathrm{a} = \gamma \kappa_\mathrm{max}, \tag{6.23}$$

where κ_max is the number of steps necessary for calculation of the entire self motion curve.

Although there is not much written in the literature about the kinematic flexibility, we must be aware that with redundant mechanisms it is of utmost importance. The problem is that usually it is not possible to estimate the kinematic flexibility when $D > 1$. With mechanisms where $D = 2$, the self motion represents a surface in an n-dimensional space of the joint coordinates, while there is a volume within the $D = 3$ mechanisms. In general these are D-parametric multidimensional spaces which in the same way as the self motion curves, split into several parts for different values of the primary coordinates. Unfortunately, they cannot be efficiently calculated, assessed, or visualized.

6.3.2 Examples of Kinematic Redundancy

When dealing with a secondary task, such as manipulability or kinematic index, which only has a single coordinate s, i.e. $l = 1$, the inverse kinematic equations are simplified. The Jacobian matrix has only a single row

$$\mathbf{J}_\mathrm{s} = \begin{bmatrix} \frac{\partial s}{\partial q_1} & \frac{\partial s}{\partial q_2} & \cdots & \frac{\partial s}{\partial q_n} \end{bmatrix},$$

its generalized inverse is written as follows

$$\mathbf{J}_\mathrm{s}^+ = \gamma \nabla_\mathrm{s},$$

where $\gamma = 1/\mathbf{J}_\mathrm{s}\mathbf{J}_\mathrm{s}^\mathrm{T}$ is a scalar and ∇_s is a column vector describing the gradient of the secondary coordinate. Also the desired displacement, expressed in the secondary coordinates, is simplified

$$ds = s - s_\mathrm{q}.$$

When we want to increase the coordinate s as much as possible, it suffices that $ds > 0$, while when we want to decrease it to a large extent, then it is sufficient that $ds < 0$. The solution of the inverse kinematics of such a redundant mechanism has the following form

$$d\mathbf{q} = \mathbf{J}_\mathrm{pA}^+ d\mathbf{p} + \gamma \mathbf{N}_\mathrm{pA} \nabla_\mathrm{s}, \tag{6.24}$$

when we want s to be as large as possible, and

$$d\mathbf{q} = \mathbf{J}_\mathrm{pA}^+ d\mathbf{p} - \gamma \mathbf{N}_\mathrm{pA} \nabla_\mathrm{s}, \tag{6.25}$$

when we desire a minimal value of the secondary coordinate s. The constant γ does not need to be calculated. It can be determined by trial, so that the convergence of the numeric approach is adequate. It is advantageous to use the weighting matrix according to formula (6.19)

$$\mathbf{A} = \nabla_\mathrm{s} \nabla_\mathrm{s}^\mathrm{T} + \varepsilon \mathbf{I}.$$

It has been demonstrated that kinematic redundancy provides an infinite number of possibilities of how to solve the primary task. The planar 3R redundant mechanism from Fig. 6.11, whose primary task is to position the gripper at point P, can by the use of self motion pass from one configuration into another and simultaneously significantly change its properties. Below are shown three configurations of the mechanism where the gripper is at the same point in the workspace. In the first configuration the manipulability ellipsoid is round, in the second it is the largest in area and in the third case the most flat. In the first case the mechanism can be evenly displaced into all directions, in the second case its translational velocities are the largest, in the third case the mechanism can more effectively resist a force in the direction of the shortest axis of the manipulability ellipsoid. Recall that the manipulability index of a non-redundant mechanism is only altered when the position of

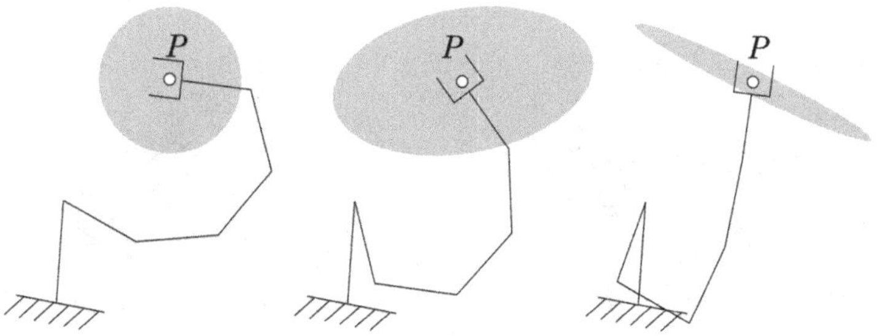

Fig. 6.11 Different manipulability ellipsoids of a redundant mechanism in the same point of a workspace

the gripper is changed, while the redundant mechanism has the ability of adaptation in the same point of the workspace.

We can solve various combinations of primary and secondary tasks by the use of the mathematical tools introduced in this section. A redundant mechanism can avoid an obstacle in the workspace while executing a primary task. This can be achieved by selecting the minimal distance between the obstacle and the mechanism as the coordinate of the secondary task [56]. A redundant mechanism will avoid the kinematic singularity, when we shall as a secondary task increase the kinematic index and in this way aim at more isotropic configurations. As the secondary task we can minimize the sensitivity to possible motor failures [71] or minimize the joint torques [27]. To resolve the redundancy, the most widely used method is the task priority method, but it is also important to mention the extended Jacobian method which is conceptually different but gives equivalent results [37].

Kinematic redundancy is also common in living organisms. The human arm has a multiple of ten muscles, while only six pairs of muscles would be sufficient for accomplishing its primary task, which is positioning and orienting of the hand. The self motion of the arm is expressed as the rotation of the center of the elbow around the axis connecting the centers of the shoulder and the wrist joint (Fig. 6.12). Despite a rather limited amount of this rotation, the arm can drastically change its properties in the center of the workspace. When the elbow is aligned with the trunk, the manipulability ellipsoid is flat so that the shorter axis is approximately vertical, while both longer axes are horizontal. In this configuration the arm can most efficiently resist the vertical force. When the elbow is lifted to the height of the shoulder, the manipulability ellipsoid becomes completely round. In this configuration a person can most efficiently execute the task requiring constant translational velocities. When calculating the manipulability indices from Fig. 6.12, the anthropomorphic mechanism with three rotations in the shoulder and one rotation in the elbow joint was used.

Some investigations [35] have demonstrated that human movements can be modeled as the movements of a redundant mechanism, where the secondary task is represented by minimizing the effort which can be approximated by the joint torques. Human beings, in their motions, often take configurations which are close to kinematic singularities. For example, during weight lifting, the weights are first lifted upwards and caught at the height of the shoulders. This is the pose when the elbows

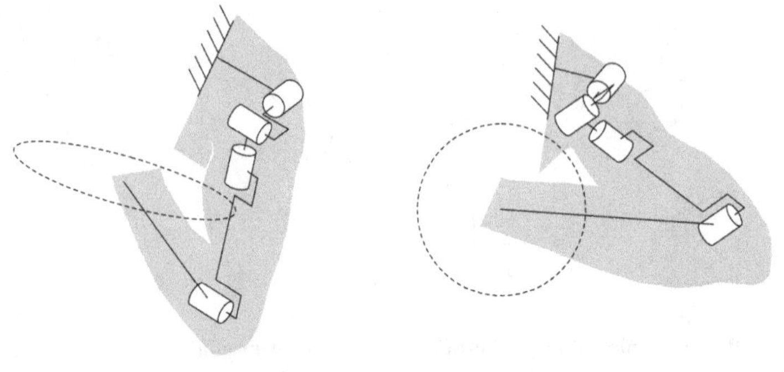

Fig. 6.12 The most round and the most flat manipulability ellipsoids of an anthropomorphic mechanism in the center of the workspace

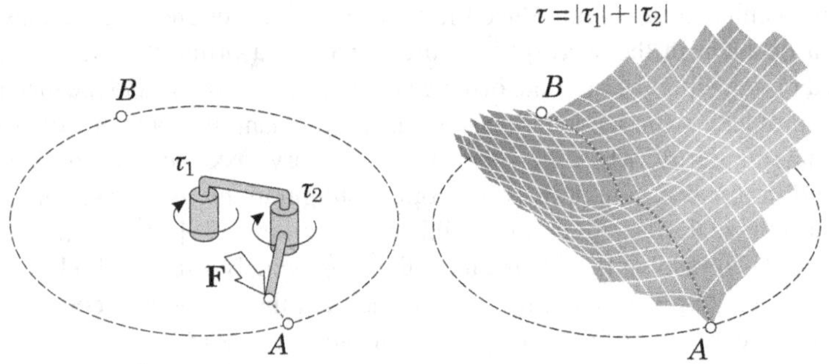

Fig. 6.13 The collective torque τ in the joints of a 2R mechanism under the influence of the external force **F** and the most efficient trajectory from point A to B

are most downwards in vicinity of the kinematic singularity shown in Fig. 6.12. Afterwards the weights are pushed by a stroke upwards and retained by the extended arms, which is again a kinematically singular configuration.

In a simplified way, weight lifting can be shown in the sagittal plane with a 2R mechanism under the influence of an external force **F** (Fig. 6.13). The variable τ_1 corresponds to the shoulder torque, while τ_2 is the torque in the elbow. When ignoring the gravity of the arms and while assessing the joint torques

$$\tau = |\tau_1| + |\tau_2|$$

only as a consequence of the force **F**, it can be shown that the maximal shoulder torque is significantly smaller than the elbow torque. During movement from point A to point B, where the torque τ is minimal, we have

$$\tau \approx |\tau_2|.$$

Here, we confront a risky finding which is controversial to the standard approaches in mechanical engineering. The elbow, which in the kinematic chain of the arm is placed after the shoulder, should be stronger than the shoulder. It is difficult to believe that an experienced constructor would design such a mechanism, however this is the case with the human arm.

6.3.3 Inverse Kinematic Solution of a Non-redundant Mechanism

Consider a mechanism with $n = 6$ degrees of freedom. Its task is first, to position the gripper with $m = 3$ degrees of freedom and second, to orient the gripper with another $l = 3$ degrees of freedom. The position and orientation of the gripper are defined by the vectors $\mathbf{p}$ and $\mathbf{s}$ respectively, each containing three coordinates. When considering both tasks as single collective task, we are dealing with a common non-redundant mechanism. The kinematic equations are written in the following form

$$\begin{bmatrix} d\mathbf{p} \\ d\mathbf{s} \end{bmatrix} = \begin{bmatrix} \mathbf{J}_{\mathrm{p}} \\ \mathbf{J}_{\mathrm{s}} \end{bmatrix} = \mathbf{J}d\mathbf{q}.$$

As the matrices $\mathbf{J}_{\mathrm{p}}$ and $\mathbf{J}_{\mathrm{s}}$ have the dimension 3×6, the Jacobian matrix $\mathbf{J}$ has the dimension 6×6 and can be inverted as with any square matrix. The solution of the inverse kinematics is therefore

$$d\mathbf{q} = \mathbf{J}^{-1} \begin{bmatrix} d\mathbf{p} \\ d\mathbf{s} \end{bmatrix}, \tag{6.26}$$

when the matrix $\mathbf{J}$ is not singular.

We write the solution in such a way that positioning will be assigned as the primary task and orienting as the secondary task which is subordinated to the primary task. We use the basic form of the task priority method

$$d\mathbf{q} = \mathbf{J}_{\mathrm{pA}}^{+} d\mathbf{p} + \mathbf{N}_{\mathrm{pA}} \mathbf{J}_{\mathrm{s}}^{+} d\mathbf{s}$$

and we rewrite the equation as follows

$$d\mathbf{q} = \begin{bmatrix} \mathbf{J}_{\mathrm{pA}}^{+} & \mathbf{N}_{\mathrm{pA}} \mathbf{J}_{\mathrm{s}}^{+} \end{bmatrix} \begin{bmatrix} d\mathbf{p} \\ d\mathbf{s} \end{bmatrix}. \tag{6.27}$$

In order to be able to equate the solutions (6.26) and (6.27), we multiply the expression (6.27) with the matrix $\mathbf{J}$ and write

$$\begin{bmatrix} \mathbf{J}_{\mathrm{p}} \\ \mathbf{J}_{\mathrm{s}} \end{bmatrix} \begin{bmatrix} \mathbf{J}_{\mathrm{pA}}^{+} & \mathbf{N}_{\mathrm{pA}} \mathbf{J}_{\mathrm{s}}^{+} \end{bmatrix} = \mathbf{I},$$

from which the following conditions can be derived for each component

$$\mathbf{J}_{\mathrm{p}} \mathbf{J}_{\mathrm{pA}}^{+} = \mathbf{I},$$

$$\mathbf{J}_{\mathrm{p}}\mathbf{N}_{\mathrm{pA}}\mathbf{J}_{\mathrm{s}}^{+} = \mathbf{0},$$
$$\mathbf{J}_{\mathrm{s}}\mathbf{J}_{\mathrm{pA}}^{+} = \mathbf{0},$$
$$\mathbf{J}_{\mathrm{s}}\mathbf{N}_{\mathrm{pA}}\mathbf{J}_{\mathrm{s}}^{+} = \mathbf{I}.$$

The first two conditions are valid. To meet the second two conditions we assume

$$\mathbf{A}^{-1} = \frac{1}{\varepsilon}\mathbf{N}_{\mathrm{s}},$$

where

$$\mathbf{N}_{\mathrm{s}} = \mathbf{I} - \mathbf{J}_{\mathrm{s}}^{+}\mathbf{J}_{\mathrm{s}},$$

and ε is an arbitrary nonzero scalar. This assumption finds its mathematical background in the limit definition of the generalized inverse

$$\mathbf{J}_{\mathrm{s}}^{+} = \lim_{\varepsilon\to 0}\left(\mathbf{J}_{\mathrm{s}}^{\mathrm{T}}\mathbf{J}_{\mathrm{s}} + \varepsilon\mathbf{I}\right)^{-1}\mathbf{J}_{\mathrm{s}}^{\mathrm{T}},$$

from where it follows

$$\mathbf{N}_{\mathrm{s}} = \lim_{\varepsilon\to 0}\left(\frac{1}{\varepsilon}\mathbf{J}_{\mathrm{s}}^{\mathrm{T}}\mathbf{J}_{\mathrm{s}} + \mathbf{I}\right)^{-1} = \lim_{\varepsilon\to 0}\varepsilon\left(\mathbf{J}_{\mathrm{s}}^{\mathrm{T}}\mathbf{J}_{\mathrm{s}} + \varepsilon\mathbf{I}\right)^{-1}.$$

Here we can recognize the approximation of the best weighting matrix as developed in one of the previous sections so that

$$\mathbf{N}_{\mathrm{s}} = \lim_{\varepsilon\to 0}\varepsilon\mathbf{A}^{-1}.$$

As $\mathbf{J}_{\mathrm{s}}\mathbf{N}_{\mathrm{s}} = \mathbf{0}$, the conditions $\mathbf{J}_{\mathrm{s}}\mathbf{J}_{\mathrm{pA}}^{+} = \mathbf{0}$ and $\mathbf{J}_{\mathrm{s}}\mathbf{N}_{\mathrm{pA}}\mathbf{J}_{\mathrm{s}}^{+} = \mathbf{I}$ are fulfilled

$$\mathbf{J}_{\mathrm{s}}\mathbf{J}_{\mathrm{pA}}^{+} = \mathbf{J}_{\mathrm{s}}\mathbf{N}_{\mathrm{s}}\mathbf{J}_{\mathrm{p}}^{\mathrm{T}}\left(\mathbf{J}_{\mathrm{p}}\mathbf{N}_{\mathrm{s}}\mathbf{J}_{\mathrm{p}}^{\mathrm{T}}\right)^{-1} = \mathbf{0}$$

and

$$\mathbf{J}_{\mathrm{s}}\mathbf{N}_{\mathrm{pA}}\mathbf{J}_{\mathrm{s}}^{+} = \mathbf{J}_{\mathrm{s}}\left(\mathbf{I} - \mathbf{N}_{\mathrm{s}}\mathbf{J}_{\mathrm{p}}^{\mathrm{T}}\left(\mathbf{J}_{\mathrm{p}}\mathbf{N}_{\mathrm{s}}\mathbf{J}_{\mathrm{p}}^{\mathrm{T}}\right)^{-1}\mathbf{J}_{\mathrm{p}}\right)\mathbf{J}_{\mathrm{s}}^{+} = \mathbf{J}_{\mathrm{s}}\mathbf{J}_{\mathrm{s}}^{+} = \mathbf{I}.$$

Now we can rewrite (6.27) as follows

$$d\mathbf{q} = \mathbf{N}_{\mathrm{s}}\mathbf{J}_{\mathrm{p}}^{\mathrm{T}}\left(\mathbf{J}_{\mathrm{p}}\mathbf{N}_{\mathrm{s}}\mathbf{J}_{\mathrm{p}}^{\mathrm{T}}\right)^{-1}d\mathbf{p} + \mathbf{N}_{\mathrm{pA}}\mathbf{J}_{\mathrm{s}}^{+}d\mathbf{s}.$$

It appears that because of the matrix $\mathbf{N}_{\mathrm{pA}}$, the primary task is superior to the secondary task. At the same time it appears like the secondary task is superior to the primary because of the matrix $\mathbf{N}_{\mathrm{s}}$. From a mathematical point of view the solutions (6.26) and (6.27) are in the present form equivalent. Also the calculation of the inverse matrix $\mathbf{J}^{-1}$ requires approximately the same number of arithmetic operations as calculation of the matrices $\mathbf{J}_{\mathrm{pA}}^{+}$, $\mathbf{N}_{\mathrm{pA}}$, $\mathbf{J}_{\mathrm{s}}^{+}$, and $\mathbf{N}_{\mathrm{s}}$ all together. The course of numerical iterations is the same when we use either of the two solutions for calculating the inverse kinematics of a non-redundant mechanism.

The problem can be solved in a slightly different way. As the weighting matrix we apply the approximation of the matrix $\mathbf{N}_s$ as follows

$$\mathbf{A}^{-1} = \mathbf{I} - \mathbf{J}_s^{+}\mathbf{J}_s + \varepsilon\mathbf{I} = \mathbf{I}(1+\varepsilon) - \mathbf{J}_s^{+}\mathbf{J}_s, \tag{6.28}$$

where ε is a small value. The difference appears in the following way. Suppose that the tasks of positioning and orienting are part of a welding process. The technology of welding requires that the mechanism follows position with high accuracy, while the accuracy of the gripper/torch orientation is not as significant. When everything works fine there is no difference between the methods (6.26) and (6.27). Suppose that at some instant a malfunction of the mechanism occurs, such as the loss of a joint actuator. When solving both tasks by using expression (6.26), considering the mechanism as non-redundant, error will arise in both position and orientation. The welding process will be compromised and because of the positional error there will either be a collision of the welded object and the gripper/torch (likely resulting in some damage), or there will be too much of a gap between the workpiece and the torch, resulting in an inferior weld. When solving the problem by the use of the task priority method (6.27) and inverse weighting matrix (6.28), the primary task is superior to the secondary task and all error is manifested in the secondary task, i.e. in orienting of the welding gun. The welded object and the mechanism will remain unharmed and the manufacturing process may remain within some acceptable limitations. There are numerous similar practical examples. By appropriate division of the external coordinates into primary and secondary the functioning of the mechanism can become more reliable [49].

6.3.4 Hyperredundancy

A mechanism is called hyperredundant when it has numerous superfluous degrees of freedom with respect to the requirements of the task to be accomplished. In the language of mathematics, this can be written as follows

$$n \gg m.$$

The hyperredundant mechanisms usually have the form of a snake, tentacle or elephant's trunk and are even more adaptable than the ordinary redundant mechanisms. Although these mechanisms are well known from a theoretical point of view, they are almost nonexistent in practical situations, except in some experimental laboratory realizations. The reason is that the design of the hyperredundant mechanisms with classical constructional approaches is not possible. Also, the development of useful control schemes are far from realization. We can only imagine what an enormous quantity of arithmetic operations would be introduced in order to solve the inverse kinematics problem with the Newton-Raphson method, for example.

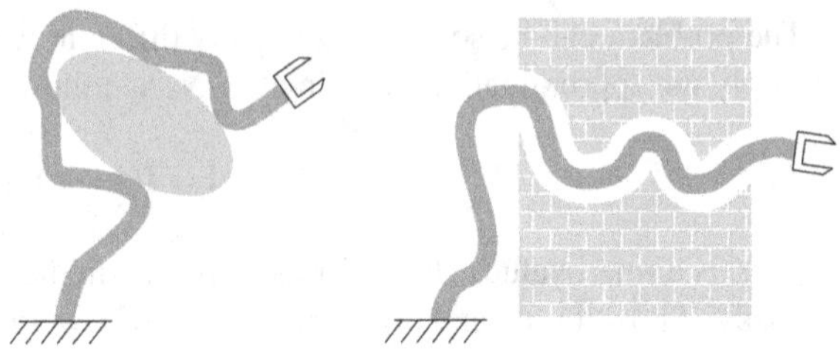

Fig. 6.14 Hyperredundant mechanisms can efficiently adapt to various tasks

We can expect that regardless of current problems, sooner or later hyperredundant mechanisms will find their way into practical use. In particular they are expected to be useful in special situations, e.g. in space or under water, where dynamics and gravity do not play such a crucial role. A hyperredundant mechanisms can wrap itself around an object, it can grasp, transfer, or orient the object. A hyperredundant mechanism can avoid obstacles in the environment and can twist into a pipe or execute a task on the other side of a hole (Fig. 6.14).

The inverse kinematics problem of a hyperredundant mechanism can be split into two parts. The first part encompasses the calculation of the inverse kinematics of the so called spinal curve of a mechanism, which has an infinite number of degrees of freedom. The aim of the second part, which bring us to the control of the mechanism, is to follow with the real mechanism as accurately as possible the shape of the spinal curve. This means that we assign an infinite number of degrees of freedom to the spinal curve and treat it as a continuum. Such an approach enables the use of different and more efficient mathematical tools [13]. Increasing the number of degrees of freedom of a mechanism, which can theoretically go to infinity, makes the mathematical analysis easier and not more difficult, as expected.

Consider an example of the spinal curve of a planar hyperredundant mechanism with the primary task of positioning the gripper. The equations of direct kinematics of the spinal curve are written as follows

$$p_1 = l \sum_{i=1}^{n} \cos \alpha_i ,$$

$$p_2 = l \sum_{i=1}^{n} \sin \alpha_i .$$

Here $n \to \infty$ is the number of degrees of freedom, $l \to 0$ is the length of the segments with equal length, nl is the collective length of the mechanism and

$$\alpha_i = q_1 + q_2 + \cdots + q_i , \quad i = 1, 2, \ldots, n.$$

For the sake of simplicity, let it be

$$p_1, p_2 \geq 0.$$

Assume that a vertical force F acts on the mechanism's end-point. The torque produced by this force in the j-th joint is

$$\tau_j = Fl \sum_{i=j}^{n} \cos\alpha_i,$$

when taking into account that all joint angles are in the interval

$$-\frac{\pi}{2} \leq \alpha_i \leq \frac{\pi}{2}, \quad i = 1, 2, \ldots, n.$$

We are interested in a configuration of the mechanism where the sum of the joint torques

$$\sum_{j=1}^{n} \tau_j = Fl \sum_{j=1}^{n} \sum_{i=j}^{n} \cos\alpha_i = Fl \sum_{i=1}^{n} i \cos\alpha_i$$

is minimized, when the position of the gripper is determined by the coordinates p_1, p_2. We could approach the problem by using standard methods of linear programming, however the calculation with $n \to \infty$ will be too time consuming. Because $n \to \infty$, the problem can only be solved with some calculating dexterity. The most favorable solution [48] is obviously the following

$$\cos\alpha_1 = \cos\alpha_2 = \cdots = \cos\alpha_k = 1,$$

from which it follows

$$\alpha_1, \alpha_2, \ldots, \alpha_k = 0,$$

and

$$\cos\alpha_{k+1} = \cos\alpha_{k+2} = \cdots = \cos\alpha_n = 0,$$

from which we have

$$\alpha_{k+1}, \alpha_{k+2}, \ldots, \alpha_n = \pm\frac{\pi}{2}.$$

Here, the following equalities must also hold

$$\sum_{i=1}^{k} l = kl = p_1$$

and

$$\sum_{i=k+1}^{n} \pm l = \sum_{i=k+1}^{r} l - \sum_{i=r+1}^{n} l = p_2,$$

from which

$$k = \frac{p_1}{l}$$

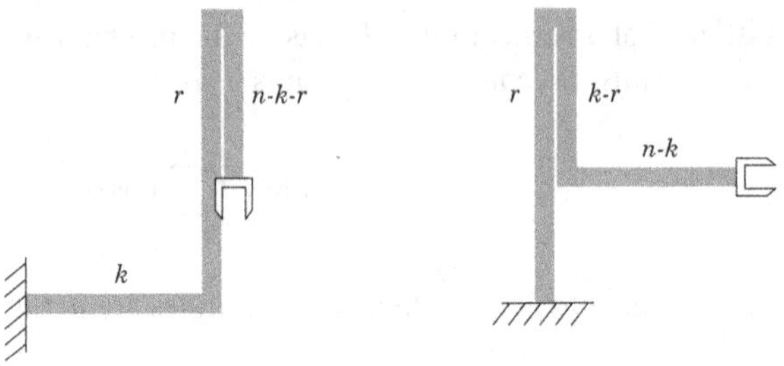

Fig. 6.15 The most favorable configurations of a mechanism, when the joint torques are the consequence of the external force (*left*) and when the joint torques are the consequence of the gravity of the segments (*right*)

and

$$r = \frac{1}{2}\left(n - k + \frac{p_2}{l}\right).$$

In the most favorable configuration of a mechanism, k represents the number of horizontal segments of a mechanism, r is the number of vertical segments turned upwards, while $n - k - r$ is the number of the downwards turned vertical segments. The solution is presented on the left side of Fig. 6.15 and is possible when

$$nl \geq p_1 + p_2.$$

In the same way we search for the most favorable configuration of the spinal curve when the joint torques are a consequence of the weight of the mechanism's links. Suppose all the segments of the mechanism are of equal weight. The gravitational force of a single segment is F_g and it acts in the center of the segment. The torque in the i-th joint is given by the following expression

$$\tau_j = F_g l\left(\left(n - j + \frac{1}{2}\right)\sum_{i=j}^{n} \cos\alpha_i - \sum_{i=j+1}^{n}\sum_{u=i}^{n} \cos\alpha_u\right).$$

With some rearrangements we can write the sum of the joint torques as follows

$$\sum_{j=1}^{n} \tau_j = F_g l \sum_{i=1}^{n} i\left(n - i + \frac{1}{2}\right)\cos\alpha_i.$$

Now, the sum of the joint torques is minimized when the minimal number of nonzero summands is placed at the mechanism's end-point. The solution has the following form

$$\cos\alpha_1 = \cos\alpha_2 = \cdots = \cos\alpha_k = 0,$$

from which

$$\alpha_1, \alpha_2, \ldots, \alpha_k = \pm\frac{\pi}{2},$$

and

$$\cos\alpha_{k+1} = \cos\alpha_{k+2} = \cdots = \cos\alpha_n = 1,$$

from which it follows

$$\alpha_{k+1}, \alpha_{k+2}, \ldots, \alpha_n = 0.$$

Here, the following two equalities must also hold

$$\sum_{i=k}^{n} l = (n-k)l = p_1$$

and

$$\sum_{i=1}^{k} \pm l = \sum_{i=1}^{r} l - \sum_{i=r+1}^{k} l = p_2,$$

from which we obtain

$$k = n - \frac{p_1}{l}$$

and

$$r = \frac{1}{2}\left(k + \frac{p_2}{l}\right).$$

In a most favorable configuration of a mechanism, k is the number of the vertical segments. There are r vertical segments which are turned upwards and $k-r$ segments that are turned downwards, while $n-k$ is the number of horizontal segments. The solution is presented on the right side of Fig. 6.15. It is interesting that the configurations depicted in Fig. 6.15 are also most favorable in the case when the joint torques are a consequence of the external force and segment weight together. It depends on the ratio of both forces as to which of the two configurations will prevail.

Chapter 7
Parallel Mechanisms

Abstract Today there is an abundance of kinematic arrangements of parallel mechanisms with different numbers of degrees of freedom available in robotics. The most commonly used are described in this chapter. In contrast to serial mechanisms, universal and spherical joints are common in parallel mechanisms, as the majority of joints in a parallel mechanism robot are passive. The complexity of parallel mechanisms becomes evident in their mathematical analysis. In parallel mechanisms, a closed-form solution to the direct kinematics problem does not, in general, exist. Difficulties are also due to the multiple solutions and to the fact that the existence of a real solution is not always guaranteed.

Parallel mechanisms have only recently emerged in industrial robotics. Their history however is extensive. In 1962, V.E. Gough and S.G. Whitehall developed parallel mechanisms for the purpose of testing automobile tires. While constructing their machine, they made use of a system of parallel actuators which provided six degrees of freedom to an attached mobile platform. In 1965, a mechanism with similar parallel kinematic structure was used by D. Stewart when designing a flight simulator. The mechanism, where the mobile platform is controlled by six actuated legs, is now called the Stewart-Gough platform.

In early eighties professor R. Clavel of École polytechnique fédérale de Lausanne in Switzerland, developed a parallel mechanism with one rotational and three translational degrees of freedom which represented the basis for a new type of industrial robot. Clavel attached three parallelogram legs to the platform resulting in a mechanism where the orientation of the platform remained unchanged while the mechanism was articulated [14]. The mechanism, which was patented in 1990 in the USA under the name Delta robot (patent no. 4,976,582), is shown in Fig. 7.1. It has exceptional kinematic and dynamic properties. The gripper, which is attached to the platform, can, in laboratory conditions, be displaced with an acceleration fifty times larger than gravitational acceleration. The original mechanism and its derivatives are now used in industrial robotics for various tasks, such as sorting and packaging. The Delta robot was also one of the first robots used in surgical procedures.

The first relevant theoretical developments in the area of parallel mechanisms appeared in the literature much later. Parallel mechanisms started to be systematically studied as late as the early eighties [30], while the amount of research increased sig-

J. Lenarčič et al., *Robot Mechanisms*,

Intelligent Systems, Control and Automation: Science and Engineering 60,
DOI 10.1007/978-94-007-4522-3_7,

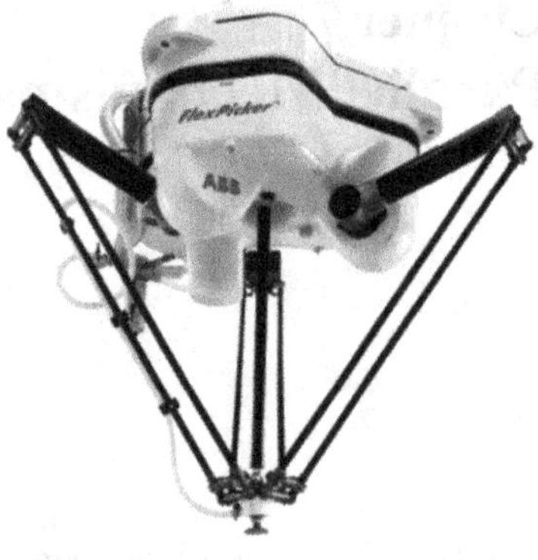

Fig. 7.1 Type of Delta robot introduced in production by ABB in 1999 under the name FlexPicker (courtesy of ABB)

nificantly towards the end of eighties and in particular into the nineties. It is interesting to note that the most important researchers from the area of parallel mechanisms gathered at the symposia *Advances in Robot Kinematics*. The first symposium was organized in 1988 in Ljubljana (Slovenia).

Today there is an abundance of kinematic arrangements of parallel mechanisms with different numbers of degrees of freedom. There are virtually no important producers of industrial robots who do not include at least one parallel mechanism in their production program. Parallel mechanisms can be found in advanced computer driven machine tools, such as in five axis CNC milling machines. When designing serial robot mechanisms, we usually make use of translational and rotational joints with a single degree of freedom. The reason is that all joints of a serial mechanism must be actuated. Joints with several degrees of freedom, such as universal and spherical joints, are rare in serial mechanisms as their construction and actuation is rather demanding. Contrary to this, universal and spherical joints are common in parallel mechanisms, as these joints are passive. We will see in this section that the majority of joints in a parallel mechanism robot are passive.

7.1 Characteristics of Parallel Mechanisms

Kinematically, the basic characteristic property of a parallel mechanism is that it consists of one or more closed loop kinematic chains. The number of degrees of freedom of a parallel mechanism is therefore less than the overall number of degrees of freedom contributed by the robot joints, unlike a serial mechanism. The number of degrees of freedom of the planar 2RR mechanism, shown in Fig. 7.2, is calculated by Grübler's formula (2.2), by entering into the equation the number of mobile segments $N = 3$, the number of joints $n = 4$ and $\lambda = 3$. As all joints have one degree of freedom, $f_i = 1, i = 1, 2, \ldots, 4$, we have

$$F = 3(3 - 4) + 4 = 1.$$

The mechanism is controlled by a single variable e.g. q, which can be related to any of four rotations [59]. The pose of the mechanism is completely determined by the

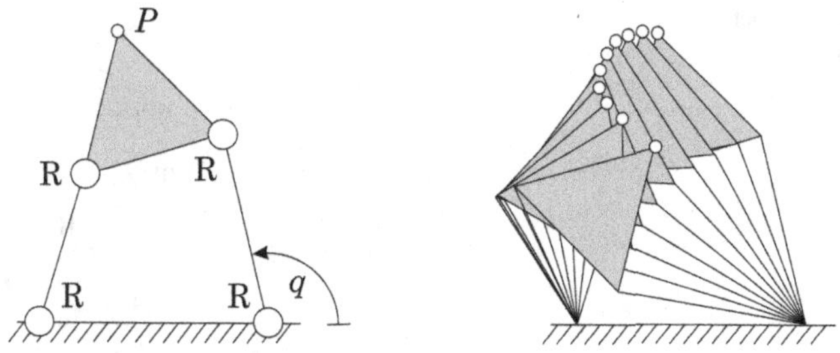

Fig. 7.2 Parallel planar 2RR mechanism

value of this variable, therefore the position of point P depends on the variable q and travels along a curve which is determined by the parameters of the mechanism i.e. the segment lengths (Fig. 7.2). We call a 2RR mechanism a parallelogram mechanism, when the two opposite segments are of equal length. With a parallelogram mechanism the orientation of the platform remains constant.

7.1.1 Components of Parallel Mechanisms

Parallel robot mechanisms are characterized by kinematic structures where two rigid segments are connected by several parallel kinematic chains (Fig. 7.3). One of the segments represents the fixed base of the parallel mechanism. The other segment is mobile and is called the platform. The kinematic chains connecting the platform with the base are called legs.

Typically, the platform carries a gripper or some type of tool. The mobility of the platform depends on the number of, and the kinematic structure of, the legs. With serial mechanisms all degrees of freedom within the joints must be controlled, while with parallel mechanism we need to control only as many joint variables as there are degrees of freedom. The most general kinematic structure of a parallel mechanism is the Stewart-Gough platform which has six degrees of freedom. With the original design of a Stewart-Gough platform the legs are serial mechanisms, however this is not a rule. With the Clavel Delta robot mechanism, the legs are hybrid mechanisms which incorporate a parallelogram portion.

Usually a parallel mechanism is referred to by the type of kinematic chains representing the legs. Thus, the mechanism in Fig. 7.3 is referred to as the SRT-USU-SS mechanism. When legs of the same type are repeated, e.g. three legs of the RRR type, the designation is a 3RRR mechanism. Several joints in contact with either the base or the platform can be placed at the same point. Therefore, often the number of contact points on the base and on the platform is denoted. The (3-1)STU mechanism has three legs of the STU type, where three S joints are separately connected to the base, while three U joints are connected to the platform at the same point. Mechanisms where all legs have the same kinematic structure, and where the legs

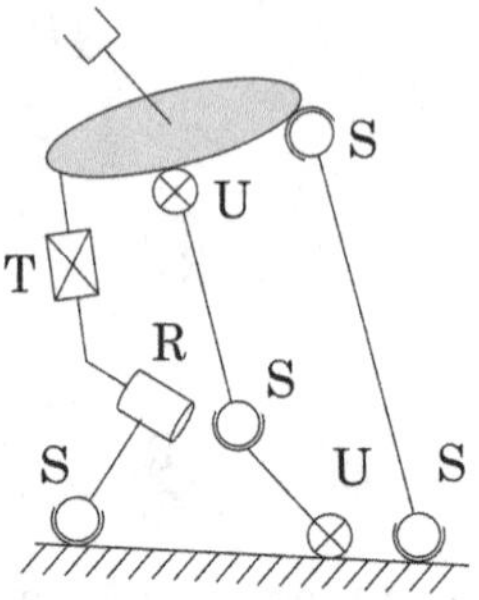

Fig. 7.3 Parallel mechanism SRT-USU-SS

are attached to the base and the platform in some symmetric pattern, are called symmetrical parallel mechanisms. Unfortunately, unique and uniform names for parallel mechanisms do not exist.

Parallel mechanisms have several advantages in comparison to serial mechanisms. These are different for various parallel mechanisms, however, in general the following properties exist:

load capacity, rigidity, and accuracy As the platform and the base are connected with several kinematic chains, the load carrying capacity of parallel mechanisms is considerably larger than that of serial mechanisms. With some types of legs, e.g. STU, the rigidity of a parallel mechanism can also be considerably greater than that of a serial mechanism. Also, the accuracy in positioning and orienting the end-effector of a parallel mechanism is several times better than in a serial mechanism.

excellent dynamic properties A parallel kinematic structure allows for all actuators and transmissions to be placed on the base and thus they are not moving. The mobile segments are therefore characterized by smaller masses and moments of inertia. Consequently, the platform can achieve high velocities and accelerations. Also the resonant frequency of a parallel mechanism is orders of magnitude higher than that of a serial mechanism making the parallel mechanism less likely to exhibit the low frequency oscillations sometimes observed in serial mechanisms.

simple construction As only the passive part of a parallel mechanism is mobile, the construction of the mechanism is simpler and less expensive. When building parallel mechanisms standard bearings, spindles, and other machine elements can be used.

However, it is also a fact that the applicability of parallel mechanisms is limited. Parallel mechanisms usually cannot avoid obstacles in their workspace and the workpiece can only be approached from a single direction. Other disadvantages are:

small workspace Parallel mechanisms have considerably smaller workspaces than serial mechanisms of a comparable size. The workspace of a parallel mechanism is intersection of the workspaces of the particular legs and is thus the reduced workspace of a serial mechanism. Their workspace may be further reduced since during motion of the platform the legs may become entangled and interfere with each other.

complex kinematics Aspects of the kinematics of parallel mechanisms become complex and lengthy. In contrast to serial mechanisms, where difficulty arises when solving the inverse kinematics problem, in parallel mechanisms difficulty arises in solving the direct kinematics problem. As well, in parallel mechanisms with less than 6 degrees of freedom, a complex coupling exists among the external coordinates.

unsurmountable kinematic singularities Serial mechanisms in kinematically singular poses lose mobility, restricting the motion of their end effector. Parallel mechanisms in singular poses gain degrees of freedom, which cannot be controlled. While singular poses in serial manipulators may be a nuisance, singular poses of parallel mechanisms result in uncontrolled motion and may be catastrophic.

7.1.2 Stewart-Gough Platform

A general Stewart-Gough platform [26, 80] is shown on the left side of Fig. 7.4. It has six legs of the UTS type. Each leg has 2 segments and 3 joints. The number of degrees of freedom of the mechanism is calculated by the formula (2.2)

$$F = 6(13-18)+36 = 6.$$

Thus, the platform of this mechanism can be spatially positioned and oriented under the control of six joint variables, which can be arbitrarily selected from the entire set of joint variables. Typically, the six translational variables are controlled, i.e. the platform is moved into a desired pose by changing the lengths of the legs. A special advantage of the Stewart-Gough platform with the UTS legs is that in the kinetostatic realm, the individual legs are two-force members. Thus loads acting on the platform are transferred to each particular leg in the form of a longitudinal force in the direction of the leg extensions and there is no transverse loading on the legs.

Solving the forward kinematics problem of the general Stewart-Gough platform is extremely complex. The general Stewart-Gough platform does not have a closed-form solution except in special cases. One such case is the (3-3)UTS platform, where pairs of spherical joints at the platform and pairs of universal joints at the base are coincident, while the legs are interconnected as shown on the right side of Fig. 7.4.

7.1.3 Delta Mechanism

The kinematic structure of the Delta robot mechanism [14] is shown in Fig. 7.5. The mechanism has three external legs and a middle leg which has no influence on the platform pose. The external legs are attached to the upper side of the base and

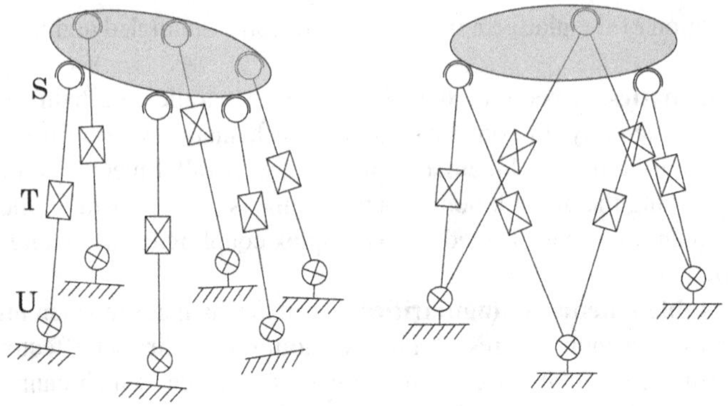

Fig. 7.4 General 6UTS and (3-3)UTS Stewart-Gough platform

Fig. 7.5 Kinematic structure of the Delta robot

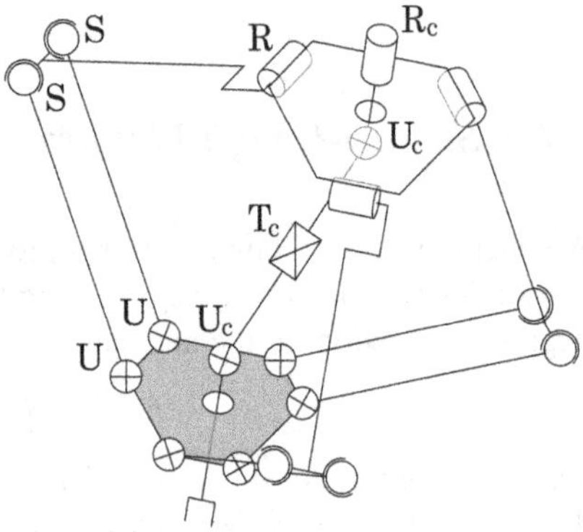

to the lower side of the platform. The three joints of the platform are placed along the circumference of a circle with an angle of $2\pi/3$ between them. The external leg is attached to the base through rotation R. There is a parallelogram mechanism between the middle of the leg and the base. It consists of two spherical joints S and two universal joints U. Thus each leg has 3 segments and 5 joints and all together 11 degrees of freedom. When not considering the middle leg, the number of degrees of freedom of the mechanism is (2.2)

$$F = 6(10 - 15) + 33 = 3.$$

The pose of the platform is determined by only three variables. In the original version of the Delta robot the three rotation angles R are the controlled variables. Due to the parallelogram structure of the legs, the platform executes translation and is always parallel to the base.

The purpose of the middle leg is to transfer the rotation R_c across the platform to the gripper at the end-point of the mechanism. It acts as a telescoping driveshaft for rotating the gripper. This leg is a cardan joint with two universal joints U_c separated by a translational joint T_c. In all, the mechanism has four degrees of freedom: three translational, enabling the spatial position of the gripper and one rotational, enabling rotation of the gripper about an axis perpendicular to the platform. All actuators of the Delta mechanism are placed on the base and are not moving. Therefore the mechanism is extremely lightweight and the platform can move with high velocities and accelerations.

7.2 Connectivity of Legs and Degrees of Freedom

Consider a parallel mechanism which has n joints and N moving segments. If F is the number of degrees of freedom of a mechanism and f is the number of degrees of freedom of the platform. It is not difficult to realize that there is the following relation between the quantities f and F

$$f \leq F = \lambda(N - n) + \sum_{i=1}^{n} f_i. \tag{7.1}$$

Grübler's formula (2.2) was used here, where f_i represents the number of degrees of freedom of the i-th mechanism joint, while $\lambda = 3$ for planar motion and $\lambda = 6$ spatial motion. In practical applications we make use of kinematic structures where the number of degrees of freedom of the platform equals the number of degrees of freedom of the mechanism

$$f = F.$$

Suppose that a mechanism has $k = 1, 2, \ldots, K$ legs, while a leg k has n_k joints and N_k segments. The number of mechanism joints equals the sum of the leg joints

$$n = \sum_{k=1}^{K} n_k, \tag{7.2}$$

while the number of mechanism segments equals the sum of the leg segments plus the platform segment

$$N = 1 + \sum_{k=1}^{K} N_k. \tag{7.3}$$

The number of degrees of freedom of leg k is ν_k and can be calculated by summing up all degrees of freedom f_j in the joints of the corresponding leg

$$\nu_k = \sum_{j=1}^{n_k} f_j, \tag{7.4}$$

therefore it should hold

$$\sum_{k=1}^{K} \nu_k = \sum_{i=1}^{n} f_i. \tag{7.5}$$

When a leg is a serial mechanism, the joints and segments occur in series in such a way that there is a joint at the beginning and at the end of the leg, therefore

$$n_k = N_k + 1, \tag{7.6}$$

from which

$$\sum_{k=1}^{K} n_k = K + \sum_{k=1}^{K} N_k$$

and by considering (7.2) and (7.3)

$$n = K + N - 1. \tag{7.7}$$

When inserting the last expression into Grübler's formula (2.2) and adding the relation (7.5) we obtain

$$F = \lambda(1 - K) + \sum_{k=1}^{K} \nu_k. \tag{7.8}$$

Through some equation manipulation we obtain an expression which relates the number of degrees of freedom of a mechanism to the number of legs and collective number of degrees of freedom in the mechanism's joints. The equation can be easily understood after arranging into the following form

$$F = \lambda - \sum_{k=1}^{K} (\lambda - \nu_k),$$

where we can notice that each leg $k = 1, 2, \ldots, K$ takes away from the mechanism $\lambda - \nu_k$ degrees of freedom.

The number of degrees of freedom of the platform is constrained by the inequality (7.1), however it also depends on the number of degrees of freedom appertaining to particular legs. This can be best understood when considering the leg as a system constraining the movement of the platform. The number of constraining degrees of freedom equals $\lambda - f$ in accordance with (2.1), the number of constraining degrees of leg k is $\lambda - \nu_k$. It is clear that the number of constraining degrees of the platform can only be greater than or equal to the highest number of constraining degrees of the legs

$$\lambda - f \geq \max_{k=1,\ldots,K} (\lambda - \nu_k),$$

therefore the inverse relation is valid

$$f \leq \min_{k=1,\ldots,K} (\nu_k).$$

The number of degrees of freedom of the platform can only be smaller than the lowest number of the degrees of freedom of any of the mechanism's legs.

The leg must therefore have at least as many degrees of freedom as the platform. When we require that the leg does not constrain the platform's movement, the leg must have at least λ degrees of freedom. We say that the leg is connective [84], when the following inequality is valid

$$f \leq \nu_k \leq \lambda, \tag{7.9}$$

$k = 1, 2, \ldots, K$. The connectivity of all legs is the basic condition for development of a parallel mechanism.

In theory it is possible to make use of legs which have more degrees of freedom than necessary. When

$$\nu_k > \lambda,$$

the leg k is kinematically redundant. It is not difficult to predict the disadvantages and advantages of parallel mechanisms with legs that have redundant degrees of freedom. However, this problem has not as yet been investigated.

7.2.1 Mechanisms with $\nu_1 = \nu_2 = \cdots = \nu_K = \lambda$

The following connectivity condition must be valid in order that the platform has λ degrees of freedom

$$\lambda \leq \nu_k \leq \lambda.$$

From here we have the requirement

$$\nu_k = \lambda,$$

$k = 1, 2, \ldots, K$. The Stewart-Gough platform is an example of such a mechanism, where the legs are kinematically identical and each leg has six degrees of freedom.

Consider the mobility of the platform when each leg has the highest number of degrees of freedom. By inserting $\nu_k = \lambda, k = 1, 2, \ldots, K$ into expression (7.8), we obtain

$$F = K\lambda - K\lambda + \lambda \quad \Longrightarrow \quad F = \lambda.$$

This statement is important as such types of mechanisms have λ degrees of freedom independent of the number of legs K. An example of a planar mechanism of this type [25] is shown in Fig. 7.6. The planar 3RRR mechanism ($\lambda = 3$) has three equal legs of RRR type, $\nu_1 = \nu_2 = \nu_3 = 3$. The number of degrees of freedom of the mechanism is in accordance with (7.8)

$$F = 3(1 - 3) + 9 = 3$$

and does not change when adding or taking a leg away. The platform also has $f = 3$ degrees of freedom and can therefore be arbitrarily positioned and oriented in the plane.

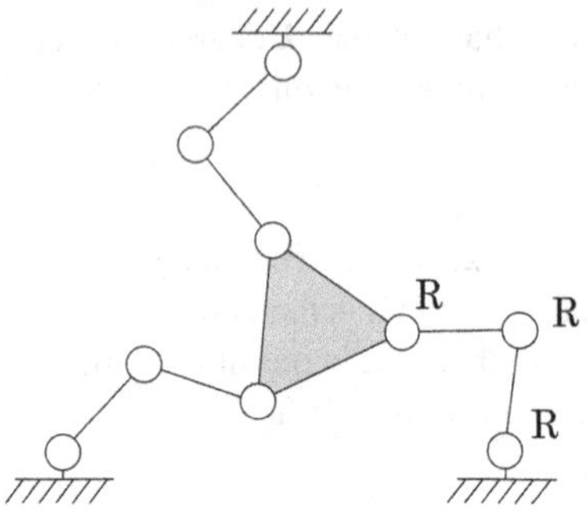

Fig. 7.6 The platform can be with planar 3RRR mechanism arbitrarily positioned and oriented

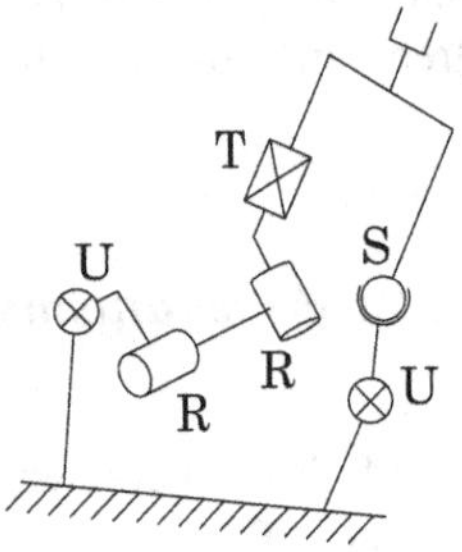

Fig. 7.7 Each leg takes away from the platform $\lambda - \nu_k$ degrees of freedom. As there is $\nu_1 = 5$ and $\nu_2 = 5$, the platform has $f = 4$ degrees of freedom

7.2.2 Mechanisms with $\nu_1 = \nu_2 = \cdots = \nu_K = F$

Consider when all the legs have the minimal permitted number of degrees of freedom. Inserting $\nu_k = F, k = 1, 2, \ldots, K$ into (7.8), it follows

$$(F - \lambda)(K - 1) = 0.$$

This equation is valid when

$$F = \lambda.$$

This is in accordance with the previous example. The equation is also valid for $F < \lambda$ when

$$K = 1.$$

The mechanism is functional at $F < \lambda$, when it has only a single leg regardless of the number of degrees of freedom F and the value λ. The mechanism in Fig. 7.7 has two legs with five degrees of freedom, therefore the number of degrees of freedom of the platform can only be less than five.

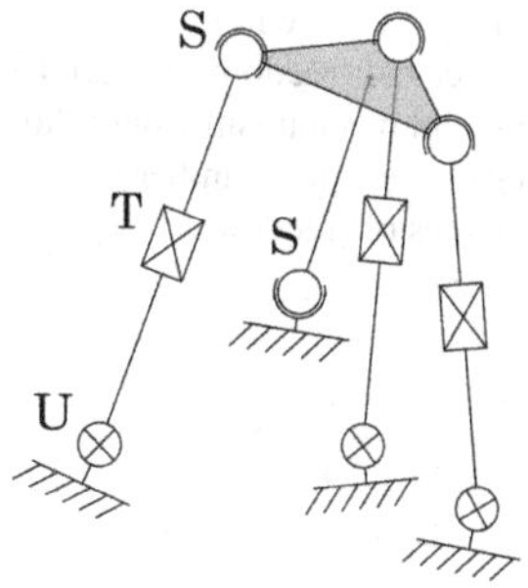

Fig. 7.8 Parallel spatial mechanism S-3UTS, where the platform is displaced by changing its pose around the spherical joint of the middle leg

7.2.3 Mechanisms with $\nu_1 = F$ and $\nu_2 = \nu_3 = \cdots = \nu_K = \lambda$

Here only one of the legs of the mechanism has the minimal permitted number of degrees of freedom, $\nu_1 = F$. For the other legs there is $\nu_k = \lambda, k = 2, \ldots, K$. The following expression must be inserted into (7.8)

$$\sum_{k=1}^{K} \nu_k = F + (K-1)\lambda,$$

which gives

$$F = F + (K-1)\lambda - (K-1)\lambda.$$

This equation is always valid and such a mechanism is functional irregardless of the number of legs and number of degrees of freedom. An example of such a mechanism is the S-3UTS mechanism shown in Fig. 7.8. The mechanism ($\lambda = 6$) has a middle leg of S type with three degrees of freedom ($\nu_1 = 3$) and three equal external legs of the UTS type with six degrees of freedom ($\nu_2 = \nu_3 = \nu_4 = 6$). The number of degrees of freedom of the mechanism is calculated by the formula (7.8)

$$F = 6(1-4) + 21 = 3.$$

The platform has the same number of degrees of freedom.

7.2.4 Mechanisms with $\nu_1 = \nu_2 = \cdots = \nu_K$

Finally, we examine the more general properties of parallel mechanisms, where all legs have an equal number of degrees of freedom. The number of degrees of freedom of such spatial parallel mechanisms is presented in Table 7.1 for different numbers of legs K with the same number of degrees of freedom in each leg $\nu_k, k = 1, 2, \ldots, K$. Only a small number of such mechanisms are movable. Most are structures with zero or fewer degrees of freedom. Mechanisms with a negative number of degrees of freedom are overconstrained.

Suppose that we want to build a mechanism with a single actuator in each leg, as is the case with the Stewart-Gough platform. Such a mechanism must have $K = F$

Table 7.1 The number of degrees of freedom of spatial parallel mechanism where the legs have equal number of degrees of freedom

ν_k	1	2	3	4	5	6
$K=1$	1	2	3	4	5	6
$K=2$	−4	−2	0	2	4	6
$K=3$	−9	−6	−3	0	3	6
$K=4$	−14	−10	−6	−2	2	6
$K=5$	−19	−14	−9	−4	1	6
$K=6$	−24	−18	−12	−6	0	6

Table 7.2 The number of degrees of freedom of a planar parallel mechanism where the legs have equal number of degrees of freedom

ν_k	1	2	3
$K=1$	1	2	3
$K=2$	−1	1	3
$K=3$	−3	0	3

legs, as a mechanism with $K < F$ legs cannot be controlled, while with $K > F$ the actuated legs counteract. Table 7.1 shows that there are only a few possibilities. There exist the following combinations

$$\begin{aligned} K &= 1, & \nu_1 &= 1, \\ K &= 2, & \nu_1 &= \nu_2 = 4, \\ K &= 3, & \nu_1 &= \nu_2 = \nu_3 = 5, \\ K &= 6, & \nu_1 &= \nu_2 = \cdots = \nu_6 = 6. \end{aligned}$$

There are no mechanisms of this type with four or five degrees of freedom [30].

Table 7.2 shows the numbers of degrees of freedom of planar parallel mechanisms as related to the number of legs K and the number of degrees of freedom in the legs ν_k, $k = 1, 2, \ldots, K$. There are only a few mechanisms which can move, the others being rigid structures where the number of degrees of freedom is equal to or less than zero. When trying to build a mechanism with a single actuator in each leg, there are only two possibilities

$$\begin{aligned} K &= 1, & \nu_1 &= 1, \\ K &= 3, & \nu_1 &= \nu_2 = \nu_3 = 3. \end{aligned}$$

There is no such mechanism with two degrees of freedom.

7.3 Kinematic Equations

There is no general procedure for calculating the inverse or direct kinematics of parallel mechanisms, as their kinematic structures are very diverse. In contrast to

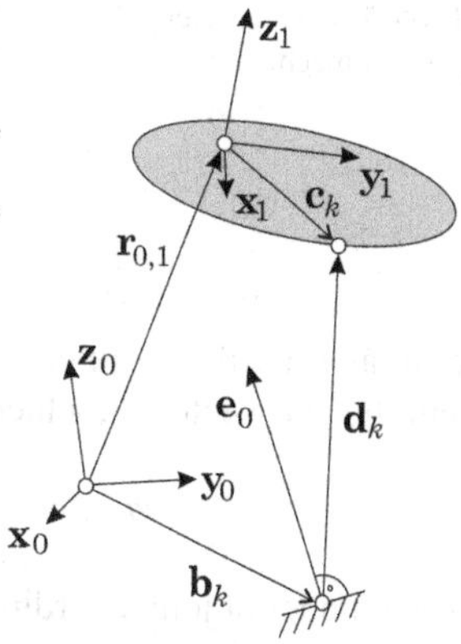

Fig. 7.9 Parameters of parallel mechanism

serial mechanisms, solving the direct kinematics problem of parallel mechanisms is a complex problem, whereas the inverse kinematics problem is simpler [60]. As a rule, the inverse kinematics problem becomes simpler as the platform has more degrees of freedom and is simplest when $F = \lambda$.

7.3.1 Parameters and Variables of a Parallel Mechanism

The coordinate frame $\mathbf{x}_1, \mathbf{y}_1, \mathbf{z}_1$ is attached to the platform of a parallel mechanism, while the frame $\mathbf{x}_0, \mathbf{y}_0, \mathbf{z}_0$ is fixed to its base (Fig. 7.9). The position of the platform is given with respect to fixed coordinate frame by the vector $\mathbf{r}_{0,1}$, while the orientation of the platform can be described by an arbitrary triple of the orientation angles ψ, θ, ϕ occurring between both frames.

The parameters of a parallel mechanism are vectors representing the attachments of the legs $k = 1, 2, \ldots, K$ to the base and the platform. In this way the vector $\mathbf{b}_k$ defines the attachment of the leg k to the base expressed in the frame $\mathbf{x}_0$, $\mathbf{y}_0$, $\mathbf{z}_0$, while the vector $\mathbf{c}_k$ defines the attachment of the leg on the platform in the frame $\mathbf{x}_1$, $\mathbf{y}_1$, $\mathbf{z}_1$.

The vectors

$$\mathbf{d}_k = \mathbf{r}_{0,1} + \mathbf{c}_k - \mathbf{b}_k, \quad k = 1, 2, \ldots, K, \tag{7.10}$$

describe the directions and the lengths of the legs. The joint coordinates are expressed in various ways, taking into account the kinematic structure of the legs and considering which joint variables are to be controlled. With the Stewart-Gough platform (Fig. 7.4) and with the S-3UTS mechanism (Fig. 7.8) usually the length of the legs is controlled, therefore the joint coordinates are given by the following norms

$$q_k = \|\mathbf{d}_k\|, \quad k = 1, 2, \ldots, K. \tag{7.11}$$

With the Delta robot mechanism (Fig. 7.5) and the planar 3RRR mechanism (Fig. 7.6) the joint rotations at the base of the legs are controlled. The joint co-

Table 7.3 Parameters of parallel mechanism

k	1	2	...	K
$\mathbf{b}_k^{(0)}$	$\mathbf{b}_1^{(0)}$	$\mathbf{b}_2^{(0)}$	...	$\mathbf{b}_K^{(0)}$
$\mathbf{c}_k^{(1)}$	$\mathbf{c}_1^{(1)}$	$\mathbf{c}_2^{(1)}$	...	$\mathbf{c}_K^{(1)}$

ordinates are therefore calculated as the inclination angles between the leg vectors and the unit vector $\mathbf{e}_0$, which is perpendicular to the base,

$$q_k = \arccos \frac{\mathbf{d}_k \cdot \mathbf{e}_0}{\|\mathbf{d}_k\|}, \quad k = 1, 2, \ldots, K. \tag{7.12}$$

The number of joint coordinates equals the number of degrees of freedom of a mechanism. When a single variable is controlled with each leg, we have

$$K = F. \tag{7.13}$$

This is the case with most parallel mechanisms in use today. We will assume this equation as valid unless stated otherwise. An exception for example is the S-3UTS mechanism (Fig. 7.8) which is controlled only through the translations of the external legs, while the middle leg is passive.

The parameters of the mechanism are the vectors $\mathbf{b}_k$, $k = 1, 2, \ldots, K$ expressed in the frame $\mathbf{x}_0, \mathbf{y}_0, \mathbf{z}_0$ and the vectors $\mathbf{c}_k$ expressed in the frame $\mathbf{x}_1, \mathbf{y}_1, \mathbf{z}_1$, as shown in Table 7.3.

7.3.2 Inverse Kinematics

The external coordinates of a parallel mechanism are represented by the position and the orientation coordinates of the platform. With a spatial mechanism where $\lambda = 6$, the vector of external coordinates is given as follows

$$\mathbf{p} = (r_x, r_y, r_z, \psi, \theta, \phi)^T,$$

while its time derivatives are

$$\dot{\mathbf{p}} = (v_x, v_y, v_z, \dot{\psi}, \dot{\theta}, \dot{\phi})^T, \qquad \ddot{\mathbf{p}} = (a_x, a_y, a_z, \ddot{\psi}, \ddot{\theta}, \ddot{\phi})^T.$$

Here

$$\mathbf{r}_{0,1}^{(0)} = (r_x, r_y, r_z)^T$$

describes the position of the platform, while ψ, θ, ϕ are the orientation angles. The translational velocity of the platform is given by

$$\mathbf{v}_{0,1}^{(0)} = \dot{\mathbf{r}}_{0,1}^{(0)} = (v_x, v_y, v_z)^T,$$

while $\dot{\psi}, \dot{\theta}, \dot{\phi}$ are time derivatives of the orientation angles. Likewise

$$\mathbf{a}_{0,1}^{(0)} = \dot{\mathbf{v}}_{0,1}^{(0)} = (a_x, a_y, a_z)^T$$

is the translational acceleration of the platform, while $\ddot{\psi}$, $\ddot{\theta}$, $\ddot{\phi}$ are the second time derivatives of the orientational angles. With planar mechanisms, where $\lambda = 3$, we have

$$\mathbf{p} = (r_{\mathrm{x}}, r_{\mathrm{y}}, \psi)^{\mathrm{T}}, \qquad \dot{\mathbf{p}} = (v_{\mathrm{x}}, v_{\mathrm{y}}, \dot{\psi})^{\mathrm{T}}, \qquad \ddot{\mathbf{p}} = (a_{\mathrm{x}}, a_{\mathrm{y}}, \ddot{\psi})^{\mathrm{T}}.$$

The first and second time derivatives of the orientational angles can be related to the angular velocity and angular acceleration of the platform as follows

$$\boldsymbol{\omega}_{0,1}^{(0)} = \mathbf{W} \begin{bmatrix} \dot{\psi} \\ \dot{\theta} \\ \dot{\phi} \end{bmatrix},$$

$$\mathbf{u}_{0,1}^{(0)} = \mathbf{W}_0 \begin{bmatrix} \dot{\psi}\dot{\theta} \\ \dot{\psi}\dot{\phi} \\ \dot{\theta}\dot{\phi} \end{bmatrix} + \mathbf{W} \begin{bmatrix} \ddot{\psi} \\ \ddot{\theta} \\ \ddot{\phi} \end{bmatrix}.$$

The matrices $\mathbf{W}_0$, $\mathbf{W}$ are 3×3 matrices defined in Sect. 1.6.

The orientation of the platform, expressed by the use of orientation angles ψ, θ, ϕ, can be converted into a rotation matrix $\mathbf{A}_{0,1}$. Afterwards the leg vectors are calculated as follows

$$\mathbf{d}_k^{(0)} = \mathbf{r}_{0,1}^{(0)} + \mathbf{A}_{0,1}\mathbf{c}_k^{(1)} - \mathbf{b}_k^{(0)}, \quad k = 1, 2, \ldots, K. \tag{7.14}$$

As a rule the legs are represented by simple mechanisms. Therefore, the joint coordinates can be expressed as scalar functions of the leg vectors

$$q_k = \xi_k\big(\mathbf{d}_k^{(0)}\big), \quad k = 1, 2, \ldots, K, \tag{7.15}$$

already introduced in (7.11) and (7.12). The joint coordinates of a parallel mechanism can be collected into a K-dimensional vector $\mathbf{q}$.

The inverse kinematics problem requires calculation of the joint coordinates $\mathbf{q}$ from the given values of the external coordinates $\mathbf{p}$. This is a simple problem for parallel mechanisms, if we know the values of all λ external coordinates $\mathbf{p}$. The joint coordinates can be expressed explicitly in terms of the external coordinates

$$\mathbf{q} = \mathbf{q}_{\mathrm{P}}(\mathbf{p}),$$

where $\mathbf{q}_{\mathrm{P}}$ is a corresponding K-dimensional vector function. Usually, a single solution of the joint coordinates vector $\mathbf{q}$ exists for given values of external coordinates $\mathbf{p}$. However, this depends on the kinematic structure of the legs.

When $K = \lambda$ all external coordinates $\mathbf{p}$ are independent and can be arbitrarily selected within the mechanism workspace. Things become complicated when $K < \lambda$. Then only K external coordinates are independent, while we cannot arbitrarily choose the values of $\lambda - K$ external coordinates. With mechanisms where $K < \lambda$, the external coordinates are related to each other through the kinematics of the mechanism. It constrains the number of poses of the platform, which is to be expected as $K < \lambda$. This was observed while displacing the 2RR mechanism in Fig. 7.2.

Equation (7.14) represents a sequence of translation and rotation. First find the time derivatives of the leg vectors

$$\dot{\mathbf{d}}_k^{(0)} = \dot{\mathbf{r}}_{0,1}^{(0)} + \dot{\mathbf{A}}_{0,1}\mathbf{c}_k^{(1)}, \quad k = 1, 2, \ldots, K.$$

As the rotation matrix is orthogonal and $\mathbf{A}_{0,1}^{\mathrm{T}}\mathbf{A}_{0,1} = \mathbf{I}$, we can write

$$\dot{\mathbf{d}}_k^{(0)} = \dot{\mathbf{r}}_{0,1}^{(0)} + \dot{\mathbf{A}}_{0,1}\mathbf{A}_{0,1}^{\mathrm{T}}\mathbf{A}_{0,1}\mathbf{c}_k^{(1)}, \quad k = 1, 2, \ldots, K.$$

In the above expression we can recognize the angular velocity matrix of the platform

$$\boldsymbol{\Omega}_{0,1} = \dot{\mathbf{A}}_{0,1}\mathbf{A}_{0,1}^{\mathrm{T}} = \boldsymbol{\omega}_{0,1}^{(0)} \otimes \mathbf{I},$$

therefore

$$\dot{\mathbf{d}}_k^{(0)} = \mathbf{v}_{0,1}^{(0)} + \boldsymbol{\Omega}_{0,1}\mathbf{A}_{0,1}\mathbf{c}_k^{(1)}. \tag{7.16}$$

The following form can also be used

$$\dot{\mathbf{d}}_k^{(0)} = \mathbf{v}_{0,1}^{(0)} + \boldsymbol{\omega}_{0,1}^{(0)} \times \left(\mathbf{A}_{0,1}\mathbf{c}_k^{(1)}\right), \quad k = 1, 2, \ldots, K. \tag{7.17}$$

Taking the time derivative of (7.16)

$$\ddot{\mathbf{d}}_k^{(0)} = \dot{\mathbf{v}}_{0,1}^{(0)} + \dot{\boldsymbol{\Omega}}_{0,1}\mathbf{A}_{0,1}\mathbf{c}_k^{(1)} + \boldsymbol{\Omega}_{0,1}\dot{\mathbf{A}}_{0,1}\mathbf{c}_k^{(1)}, \quad k = 1, 2, \ldots, K,$$

and considering the orthogonality of the rotation matrix we have

$$\ddot{\mathbf{d}}_k^{(0)} = \mathbf{a}_{0,1}^{(0)} + \left(\dot{\boldsymbol{\Omega}} + \boldsymbol{\Omega}_{0,1}^2\right)\mathbf{A}_{0,1}\mathbf{c}_k^{(1)}, \quad k = 1, 2, \ldots, K.$$

In the above expression we have the angular acceleration matrix

$$\boldsymbol{\Psi}_{0,1} = \dot{\boldsymbol{\Omega}}_{0,1} + \boldsymbol{\Omega}_{0,1}^2 = \mathbf{u}_{0,1}^{(0)} \otimes \mathbf{I}.$$

Which gives the accelerations of the leg vectors as

$$\ddot{\mathbf{d}}_k^{(0)} = \mathbf{a}_{0,1}^{(0)} + \boldsymbol{\Psi}_{0,1}\mathbf{A}_{0,1}\mathbf{c}_k^{(1)} \tag{7.18}$$

or in vector form

$$\ddot{\mathbf{d}}_k^{(0)} = \mathbf{a}_{0,1}^{(0)} + \mathbf{u}_{0,1}^{(0)} \times \left(\mathbf{A}_{0,1}\mathbf{c}_k^{(1)}\right), \quad k = 1, 2, \ldots, K. \tag{7.19}$$

Solving the complete inverse kinematics problem of a parallel mechanism requires relating the velocities and accelerations of the joint coordinates to the velocities and accelerations of the leg vectors. Preserving the generality of the expression (7.15), we can only write the formal relations. For the velocities of the joint coordinates we have

$$\dot{q}_k = \nabla_k^{\mathrm{T}}\dot{\mathbf{d}}_k^{(0)}, \quad k = 1, 2, \ldots, K. \tag{7.20}$$

The vector ∇_k is the gradient of the function ξ_k with regard to the components of the vector $\mathbf{d}_k^{(0)}$

$$\nabla_k^{\mathrm{T}} = \left(\frac{\partial \xi_k}{\partial d_{kx}}, \frac{\partial \xi_k}{\partial d_{ky}}, \frac{\partial \xi_k}{\partial d_{kz}}\right), \quad k = 1, 2, \ldots, K.$$

For the accelerations of the joint coordinates we obtain

$$\ddot{q}_k = \nabla_k^{\mathrm{T}} \ddot{\mathbf{d}}_k^{(0)} + \dot{\mathbf{d}}_k^{\mathrm{T}(0)} \begin{bmatrix} \dfrac{\partial \nabla_k^{\mathrm{T}}}{\partial d_{k\mathrm{x}}} \\ \dfrac{\partial \nabla_k^{\mathrm{T}}}{\partial d_{k\mathrm{y}}} \\ \dfrac{\partial \nabla_k^{\mathrm{T}}}{\partial d_{k\mathrm{z}}} \end{bmatrix} \dot{\mathbf{d}}_k^{(0)}, \quad k = 1, 2, \ldots, K. \tag{7.21}$$

It is obvious that the velocities and accelerations of the external and joint coordinates can be exposed in the last two expressions. In this way the Jacobian and Hessian formulations are obtained, similar to kinematics of serial mechanisms

$$\dot{\mathbf{q}} = \mathbf{Y}\dot{\mathbf{p}} \tag{7.22}$$

and

$$\ddot{\mathbf{q}} = \mathbf{Y}\ddot{\mathbf{p}} + \dot{\mathbf{p}}^{\mathrm{T}}\mathbf{X}\dot{\mathbf{p}}, \tag{7.23}$$

where

$$\mathbf{Y} = \begin{bmatrix} \dfrac{\partial q_{\mathrm{P}1}}{\partial p_1} & \dfrac{\partial q_{\mathrm{P}1}}{\partial p_2} & \cdots & \dfrac{\partial q_{\mathrm{P}1}}{\partial p_\lambda} \\ \dfrac{\partial q_{\mathrm{P}2}}{\partial p_1} & \dfrac{\partial p_{\mathrm{P}2}}{\partial p_2} & \cdots & \dfrac{\partial q_{\mathrm{P}2}}{\partial p_\lambda} \\ \vdots & \vdots & & \vdots \\ \dfrac{\partial q_{\mathrm{P}K}}{\partial p_1} & \dfrac{\partial q_{\mathrm{P}K}}{\partial p_2} & \cdots & \dfrac{\partial q_{\mathrm{P}K}}{\partial p_\lambda} \end{bmatrix} \tag{7.24}$$

is the $K \times \lambda$-dimensional Jacobian matrix and

$$\mathbf{X} = \begin{bmatrix} \dfrac{\partial \mathbf{Y}}{\partial p_1} \\ \dfrac{\partial \mathbf{Y}}{\partial p_2} \\ \vdots \\ \dfrac{\partial \mathbf{Y}}{\partial p_\lambda} \end{bmatrix}$$

is the $\lambda \times (K \times \lambda)$-dimensional Hessian matrix of the parallel mechanism. We have to mention that in parallel mechanisms the Jacobian and the Hessian matrices relate the velocities and accelerations of joint and external coordinates in the opposite direction in comparison to serial mechanisms, so that here they are both functions of the external coordinates.

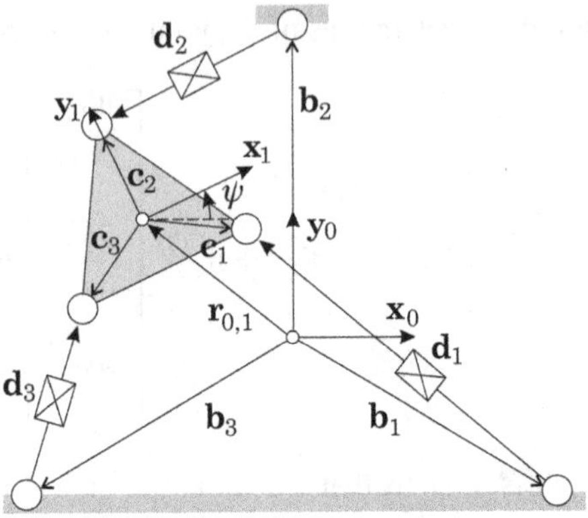

Fig. 7.10 Parallel mechanism 3RTR

As the rotation matrix $\mathbf{A}_{0,1}$ is a function of the angles ψ, θ and ϕ describing the orientation of the platform, its time derivative can be expressed in the following way

$$\dot{\mathbf{A}}_{0,1} = \frac{\partial \mathbf{A}_{0,1}}{\partial \psi}\dot{\psi} + \frac{\partial \mathbf{A}_{0,1}}{\partial \theta}\dot{\theta} + \frac{\partial \mathbf{A}_{0,1}}{\partial \phi}\dot{\phi}.$$

We obtain

$$\dot{\mathbf{d}}_k^{(0)} = \mathbf{v}_{0,1}^{(0)} + \left(\frac{\partial \mathbf{A}_{0,1}}{\partial \psi}\dot{\psi} + \frac{\partial \mathbf{A}_{0,1}}{\partial \theta}\dot{\theta} + \frac{\partial \mathbf{A}_{0,1}}{\partial \phi}\dot{\phi}\right)\mathbf{c}_k^{(1)}$$

or in matrix form

$$\dot{\mathbf{d}}_k^{(0)} = \left[\mathbf{I} \quad \frac{\partial \mathbf{A}_{0,1}}{\partial \psi}\mathbf{c}_k^{(1)} \quad \frac{\partial \mathbf{A}_{0,1}}{\partial \theta}\mathbf{c}_k^{(1)} \quad \frac{\partial \mathbf{A}_{0,1}}{\partial \phi}\mathbf{c}_k^{(1)}\right]\dot{\mathbf{p}}, \quad k = 1, 2, \ldots, K.$$

It follows

$$\dot{q}_k = \left[\nabla_k^{\mathrm{T}} \quad \nabla_k^{\mathrm{T}}\frac{\partial \mathbf{A}_{0,1}}{\partial \psi}\mathbf{c}_k^{(1)} \quad \nabla_k^{\mathrm{T}}\frac{\partial \mathbf{A}_{0,1}}{\partial \theta}\mathbf{c}_k^{(1)} \quad \nabla_k^{\mathrm{T}}\frac{\partial \mathbf{A}_{0,1}}{\partial \phi}\mathbf{c}_k^{(1)}\right]\dot{\mathbf{p}},$$

$k = 1, 2, \ldots, K$. In this way we obtain the Jacobian matrix of a parallel mechanism

$$\mathbf{Y} = \begin{bmatrix} \nabla_1^{\mathrm{T}} & \nabla_1^{\mathrm{T}}\frac{\partial \mathbf{A}_{0,1}}{\partial \psi}\mathbf{c}_1^{(1)} & \nabla_1^{\mathrm{T}}\frac{\partial \mathbf{A}_{0,1}}{\partial \theta}\mathbf{c}_1^{(1)} & \nabla_1^{\mathrm{T}}\frac{\partial \mathbf{A}_{0,1}}{\partial \phi}\mathbf{c}_1^{(1)} \\ \nabla_2^{\mathrm{T}} & \nabla_2^{\mathrm{T}}\frac{\partial \mathbf{A}_{0,1}}{\partial \psi}\mathbf{c}_2^{(1)} & \nabla_2^{\mathrm{T}}\frac{\partial \mathbf{A}_{0,1}}{\partial \theta}\mathbf{c}_2^{(1)} & \nabla_2^{\mathrm{T}}\frac{\partial \mathbf{A}_{0,1}}{\partial \phi}\mathbf{c}_2^{(1)} \\ \vdots & \vdots & \vdots & \vdots \\ \nabla_K^{\mathrm{T}} & \nabla_K^{\mathrm{T}}\frac{\partial \mathbf{A}_{0,1}}{\partial \psi}\mathbf{c}_K^{(1)} & \nabla_K^{\mathrm{T}}\frac{\partial \mathbf{A}_{0,1}}{\partial \theta}\mathbf{c}_K^{(1)} & \nabla_K^{\mathrm{T}}\frac{\partial \mathbf{A}_{0,1}}{\partial \phi}\mathbf{c}_K^{(1)} \end{bmatrix}. \tag{7.25}$$

As an example, consider the planar 3RTR mechanism shown in Fig. 7.10. Its parameters are presented in Table 7.4. The joint coordinates are the leg lengths q_1, q_2 and q_3. The first two external coordinates, r_x and r_y, represent the position of the platform. The third external coordinate is the angle ψ. The rotation matrix, describ-

Table 7.4 Parameters of the parallel mechanism 3RTR

k	1	2	3
$\mathbf{b}_k^{(0)}$	$b\sqrt{3}/2$	0	$-b\sqrt{3}/2$
	$-b/2$	b	$-b/2$
	0	0	0
$\mathbf{c}_k^{(1)}$	$c\sqrt{3}/2$	0	$-c\sqrt{3}/2$
	$-c/2$	c	$-c/2$
	0	0	0

ing the orientation of the platform, corresponds to a rotation about an axis perpendicular to the plane in which the mechanism moves. Therefore

$$\mathbf{A}_{0,1} = \begin{bmatrix} \cos\psi & -\sin\psi & 0 \\ \sin\psi & \cos\psi & 0 \\ 0 & 0 & 1 \end{bmatrix}.$$

The leg vectors are calculated as follows

$$\mathbf{d}_1^{(0)} = \begin{bmatrix} r_\mathrm{x} + \dfrac{c\sqrt{3}}{2}\cos\psi + \dfrac{c}{2}\sin\psi - \dfrac{b\sqrt{3}}{2} \\ r_\mathrm{y} + \dfrac{c\sqrt{3}}{2}\sin\psi - \dfrac{c}{2}\cos\psi + \dfrac{b}{2} \\ 0 \end{bmatrix},$$

$$\mathbf{d}_2^{(0)} = \begin{bmatrix} r_\mathrm{x} - c\sin\psi \\ r_\mathrm{y} + c\cos\psi - b \\ 0 \end{bmatrix},$$

$$\mathbf{d}_3^{(0)} = \begin{bmatrix} r_\mathrm{x} - \dfrac{c\sqrt{3}}{2}\cos\psi + \dfrac{c}{2}\sin\psi + \dfrac{b\sqrt{3}}{2} \\ r_\mathrm{y} - \dfrac{c\sqrt{3}}{2}\sin\psi - \dfrac{c}{2}\cos\psi + \dfrac{b}{2} \\ 0 \end{bmatrix}.$$

The joint coordinates are

$$q_1 = \sqrt{d_{1\mathrm{x}}^2 + d_{1\mathrm{y}}^2}, \qquad q_2 = \sqrt{d_{2\mathrm{x}}^2 + d_{2\mathrm{y}}^2}, \qquad q_3 = \sqrt{d_{3\mathrm{x}}^2 + d_{3\mathrm{y}}^2}.$$

In this way the inverse kinematics problem of the mechanism of Fig. 7.10 is solved. By assigning values to the external coordinates r_x, r_y, ψ, there is only one solution for the joint coordinates q_1, q_2, q_3.

To calculate the Jacobian matrix of mechanism, we first determine the following derivative

$$\frac{\partial \mathbf{A}_{0,1}}{\partial \psi} = \begin{bmatrix} -\sin\psi & -\cos\psi & 0 \\ \cos\psi & -\sin\psi & 0 \\ 0 & 0 & 0 \end{bmatrix},$$

and also the gradients

$$\nabla_1^{\mathrm{T}} = \frac{1}{q_1}(d_{1\mathrm{x}}, d_{1\mathrm{y}}, 0), \qquad \nabla_2^{\mathrm{T}} = \frac{1}{q_2}(d_{2\mathrm{x}}, d_{2\mathrm{y}}, 0), \qquad \nabla_3^{\mathrm{T}} = \frac{1}{q_3}(d_{3\mathrm{x}}, d_{3\mathrm{y}}, 0)$$

and substitute them into formula (7.25). The resulting Jacobian matrix **Y** has dimension 3 × 3, because we include only the columns which are related to the chosen external coordinates.

$$\mathbf{Y} = \begin{bmatrix} \frac{d_{1\mathrm{x}}}{q_1} & \frac{d_{1\mathrm{y}}}{q_1} & \frac{c}{2q_1}\left((-\sqrt{3}d_{1\mathrm{x}} + d_{1\mathrm{y}})\sin\psi + (d_{1\mathrm{x}} + \sqrt{3}d_{1\mathrm{y}})\cos\psi\right) \\ \frac{d_{2\mathrm{x}}}{q_2} & \frac{d_{2\mathrm{y}}}{q_2} & \frac{c}{2q_2}(-d_{2\mathrm{y}}\sin\psi - d_{2\mathrm{x}}\cos\psi) \\ \frac{d_{3\mathrm{x}}}{q_3} & \frac{d_{3\mathrm{y}}}{q_3} & \frac{c}{2q_3}\left((\sqrt{3}d_{3\mathrm{x}} + d_{3\mathrm{y}})\sin\psi + (d_{3\mathrm{x}} - \sqrt{3}d_{3\mathrm{y}})\cos\psi\right) \end{bmatrix}.$$

7.3.3 Direct Kinematics

The direct kinematics problem of parallel mechanisms requires determining the external coordinates **p**, which represent the position and orientation of the platform, from a given set of joint coordinates **q**, which are typically the leg lengths.

For example, consider solving the direct kinematics problem for the 2RR planar mechanism which was introduced in Fig. 7.2. The mechanism has one degree of freedom, which is controlled through the joint coordinate q. The end-point of the mechanism P can move along a curve whose shape depends on the segment lengths. The position of point P on the curve is related to the coordinate q. We can easily realize, that there exists a real solution of the direct kinematics problem only in the case when no segment of the mechanism is longer than the sum of the lengths of the other three segments. This can be written mathematically as follows

$$l_i < \frac{1}{2}\sum_{j=1}^{4} l_j, \quad i = 1, 2, \ldots, 4.$$

When the inequality is valid and when

$$l_3 + l_4 < l_1 + l_2,$$

a real solution is obtained within the interval

$$q_{\min} \leq q \leq q_{\max},$$

whose limits are

$$q_{\max} = \pi + \arccos\frac{l_1^2 + l_2^2 - (l_3 + l_4)^2}{2l_1 l_2},$$

$$q_{\min} = \pi - \arccos\frac{l_1^2 + l_2^2 - (l_3 + l_4)^2}{2l_1 l_2}.$$

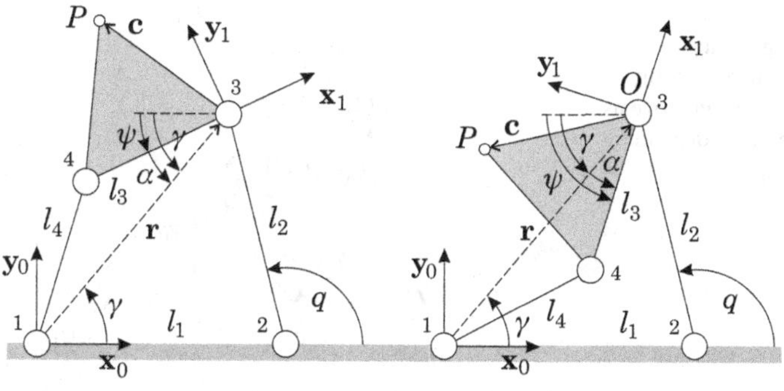

Fig. 7.11 Two poses of 2RR mechanism appertaining to the same angle q

In Fig. 7.11 the coordinate frame attached to the platform, is placed at the center of joint 3, while the fixed reference frame is at the center of joint 1. The position of the center of joint 3 with respect to the reference frame is obtained as a function of the joint variable q as follows

$$r_x = l_1 + l_2 \cos q,$$
$$r_y = l_2 \sin q.$$

If we want to know the position of the end-point on the platform P, we must know the orientation of the platform, which is defined by the angle ψ. First we find the angle γ

$$\gamma = \arctan_2 \frac{r_y}{r_x},$$

and then the positive value of the angle α

$$\alpha = \arccos \frac{r^2 + l_3^2 - l_1^2}{2rl_3},$$

where

$$r = \sqrt{r_x^2 + r_y^2}.$$

If $q_{min} < q < q_{max}$, two orientations are possible (Fig. 7.11)

$$\psi = \gamma \pm \alpha.$$

When $q = q_{min}$ or $q = q_{max}$ only one solution exists, as $\alpha = 0$. Now we can determine the two components of the position of point P with respect to the reference coordinate frame as follows

$$p_x = r_x + c_x \cos \psi - c_y \sin \psi,$$
$$p_y = r_y + c_x \sin \psi + c_y \cos \psi.$$

There exist two combinations of p_x and p_y, as we have two values of the orientation angle ψ. Here, c_x and c_y are the elements of the vector $\mathbf{c}$, expressed in the frame attached to the platform.

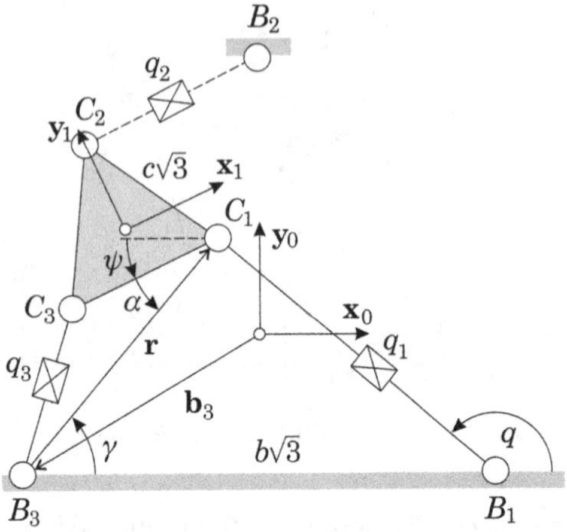

Fig. 7.12 When disconnecting the leg of a 3RTR mechanism at point C_2, the platform can move as a 2RR mechanism depending on the angle q

In this way the direct kinematics problem of a simple parallel 2RR mechanism is solved in a closed-form. We can see that the motion of the parallel 2RR mechanism is identical to the self motion of a redundant serial 3R mechanism, with the joints 1, 4 and 3 and segments l_4, l_3 and l_2, whose end-point is located at the center of joint 2. The various types of these motions are delineated in the same way as they are in the self motion of the 3R mechanism.

Consider determining the position and orientation of the platform of the 3RTR mechanism introduced in Fig. 7.12. The parameters of the mechanism are given in Table 7.4. The mechanism has three degrees of freedom. Suppose that we know the values of the joint coordinates represented by the leg lengths q_1, q_2 and q_3. We approach the problem by virtually disconnecting the upper leg at point C_2. The rest of the mechanism can move with one degree of freedom in the same way as a 2RR mechanism. The pose of the platform can be expressed as a function of the angle q. First we determine the vector

$$r_x = b\sqrt{3} + q_1 \cos q,$$

$$r_y = q_1 \sin q,$$

and afterwards the orientation angle ψ

$$\psi = \gamma \pm \alpha,$$

where

$$\gamma = \arctan_2 \frac{r_y}{r_x},$$

and

$$\alpha = \arccos \frac{r^2 + 3c^2 - q_3^2}{2rc\sqrt{3}}.$$

In this way we obtain the position of point C_2 with respect to the reference frame as function of the angle q

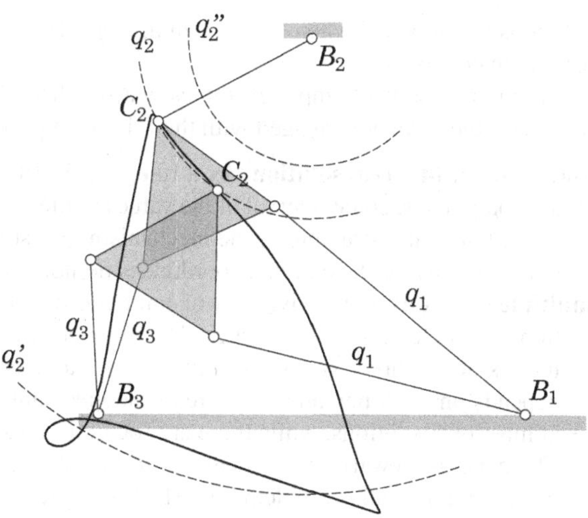

Fig. 7.13 Two poses of the 3RTR mechanism, appertaining to the same leg length q_2

$$s_x = -\frac{b\sqrt{3}}{2} + r_x - \frac{c\sqrt{3}}{2}\cos\psi - \frac{3c}{2}\sin\psi,$$
$$s_y = -\frac{b}{2} + r_y - \frac{c\sqrt{3}}{2}\sin\psi + \frac{3c}{2}\cos\psi.$$

There are two combinations of s_x, s_y since there are two possible values of ψ.

Knowing the possible positions of point C_2, we find the solution of the direct kinematics problem of the 3RTR mechanism in such a way that we determine those positions which lay on a circle with radius q_2 and center at B_2. We introduce the condition

$$s_x^2 + (s_y - b)^2 = q_2^2.$$

We see that the direct kinematics problem of the 3RTR mechanism has at most four solutions. The quadratic equation yields two solutions for both combinations of s_x and s_y. Depending on the mechanism's parameters and the values of the joint variables, the number of real solutions can be less than four or zero. The solutions can be found numerically, or by symbolic approaches which transform the above equations into a four order polynomial. In this manner we first obtain solutions for the orientation of the platform ψ, and afterwards we determine the corresponding position of the platform

$$p_x = s_x + c\sin\psi,$$
$$p_y = s_y - c\cos\psi.$$

Figure 7.13 shows the trajectory of point C_2 at given leg lengths q_1 and q_3, while sweeping through the values of the angle q in the interval between q_{min} and q_{max} and considering both solutions for the orientation angle ψ. The solutions of the direct kinematics problem can be obtained by finding the crossings of this trajectory and a circle with the radius q_2. In Fig. 7.13 two possible solutions are presented. Four solutions are obtained when the leg length is q_2'. There is no real solution when the leg length equals q_2''.

Through these two simple examples, we see that while solving the direct kinematics problem we are engaged with the following problems:

nonexistence of a real solution As a rule, the direct kinematics problem is solvable only inside some interval of the values of the joint variables, which is related to the kinematic structure of the mechanism. For some values of joint variables, real solutions for the external coordinates do not exist.

multiple solutions For a given set of joint coordinates, there exist multiple solutions for the external coordinates. The number of poses of the platform (i.e. solutions of the direct kinematics problem) for a given combination of leg lengths depends on the kinematic structure of the mechanism. The 2RR mechanism has at most two solutions, while there are four solutions with the 3RTR mechanism. The general Stewart-Gough platform in Fig. 7.4 has forty possible solutions of the direct kinematics problem [20, 31]. For a selected combination of leg lengths there exist forty different poses of the platform. Sometimes the platform can transition from one pose into another, but only with proper control of the leg lengths. Sometimes two poses of the platform cannot be transitioned between as the legs get entangled. In such cases, the platform transits from one pose into another only by dismantle the legs in the first pose and reassembling them in the new pose.

kinematic singularity When a parallel mechanism is in kinematic singularity, the platform gains one or more degrees of freedom, which cannot be controlled by the joint variables. Grübler's formula is no longer valid and the platform cannot resist a force or moment in at least one direction. Kinematic singularities are detrimental to parallel mechanisms.

nonexistence of closed-form solutions In general for a given set of joint coordinates, it is not possible to find an exact solution to the direct kinematics problem, even if a real solution exists. In such cases we resort to numerical techniques which may not necessarily converge and may not find all the solutions.

In parallel mechanisms, in general, it is not possible to express the external coordinates as explicit functions of the joint coordinates

$$\mathbf{p} = \mathbf{p}_{\mathrm{q}} \tag{7.26}$$

whereas with serial mechanisms it is quite straightforward. In parallel mechanisms the equations relating the external coordinates to the joint variables are nonlinear. Usually these are coupled trigonometric and quadratic equations which can be solved in closed-form only in special cases. There exist no rules as how to approach symbolic solutions. Some possibilities were described when solving the inverse kinematics problem of serial mechanisms, such as triangulation, or transforming the trigonometric equations into polynomial equations. Iterative numerical solutions can be successful. However they introduce other problems when the number of solutions and/or their approximate values are unknown. With gradient methods the Jacobian form of the direct kinematic equations are useful

$$\dot{\mathbf{p}} = \mathbf{Y}^{-1}\dot{\mathbf{q}}. \tag{7.27}$$

The method is valid when the Jacobian matrix $\mathbf{Y}$ is not singular. The advantageous property of this formulation is the linear relationship it provides between the velocities of the external and joint variables.

7.3.4 Kinematic Singularities

In serial mechanisms kinematic singularities have both positive and negative aspects. When a serial mechanism is in a kinematic singularity, the number of independent external coordinates is decreased, i.e. a degree of freedom is lost, which is negative for mobility. However, in the direction of decreased mobility the mechanism efficiently resists external forces or moments, with a high mechanical advantage, which is a desirable property in minimizing actuator forces or torques.

With parallel mechanisms, the kinematic singularity is strictly a negative property. Kinematic singularities offer no positive benefit to a parallel mechanism. A parallel mechanism in a kinematic singularity gains mobility. However, there is no advantage from this increased mobility as the gained degrees of freedom cannot be controlled by the joint variables. The mechanism cannot resist a force or moment associated with the gained degree of freedom [60].

With parallel mechanisms the Jacobian matrix transforms the vector of increments of external coordinates into the vector of increments of joint variables

$$d\mathbf{q} = \mathbf{Y}d\mathbf{p},$$

while the inverse equation

$$d\mathbf{p} = \mathbf{Y}^{-1}d\mathbf{q}$$

does not hold when the matrix $\mathbf{Y}$ is singular.

The matrix $\mathbf{Y}$ becomes singular due to one of the following reasons:

indefiniteness of orientational angles When the orientation of the platform is expressed with Euler orientational angles, the matrix $\mathbf{Y}$ is singular when $\theta = 0 \pm 2k\pi$, $k = 0, 1, \ldots$. When the orientation is given with the YPR angles, the matrix $\mathbf{Y}$ is singular when $\theta = \pm\pi/2 \pm 2k\pi$, $k = 0, 1, \ldots$.

kinematic singularities of legs The legs of a parallel mechanism are typically simple serial mechanisms. Analysis of the serial mechanism defines values of the leg's joint variables which correspond to singular leg configurations, where the leg has lost mobility. We try to avoid the singular leg configurations, as the parallel mechanism cannot be controlled when the legs do not have the anticipated mobility. Most parallel mechanisms are designed in such a way that the legs do not reach their singular configurations within the mechanism's workspace.

kinematic singularities of the parallel structure For some values of the external coordinates, the parallel mechanism can be kinematically singularity, even when the legs are not in singular configurations. In such a pose the platform gains one or more degrees of freedom which cannot be controlled by the joint variables. In practical applications such poses are forbidden, as the mechanism is unable to move out of them. The damage which can occur to the mechanism, or to the object being manipulated, can be catastrophic.

The indefiniteness of orientational angles is not a singularity of the mechanism but is entirely a manifestation of the mathematical representation of orientations. Such singularities can be overcome by mathematically reformulating orientation in terms of a more suitable set of three angles. True kinematic singularities of the mechanism are a consequence of the mechanical properties of the legs and above all, the consequence of the association between the legs which is due to their arrangement. These are purely mechanical problems which cannot be overcome by a mathematical reformulation.

The Jacobian matrix $\mathbf{Y}$, defined by formula (7.25), has two submatrices

$$\mathbf{Y} = \begin{bmatrix} \mathbf{Y}_{\mathrm{v}} & \mathbf{Y}_{\dot{\mathrm{g}}} \end{bmatrix}. \tag{7.28}$$

The matrix $\mathbf{Y}_{\mathrm{v}}$ is of dimension $K \times 3$ and encompasses the first three columns of the matrix $\mathbf{Y}$ and describes the translational velocity of the platform. The matrix $\mathbf{Y}_{\dot{\mathrm{g}}}$ also has dimension $K \times 3$ and includes the other three columns of the matrix $\mathbf{Y}$ and describes the first time derivatives of the orientational angles. When studying the singularities of a mechanism, instead of the matrix $\mathbf{Y}_{\dot{\mathrm{g}}}$ we rather make use of the matrix $\mathbf{Y}_{\omega}$, describing the angular velocity of the platform. In this way we avoid the indefiniteness and the singularities associated with angles describing orientation.

We proceed from the formula for the vectors of leg velocities (7.16) given in the following form

$$\dot{\mathbf{d}}_k^{(0)} = \mathbf{v}_{0,1}^{(0)} + \left(\boldsymbol{\omega}_{0,1}^{(0)} \otimes \mathbf{I}\right)\mathbf{A}_{0,1}\mathbf{c}_k^{(1)}, \quad k = 1, 2, \ldots, K.$$

This can be rewritten as follows

$$\dot{\mathbf{d}}_k^{(0)} = \mathbf{v}_{0,1}^{(0)} - \left(\mathbf{A}_{0,1}\mathbf{c}_k^{(1)} \otimes \mathbf{I}\right)\boldsymbol{\omega}_{0,1}^{(0)},$$

from which we have

$$\dot{\mathbf{d}}_k^{(0)} = \begin{bmatrix} \mathbf{I} & -\mathbf{A}_{0,1}\mathbf{c}_k^{(1)} \otimes \mathbf{I} \end{bmatrix} \begin{bmatrix} \mathbf{v}_{0,1}^{(0)} \\ \boldsymbol{\omega}_{0,1}^{(0)} \end{bmatrix}, \quad k = 1, 2, \ldots, K.$$

Now we can write

$$\mathbf{Y}_{\mathrm{v}} = \begin{bmatrix} \nabla_1^{\mathrm{T}} \\ \nabla_2^{\mathrm{T}} \\ \vdots \\ \nabla_K^{\mathrm{T}} \end{bmatrix} \tag{7.29}$$

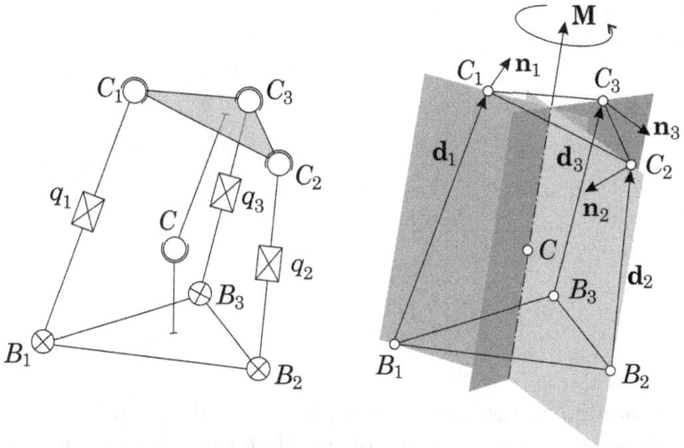

Fig. 7.14 Kinematic singularity of the S-3UTS mechanism

and

$$\mathbf{Y}_\omega = -\begin{bmatrix} \nabla_1^{\mathrm{T}}(\mathbf{A}_{0,1}\mathbf{c}_1^{(1)} \otimes \mathbf{I}) \\ \nabla_2^{\mathrm{T}}(\mathbf{A}_{0,1}\mathbf{c}_2^{(1)} \otimes \mathbf{I}) \\ \vdots \\ \nabla_K^{\mathrm{T}}(\mathbf{A}_{0,1}\mathbf{c}_K^{(1)} \otimes \mathbf{I}) \end{bmatrix}. \tag{7.30}$$

The kinematic singularities of a parallel mechanism are the singularities of the Jacobian matrix

$$\mathbf{Y} = \begin{bmatrix} \mathbf{Y}_{\mathrm{v}} & \mathbf{Y}_\omega \end{bmatrix}. \tag{7.31}$$

The S-3UTS mechanism, with three degrees of freedom, was introduced in Fig. 7.8. Figure 7.14 is used to describe the kinematic singularities of the mechanism. In general, the platform can be rotated around point C, which is at the center of the spherical joint in the middle leg. Its rotation about C is controlled by the translations within the external legs, whose lengths are q_1, q_2 and q_3. There exists at most eight real solutions of the direct kinematics problem [33].

As shown in [91], the mechanism is in a kinematic singular pose when one of the following conditions is fulfilled

$$(\mathbf{d}_1 \times \mathbf{d}_2) \cdot \mathbf{d}_3 = 0$$

or

$$(\mathbf{n}_1 \times \mathbf{n}_2) \cdot \mathbf{n}_3 = 0.$$

Here, $\mathbf{n}_1$ is normal to the plane CB_1C_1, $\mathbf{n}_2$ is normal to the plane CB_2C_2 and $\mathbf{n}_3$ is normal to the plane CB_3C_3. Both conditions are defined by a mixed (dot and cross) product of the three vectors, which is equal to zero when the three vectors are coplanar.

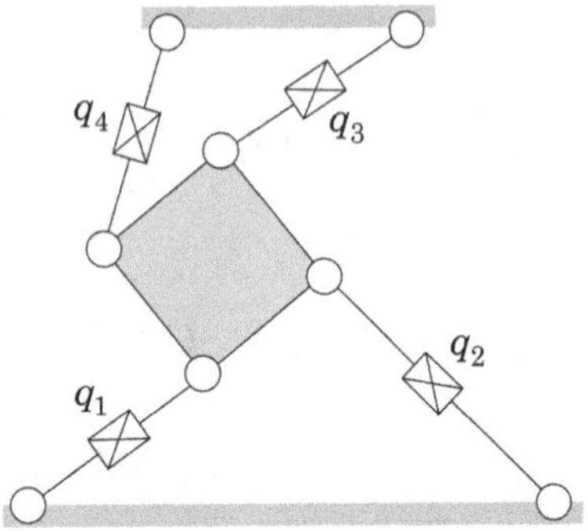

Fig. 7.15 Overconstrained parallel 4RTR mechanism

The first condition corresponds to when the external legs of the mechanism are coplanar. This occurs when the plane of the platform is coincident with the base plane, which is only possible when the length of the middle leg equals zero. The condition is also fulfilled when all three external legs are parallel. Such poses of the platform occur when the platform is congruent to the base and the corresponding corners of the platform and base are connected by equal length legs, so that the platform is also parallel to the base. We can easily avoid these kinematic singularities by selecting appropriate parameters for the mechanism.

The second condition requires that the normal vectors $\mathbf{n}_1$, $\mathbf{n}_2$ and $\mathbf{n}_3$ are coplanar. The condition is satisfied when all three planes have a common line of intersection. In this case, the mechanism cannot resist an external moment $\mathbf{M}$ about this common line. It is difficult to avoid such a singularity as it can occur in many different poses of the mechanism.

7.3.5 Overconstrained and Redundant Parallel Mechanisms

We know that we can add an arbitrary number of legs with $\nu_k = \lambda$ degrees of freedom to a parallel mechanism, without taking degrees of freedom away from the platform. Adding new controlled joint coordinates to a parallel mechanism causes the mechanism to be overconstrained. The is not the case in serial mechanisms.

The mechanism from Fig. 7.15 has $K = 4$ serial legs and twelve one degree of freedom joints. In accordance with formula (7.8) it has

$$F = 3(1 - 4) + 12 = 3$$

degrees of freedom. Therefore, it cannot be independently controlled by the variables q_1, q_2, q_3, q_4. The values of these variables are related to each other through a system of kinematic equations $\mathbf{q} = \mathbf{q}_\mathrm{p}$. The system of equations is overconstrained. It includes four equations with four joint variables and three unknown external coordinates describing the position and orientation of the platform. At first glance it appears that the additional coordinate, which is controlled, is not necessary. Nevertheless, by introducing the additional leg we can remove or even completely eliminate kinematic singularities from the mechanism's workspace [76].

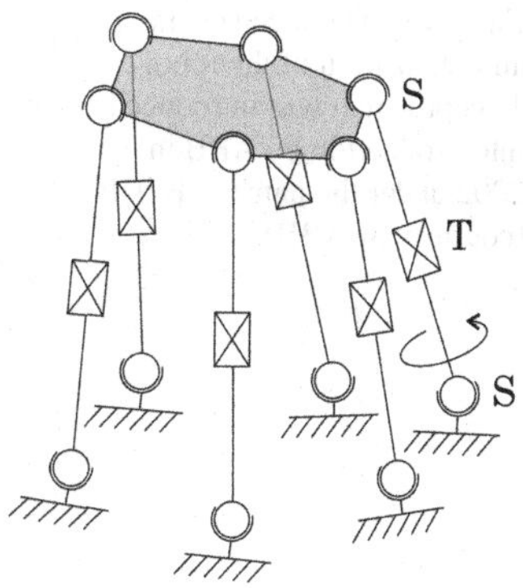

Fig. 7.16 Steward-Gough platform where the mechanism is kinematically redundant

Consider now the mechanism from Fig. 7.16, the general Stewart-Gough 6STS platform, where the legs have spherical instead of universal joints. Calculate the number of degrees of freedom of the mechanism. As the number of serial legs is $K = 6$ and there are 42 degrees of freedom in the joints, we have

$$F = 6(1 - 6) + 42 = 12.$$

Therefore the mechanism has $F = 12$, while the platform can only have $f = 6$. The position and orientation of the platform can be controlled by six variables, e.g. translations in the legs. It is not difficult to see where the other degrees of freedom are hidden. Each leg is kinematically redundant and has seven degrees of freedom, while only six are sufficient for positioning and orienting of the platform. The self motion of such legs is exhibited as rotation around the axis running along the leg as shown in Fig. 7.16. This rotation has no influence on the displacement of the platform. We can say that a parallel mechanism, where

$$F > f,$$

is kinematically redundant.

7.4 Some Examples of Parallel Mechanisms

Among industrial robots which consist of a parallel mechanism, in addition to the Delta robot which we have already seen, there is the well-known Tricept robot which was patented in 1988 by K.-E. Neumann (patent no. 4,732,525). The positional part of the robot is the parallel UT-3UTS mechanism, shown in Fig. 7.17. The middle leg of this parallel mechanism has UT structure and has three degrees of freedom. The three external legs have the UTS structure with six degrees of freedom. The UT-3UTS mechanism has $K = 4$ serial legs and 21 degrees of freedom in the joints, therefore it has (7.8)

$$F = 6(1 - 4) + 21 = 3$$

degrees of freedom, the same as the platform. The displacement of the platform is dictated by the middle leg. The platform can be rotated with regard to the base

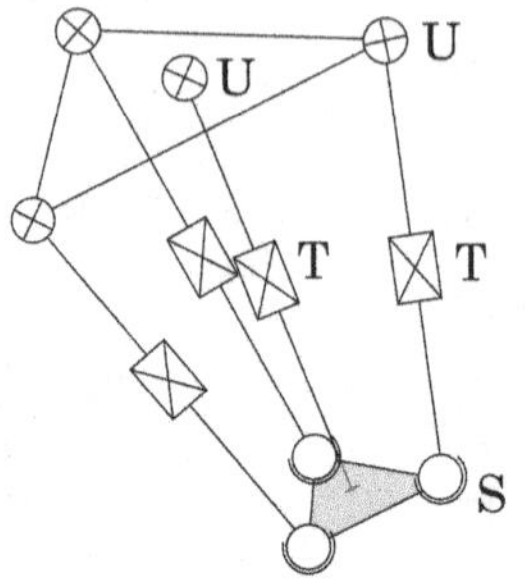

Fig. 7.17 The mechanism and realization of the robot Tricept which was introduced into production by ABB in 2002 under the name IRB 940 (courtesy of ABB)

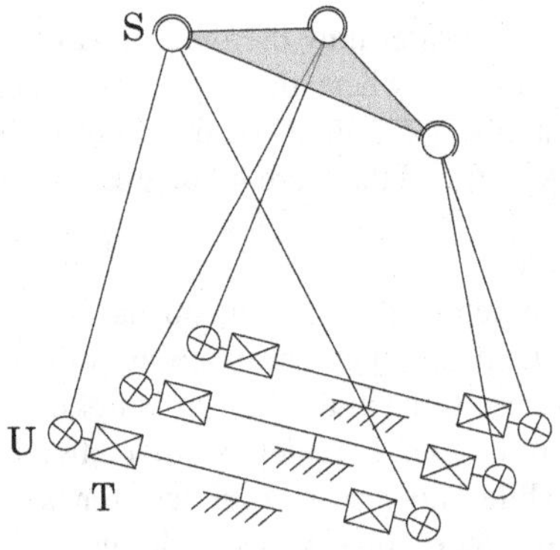

Fig. 7.18 Parallel Hexaglide mechanism

around the U joint and can be displaced from the base in the direction of the translation T. The mechanism is controlled by three translations in the external legs, while both joints of the middle leg are passive.

Hexaglide is another parallel mechanism. It was developed as a manufacturing machine aimed for milling and deburring purposes [28]. The kinematic structure of the mechanism, shown in Fig. 7.18, is 6TUS. It resembles the Stewart-Gough platform, only that the order of the joints is altered. The end-points of the legs slide over the base, while the leg lengths do not change. With the original mechanism the pairs of spherical joints are placed at the same point while a pair of translations run along the same line. In general the spherical joints are placed arbitrarily over the platform, the same as the translations along the base. The number of serial legs is $K = 6$, the number of degrees of freedom in the joints is 36, so that the number of degrees of freedom of the mechanism equals

$$F = 6(1 - 6) + 36 = 6.$$

A special category of parallel mechanisms are those whose main property is that their platform can only move in translation. For example, the mechanism 3UTU from Fig. 7.19 has $K = 3$ serial legs and 15 degrees of freedom in the joints and

$$F = 6(1 - 3) + 15 = 3$$

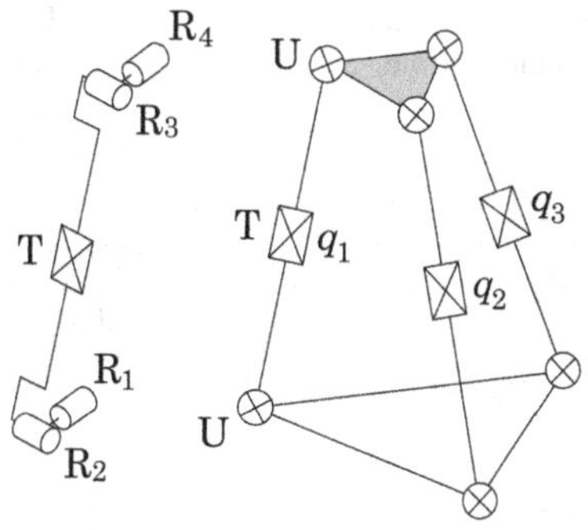

Fig. 7.19 Parallel 3UTU mechanism

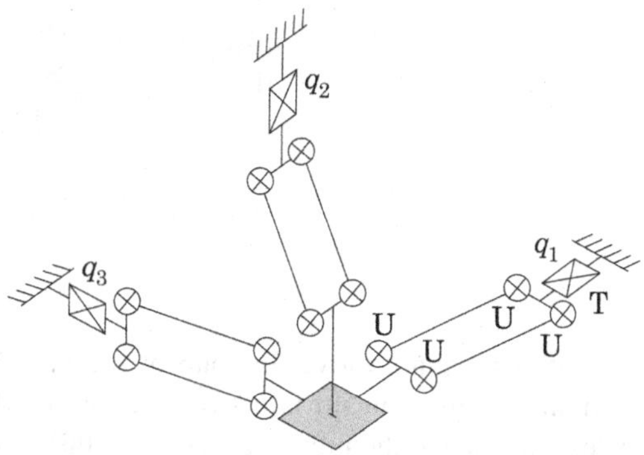

Fig. 7.20 Parallel mechanism Orthoglide

degrees of freedom. It is controlled through the translations q_1, q_2, q_3. The mechanism has a universal joint U with the rotations R_1 and R_2 at the base of the mechanism and another universal joint with the rotations R_3 and R_4 at the platform. Suppose that all rotations R_1 are coplanar as well as the rotations R_4. The platform of such a 3UTU mechanism, with the following requirements

$$R_2 \perp R_1,$$
$$R_3 \perp R_4,$$
$$R_2 \parallel R_3,$$

can only be displaced translationally [83]. One should be aware that such a mechanism is very sensitive to tolerancing and dimensional errors. If the requirements presented are not precisely fulfilled, the movement of the platform is highly distorted [18].

The mechanism called Orthoglide is also characterized by only translational displacement of its platform [88]. The number of degrees of freedom of a general realization of such mechanism, shown in Fig. 7.20, can be calculated by the Grübler's formula (2.2), as the mechanism has no serial legs. We have

$$F = 6(10 - 15) + 27 = -3,$$

as the number of mobile segments is $N = 10$ and the number of joints is $n = 15$. Grübler's formula indicates that this mechanism is in general overconstrained and

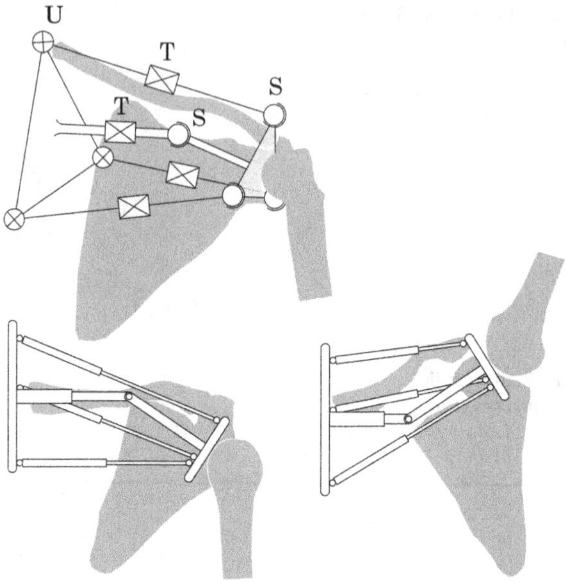

Fig. 7.21 Humanoid shoulder girdle

the platform cannot move. In spite of the fact that each leg has three degrees of freedom, the mechanism is a structure, as the displacement of one leg is not in correspondence to the movements of the other two legs.

If we desire that the Orthoglide mechanism is actually movable, we must add the requirement that the axes of the neighboring universal joints of the legs are parallel. Under these conditions, the legs do not influence the orientation of the platform an the platform can therefore be translationally displaced in all three directions. The mechanism is controlled by the variables q_1, q_2, q_3, whose pairs of axes are perpendicular.

It is no surprise that the models of parallel mechanisms can be efficiently used when studying biomechanical properties of human joints. The muscles and ligaments, which stretch over the joints, e.g. the knee joint [66], create various parallel kinematic structures. The shoulder girdle, represented by the collar bone and shoulder blade, can, as a first approximation, be modeled as having two degrees of freedom (Fig. 2.9), although in reality its movement is much more complicated. Figure 7.21 shows the kinematic model of the shoulder girdle represented by the TS-3UTS parallel mechanism [55]. The mechanism has $K = 4$ serial legs and 22 degrees of freedom in the joints, therefore it has

$$F = 6(1 - 4) + 22 = 4$$

degrees of freedom. The platform of the mechanism, which can be arbitrarily oriented or displaced from the base by controlling four translational coordinates, replicates the movement of the shoulder blade. In Fig. 7.21 two outermost poses of the mechanism are displayed in the frontal plane during abduction and adduction of the arm. The movements of the shoulder girdle are related to the movements of the

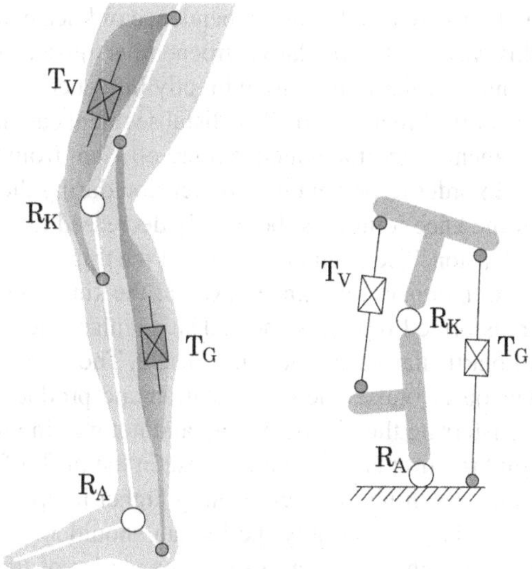

Fig. 7.22 Planar kinematic model of human leg

glenohumeral joint which is spherical joint connection the platform (shoulder blade) and the humerus (upper arm). Such coordinated movement is called the shoulder rhythm and it has been incorporated in mechanical models of the human shoulder [52].

Figure 7.22 shows a planar parallel mechanism which is a simple representation of the activity of two muscles in the human leg and their important role in vertical jumping [4]. The first is the monoarticular four-head knee extensor muscle *vastus*, while the second is the biarticular two-head ankle joint plantar flexor muscle *gastrocnemius*. In the kinematic model the knee joint movement is described by the rotation R_K, while the ankle joint movement is described by the rotation R_A. The translation T_V describes the displacement of the m. vastus and the translation T_G represents movement of the m. gastrocnemius. The attachments of the muscle to the bones were modeled by a proximal and distal rotation, which in the figure are displayed in the darker shade. Such a kinematic model of a muscle is very superficial. We must take into account that the muscle produces force only during contraction. The mechanism, as shown in the right side of the figure, includes 6 rotations, 2 translations and 6 mobile segments. The foot was considered as a fixed base. In this way the mechanism has

$$F = 3(6 - 8) + 8 = 2$$

degrees of freedom, whose functions are movements in the knee and ankle joint.

The characteristic property of the described mechanism is the biarticular muscle gastrocnemius. Biarticular muscles, in contrast to monoarticular muscles, span across two joints. The gastrocnemius muscle stretches over the knee and the ankle joint and acts simultaneously with the knee flexor and ankle extensor. During fast movement, the monoarticular vastus muscle produces knee extension, while the isometrically activated biarticular gastrocnemius muscle provides ankle plantar flexion and thus creates the interdependence of knee extension and ankle plantar flexion. In this way, the biarticular gastrocnemius muscle enables the larger proximally placed monoarticular muscle to indirectly influence the movement of the joint which is not under its direct control. The distal segment can therefore be lighter and has a smaller moment of inertia, which is more efficient from the dynamics point of view.

In order to prevent hyperextension during the vertical jump, the angular velocity of the knee joint must be briskly decreased to zero near the point of complete knee extension. The angular velocity of the thigh and shank is not decreased by the activity of monoarticular knee flexor, as the kinetic energy of both leg segments would be transformed in useless heat. The angular velocity is decreased through the activity of biarticular gastrocnemius muscle. The isometric activation of this muscle simultaneously brakes the knee rotation and produces accelerated ankle plantar flexion, transferring the kinetic energy and not wasting it. The result of such a violent ankle joint rotation is a crucial increase in the push-off velocity associated with a vertical jump, resulting in a significantly higher jump.

In this section only the human shoulder girdle and a human's lower extremity during high jump were used as examples of the parallel mechanisms found in nature. Many other parallel mechanisms in biological systems exist. Kinematic and dynamic models of these biological systems not only help us understand movements in humans and other organisms, but they also can be efficiently used when designing the mechanisms of advanced robots [5].

Chapter 8
Robot Contact

Abstract When a rigid body is in contact with another rigid body, which is fixed in the frame of reference, its motion becomes constrained (i.e., restricted) and the number of degrees of freedom associated with its motion is reduced. This restriction of the motion depends on the type of contact. When the body is in contact with several other bodies, its motion is even more restricted, and it can become immoveable. In this chapter, our goal is to present the basic mathematical description of the contacts between rigid bodies, to study the effects of several types of contacts on the constraints of the body's motion, and to determine the requirements for the modeling of a single robot contact. As the mathematical tool appropriate for a description of the contacts we use the screw representation, which relates the velocities and forces, occurring before and during the contact, with the contact normal.

By the word contact we mean a connected set of points on the surface of two bodies, where the bodies are touching each other. When dealing with two or more areas of contact, we speak about distinct contacts. A point contact is considered to be the limiting case of an infinitesimally small area contact. First, we shall study only those possible motions which are permitted in the cases of a single contact between two bodies. Constraints in the relative motion of two bodies in contact and the reduction of degrees of freedom depend on the following contact properties:

- shape of the contact area,
- relative positions of the contact areas,
- relative orientation of each contact area, and
- the effect of friction in the contact area.

At the location of a contact between two bodies, the relative motion of the two bodies is constrained to those directions where the bodies would not be deformed as a consequence of contact. The bodies are displaced in one of the following ways:

- the contact points are maintained,
- the contact points morph into a new set of points during the motion,
- the contact during the motion is broken.

J. Lenarčič et al., *Robot Mechanisms*,
Intelligent Systems, Control and Automation: Science and Engineering 60,
DOI 10.1007/978-94-007-4522-3_8, © Springer Science+Business Media Dordrecht 2013

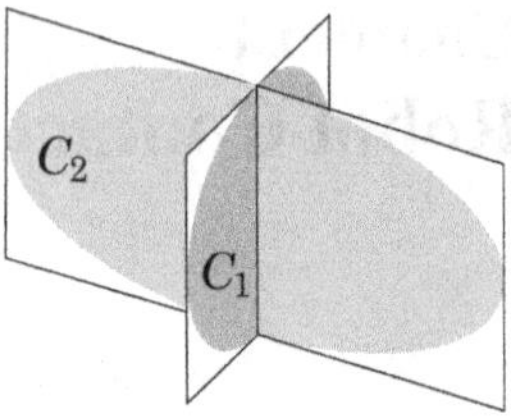

Fig. 8.1 Cross-sections and curvature radii of the contact surface

When there is sufficiently high friction between the contact areas, the motion is also constrained in the tangential direction. The contact of two bodies is further determined by the geometric properties of the contact surface, which can be described by the following parameters:

- position of the contact surface,
- orientation of the contact surface, and
- curvature of the contact surface.

In order to describe a contact, we must know the coordinates of the points belonging to the contact area, the direction of the normal to the contact surface, maximum and minimum radius of curvature of the contact surface, and the orientation of the cross-sections comprising both radii [57]. The meaning of the maximum and minimum radius of curvature and corresponding cross-sections C_1 and C_2 are shown in Fig. 8.1 for the example of an ellipsoid. With convex contact surfaces both radii are considered to be positive.

A plane contact occurs when both curvature radii are infinite. A cylinder has one positive and one infinite radius of curvature. A point contact is treated as an infinitesimally small sphere with both radii equal to zero. The contact normal for a particular surface point is normal to the plane tangent to the surface at that point and is pointing outwards from the contact area. When the radius of curvature is zero, the tangential plane cannot be uniquely defined. The surface normals of two bodies in contact have opposite directions. One of these normals is selected as the contact normal. When one of the normals is not defined, the other normal is considered as the contact normal. The direction of the contact normal will be used in the mathematical description of the contact. We shall study three types of contacts:

- point contact,
- line contact, and
- plane contact.

A contact surface represents the point contact when both curvature radii are very small or equal zero. The line contact is characterized with the first radius very small or zero, while the second radius is infinite. When both radii of curvature are infinite, we are dealing with a planar contact. With these three types of contact we can analyze the mobility of two bodies in contact. When studying single contacts between two bodies, we have nine different pairs in the case of point, line, and plane contacts:

point–point,
point–line,

point–plane,
line–point,
line–line,
line–plane,
plane–point,
plane–line,
plane–plane.

Three of the enumerated contact pairs are only transient. These are the contacts: point–point, point–line, and line–point. In these cases the contact constraints are unstable, hence these contacts will not be considered further in the text. It is not difficult to realize that from the point of view of the relative motion between two bodies, there is no difference between the contact pairs point–plane and plane–point. The same is true for the contacts line–plane and plane–line. Finally, we notice that the contact of two lines represents the same constraint as a point on a plane or a plane on a point. This is not the case for collinear lines, which represent a transient constraint which also will not be studied further. We therefore conclude that we are dealing with only three stable pairs of contacts of two bodies. These are point contact (point on plane, line on non-parallel line), line contact (line on plane), and plane contact (plane on plane).

8.1 Screw Systems

In this section a mathematical tool appropriate for description of contacts will be presented. We will introduce the screw representation [8], which relates the velocities and forces, occurring before and during the contact, with the contact normal. The relative velocity of one body approaching another body can, at any instant, be represented by a translational velocity along a unique line and angular velocity about that line (Fig. 8.2). The generalized velocity vector which encompasses both this translational and angular velocity of the body, will be called a twist.

Similarly, all the forces and moments acting on a body can be unified into a force acting along a unique line and a moment acting about that line (Fig. 8.2). The generalized force vector which encompasses both this force and moment, will be called a wrench.

Before mathematically describing twists and wrench systems, we will introduce Plücker coordinates. The Plücker coordinates of a line $\mathbf{l}$ are represented by six values. The first three values are the direction cosines of the line and will be denoted as L, M, and N. The remaining three values represent the moments of the line about the origin of the reference frame and will be denoted as P, Q, and R. The line $\mathbf{l}$ is defined by the following expression

$$\mathbf{l} = (L, M, N, P, Q, R)^{\mathrm{T}} = \begin{bmatrix} \mathbf{e} \\ \mathbf{r} \times \mathbf{e} \end{bmatrix}. \tag{8.1}$$

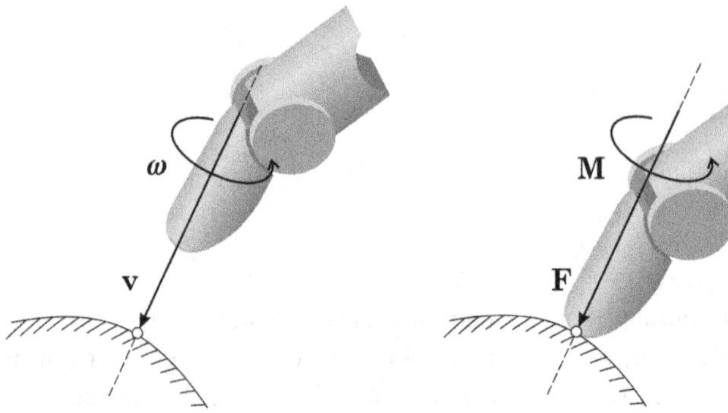

Fig. 8.2 Twist and wrench systems

Fig. 8.3 Plücker coordinates

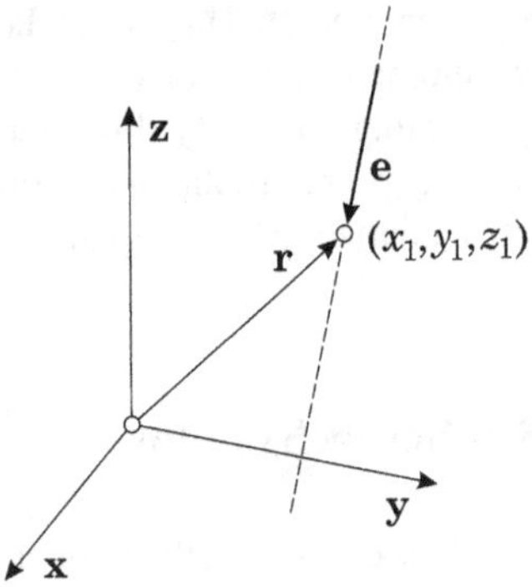

The line moments are defined as the cross product of a vector **r**, running from the origin of the coordinate frame to an arbitrary point on the line, and the unit vector **e** defined along that line (Fig. 8.3). By selecting an arbitrary point on the line with the coordinates (x_1, y_1, z_1), the line moments are calculated as follows

$$\begin{aligned} P &= y_1 N - z_1 M, \\ Q &= z_1 L - x_1 N, \\ R &= x_1 M - y_1 L. \end{aligned} \tag{8.2}$$

Four values are sufficient to define a line in the space. The definition of a line with direction cosines and line moments is therefore redundant. One of the direction cosines can be determined by knowing the other two

$$L^2 + M^2 + N^2 = 1. \tag{8.3}$$

It is also evident that the dot product of the direction cosines and line moments equals zero

$$LP + MQ + NR = \mathbf{e}^{\mathrm{T}}(\mathbf{r} \times \mathbf{e}) = 0. \tag{8.4}$$

As the magnitude of vector **e** has no meaning, all six values in expression (8.1) can be multiplied by a scalar without changing the definition of a line. As an example, let us describe with Plücker coordinates the line which runs through the points $(1, 1, 0)$ and $(1, 2, \sqrt{3})$. The line is parallel to the plane **y**, **z**, the angle of the line with respect to **x** axis is therefore 90°. The angle between the line and the **y** axis is 60° and 30° with respect to the **z** axis. We have the following direction cosines

$$\begin{aligned} L &= \cos 90^\circ = 0, \\ M &= \cos 60^\circ = \frac{1}{2}, \\ N &= \cos 30^\circ = \frac{\sqrt{3}}{2}. \end{aligned} \tag{8.5}$$

The line moments are obtained by inserting the coordinates of the first point into (8.2)

$$\begin{aligned} P &= \frac{\sqrt{3}}{2}, \\ Q &= -\frac{\sqrt{3}}{2}, \\ R &= \frac{1}{2}. \end{aligned} \tag{8.6}$$

The line running through points $(1, 1, 0)$ and $(1, 2, \sqrt{3})$ is expressed with Plücker coordinates as follows:

$$\mathbf{l} = \left(0, \frac{1}{2}, \frac{\sqrt{3}}{2}, \frac{\sqrt{3}}{2}, -\frac{\sqrt{3}}{2}, \frac{1}{2}\right)^{\mathrm{T}}. \tag{8.7}$$

We shall now study the wrench system [29]. With any set of forces and moments acting on a rigid body, there exists a unique line which will be called the wrench axis. The set of forces is equivalent to a single force acting along the wrench axis, while the set of moments is equivalent to the torque acting around this axis. The equivalent force and moment is called a wrench, **w**, and expressed as the six element vector

$$\mathbf{w} = (W_1, W_2, W_3, W_4, W_5, W_6)^{\mathrm{T}}. \tag{8.8}$$

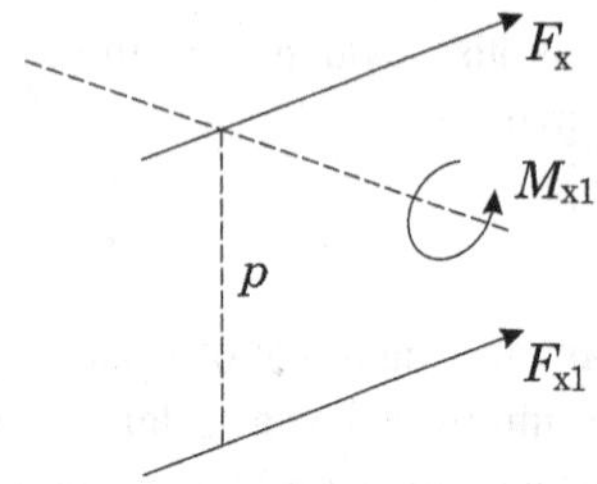

Fig. 8.4 Parallel displacement of force F_x

In (8.8) W_1, W_2, and W_3 represent the components of the force **F**, while W_4, W_5, and W_6 are the components of the moment **M**. The wrench coordinates are therefore

$$\begin{aligned} L &= W_1, \\ M &= W_2, \\ N &= W_3, \\ P &= W_4 - pW_1, \\ Q &= W_5 - pW_2, \\ R &= W_6 - pW_3. \end{aligned} \tag{8.9}$$

The above coordinates L, M, N, P, Q, R must be normalized by the expression

$$\sqrt{W_1^2 + W_2^2 + W_3^2},$$

so that the first three coordinates have the meaning of direction cosines and the other three coordinates the meaning of the line moments.

Let us examine the meaning of the line moment P from (8.9). With regard to the definition of the wrench, we can write

$$\mathbf{w} = (W_1, W_2, W_3, W_4, W_5, W_6)^T = (F_x, F_y, F_z, M_x, M_y, M_z)^T. \tag{8.10}$$

The fourth wrench component can therefore be written in the following form

$$P = M_x - pF_x. \tag{8.11}$$

A force component F_x should be displaced in such a way that the force **F** will be aligned with the wrench axis (Fig. 8.4). When displacing the force component F_x, its direction is preserved. The parallel force component is displaced by a distance p. The displaced force exerts a torque pF_{x1}, which was not present before and must be annihilated by the moment M_{x1}

$$M_{x1} = pF_{x1}. \tag{8.12}$$

We shall assume without proof that the above equation, written for one of the force components, holds also in general vector form

$$\mathbf{M} = p\mathbf{F}. \tag{8.13}$$

By multiplying the above equation with the force $\mathbf{F}$, the pitch of the screw can be expressed as

$$p = \frac{\mathbf{F}^T\mathbf{M}}{\mathbf{F}^T\mathbf{F}}. \tag{8.14}$$

The pitch, expressed in terms of the screw coordinates, has the following form

$$p = \frac{W_1W_4 + W_2W_5 + W_3W_6}{W_1^2 + W_2^2 + W_3^2}. \tag{8.15}$$

The pitch of the screw is equal to the ratio of the moment applied about a selected point on the wrench axis and the magnitude of the force along the axis. A zero pitch screw represents pure force (wrench without a moment), while infinite pitch screw represents a pure moment (wrench without a force). The magnitude of the wrench is given by

$$|\mathbf{F}| = \sqrt{W_1^2 + W_2^2 + W_3^2}. \tag{8.16}$$

We shall now study the twist system. For any motion of a rigid body in space, there exists a line about which the body rotates and along which it translates. This line is called the twist axis. The twist, $\mathbf{t}$, is also described by a six element vector

$$\mathbf{t} = (T_1, T_2, T_3, T_4, T_5, T_6)^T. \tag{8.17}$$

The elements T_1, T_2, and T_3 are the components of the body's angular velocity $\boldsymbol{\omega}$, while T_4, T_5, and T_6 are the components of the translational velocity $\mathbf{v}$ of the point in the body instantaneously coincident with the origin of the reference frame. The following are the Plücker coordinates of the twist axis

$$\begin{aligned} L &= T_1, \\ M &= T_2, \\ N &= T_3, \\ P &= T_4 - pT_1, \\ Q &= T_5 - pT_2, \\ R &= T_6 - pT_3. \end{aligned} \tag{8.18}$$

The coordinates in (8.18) must be normalized by

$$\sqrt{T_1^2 + T_2^2 + T_3^2},$$

so that the first three coordinates represent the direction cosines and the other three coordinates the line moments. The pitch of the twist is defined as

$$p = \frac{\boldsymbol{\omega}^T\mathbf{v}}{\boldsymbol{\omega}^T\boldsymbol{\omega}} = \frac{T_1T_4 + T_2T_5 + T_3T_6}{T_1^2 + T_2^2 + T_3^2}. \tag{8.19}$$

The pitch of the twist is the ratio of the magnitude of the velocity of a point on the body which lies on the twist axis and the angular velocity about the axis. A zero pitch twist corresponds to pure rotation, while an infinite pitch corresponds to pure translation. The magnitude of the twist is

$$|\omega| = \sqrt{T_1^2 + T_2^2 + T_3^2}. \tag{8.20}$$

When the magnitude of a twist or wrench equals one, we refer to it as a unit twist or a unit wrench. The unit twist or unit wrench only define the axis and pitch of either the twist or wrench. The magnitude of a twist has units of rotation per unit time, while the magnitude of a wrench has units of force. The pitch of either a twist or wrench has units of length.

The constraints imposed by a particular contact, can be described in two equivalent ways. In one description, as a consequence of a particular contact, the motion of one body relative to the other can be described by a system of twists. Alternatively, as a consequence of the contact, the forces and moments of interaction between the bodies can be described as a system of wrenches. The two systems are said to be reciprocal, meaning, that the systems of wrenches can do no work on the body with that system of twists. This is a consequence of the fundamental principle that constraint forces and moments can do no work. In the next section, the point, line, and plane contacts will be described by the use of unit twist and unit wrench systems.

8.2 Basic Contacts

The contacts between two bodies can be divided into five categories, depending on the relative number of degrees of freedom allowed by the contact. A contact with one degree of freedom means that a single parameter is sufficient to describe the relative motion of the contacting bodies. Similarly, two parameters are required to describe the possible relative motions of two bodies in a two degree of freedom contact. This continues to where we need five parameters to describe the possible relative motions of two bodies in a five degree of freedom contact.

Let us study the influence of various types of contacts on the relative mobility between two bodies. The relative number of degrees of freedom, permitted by a particular type of contact also depends on whether or not frictional forces are considered. When friction is ignored, the point contact (Fig. 8.5) has five degrees of freedom. These are three rotations, which are in figure denoted by $\mathbf{t}_1$, $\mathbf{t}_2$, and $\mathbf{t}_3$, and two translations $\mathbf{t}_4$ and $\mathbf{t}_5$.

To describe the frictionless point contact, the unit twist and unit wrench systems are used. The corresponding twist system is of dimension five and is presented by

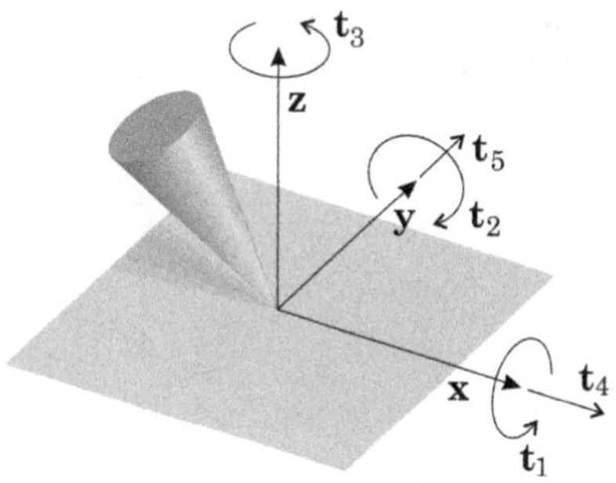

Fig. 8.5 Frictionless point contact in twist system

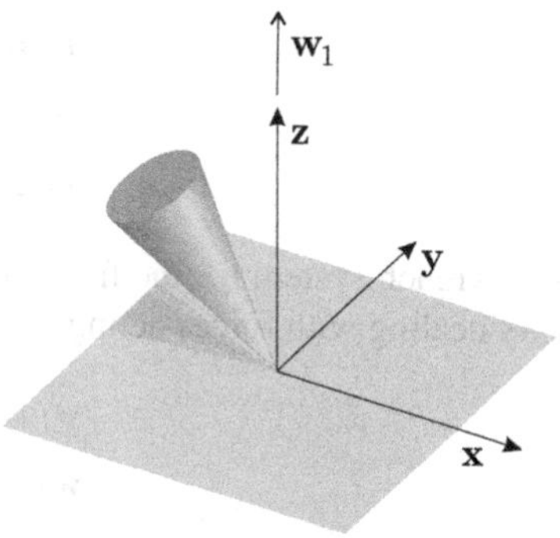

Fig. 8.6 Frictionless point contact in wrench system

the following five basic unit vectors

$$\begin{aligned}
\mathbf{t}_1 &= (1, 0, 0, 0, 0, 0)^T, \\
\mathbf{t}_2 &= (0, 1, 0, 0, 0, 0)^T, \\
\mathbf{t}_3 &= (0, 0, 1, 0, 0, 0)^T, \\
\mathbf{t}_4 &= (0, 0, 0, 1, 0, 0)^T, \\
\mathbf{t}_5 &= (0, 0, 0, 0, 1, 0)^T.
\end{aligned} \tag{8.21}$$

The wrench is reciprocal to the twist. The frictionless point contact is described by a single unit wrench. The direction of the wrench is reciprocal with respect to the permitted twist directions (Fig. 8.6) and is represented by a vertical force. The frictionless point contact is described by a single line in the wrench system

$$\mathbf{w}_1 = (0, 0, 1, 0, 0, 0)^T. \tag{8.22}$$

With friction in a point contact, sliding of a point across the plane is not possible. The point contact with friction has three degrees of freedom, all of them are

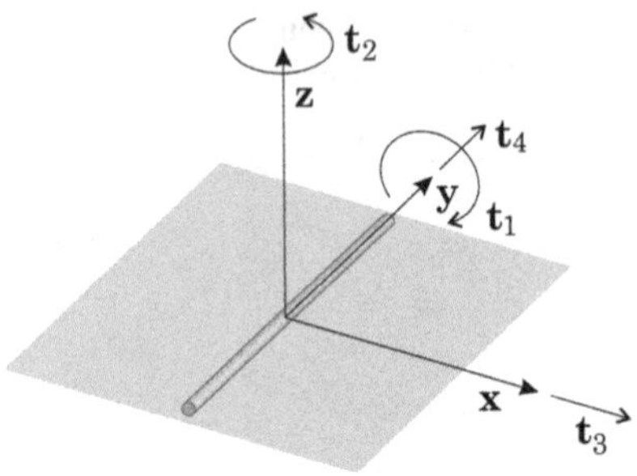

Fig. 8.7 Frictionless line contact in twist system

rotations. The corresponding twist system is of the third order

$$\begin{aligned}\mathbf{t}_1 &= (1,0,0,0,0,0)^{\mathrm{T}},\\ \mathbf{t}_2 &= (0,1,0,0,0,0)^{\mathrm{T}},\\ \mathbf{t}_3 &= (0,0,1,0,0,0)^{\mathrm{T}}.\end{aligned} \tag{8.23}$$

The wrench system for the friction point contact is of the same order. Here, we are only dealing with forces acting along the three axes of the coordinate frame

$$\begin{aligned}\mathbf{w}_1 &= (1,0,0,0,0,0)^{\mathrm{T}},\\ \mathbf{w}_2 &= (0,1,0,0,0,0)^{\mathrm{T}},\\ \mathbf{w}_3 &= (0,0,1,0,0,0)^{\mathrm{T}}.\end{aligned} \tag{8.24}$$

In the above expression the force along the **x** axis is denoted as $\mathbf{w}_1$, $\mathbf{w}_2$ belongs to the force along **y** and $\mathbf{w}_3$ along the **z** axis of the rectangular coordinate frame.

Frictionless line contact is presented in the twist system shown in Fig. 8.7. The number of degrees of freedom is reduced to four. The line can either roll on the plane or rotate about the vertical axis of the coordinate frame. Apart from these two rotational degrees of freedom $\mathbf{t}_1$ and $\mathbf{t}_2$, there are two translations, denoted in figure by $\mathbf{t}_3$ and $\mathbf{t}_4$. The frictionless line contact is described by four unit twists

$$\begin{aligned}\mathbf{t}_1 &= (0,1,0,0,0,0)^{\mathrm{T}},\\ \mathbf{t}_2 &= (0,0,1,0,0,0)^{\mathrm{T}},\\ \mathbf{t}_3 &= (0,0,0,1,0,0)^{\mathrm{T}},\\ \mathbf{t}_4 &= (0,0,0,0,1,0)^{\mathrm{T}}.\end{aligned} \tag{8.25}$$

The reciprocal wrench system (Fig. 8.8) is for the frictionless line contact two-dimensional

$$\begin{aligned}\mathbf{w}_1 &= (0,0,1,0,0,0)^{\mathrm{T}},\\ \mathbf{w}_2 &= (0,0,0,1,0,0)^{\mathrm{T}}.\end{aligned} \tag{8.26}$$

When there is friction between the line and the plane, both translations are impossible, as was the case in the previous example. Also the rotation around the vertical

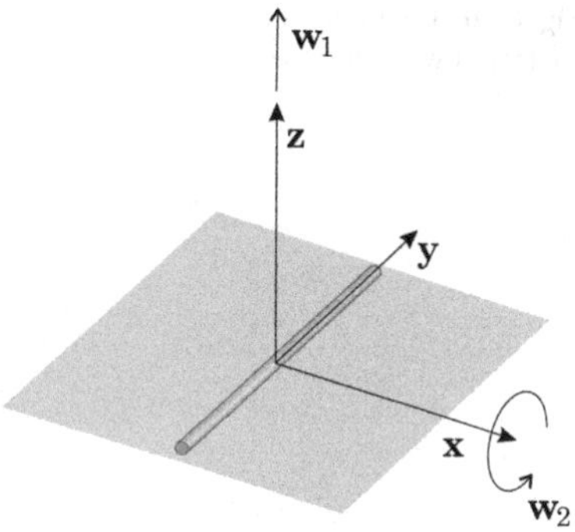

Fig. 8.8 Frictionless line contact in wrench system

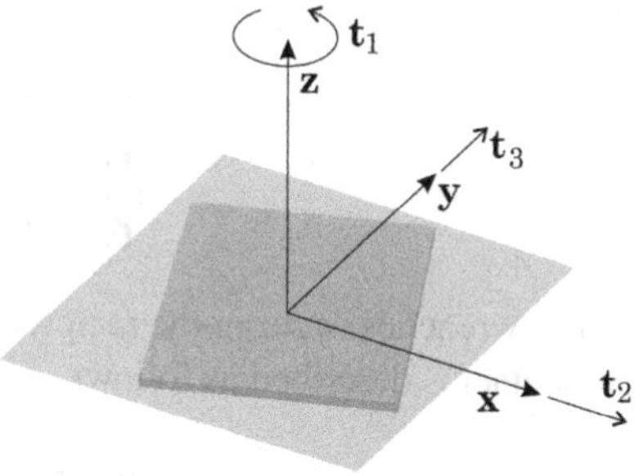

Fig. 8.9 Frictionless plane contact in twist system

axis is prevented. Only one degree of freedom remains, representing rolling of the line (rotation about **y** axis)

$$\mathbf{t}_1 = (0, 1, 0, 0, 0, 0)^{\mathrm{T}}. \tag{8.27}$$

The reciprocal wrench system consists of five lines. The first three are represented by the forces along all three axes of the rectangular coordinate frame, while the fourth and fifth component belong to the moments around the **x** and **z** axis

$$\begin{aligned}
\mathbf{w}_1 &= (1, 0, 0, 0, 0, 0)^{\mathrm{T}},\\
\mathbf{w}_2 &= (0, 1, 0, 0, 0, 0)^{\mathrm{T}},\\
\mathbf{w}_3 &= (0, 0, 1, 0, 0, 0)^{\mathrm{T}},\\
\mathbf{w}_4 &= (0, 0, 0, 1, 0, 0)^{\mathrm{T}},\\
\mathbf{w}_5 &= (0, 0, 0, 0, 0, 1)^{\mathrm{T}}.
\end{aligned} \tag{8.28}$$

In the planar contact we have three degrees of freedom. As can be seen from Fig. 8.9, we have rotation about the vertical axis $\mathbf{t}_1$ and two translations $\mathbf{t}_2$ and $\mathbf{t}_3$ in horizontal plane. The corresponding twist system consists of the following three

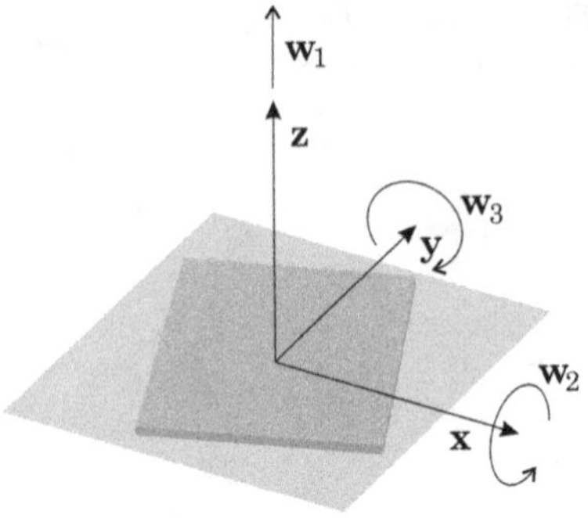

Fig. 8.10 Frictionless planar contact in wrench system

lines

$$\mathbf{t}_1 = (0, 0, 1, 0, 0, 0)^{\mathrm{T}},$$
$$\mathbf{t}_2 = (0, 0, 0, 1, 0, 0)^{\mathrm{T}}, \tag{8.29}$$
$$\mathbf{t}_3 = (0, 0, 0, 0, 1, 0)^{\mathrm{T}}.$$

The wrenches occur in the reciprocal directions (Fig. 8.10). The frictionless planar contact is described as follows

$$\mathbf{w}_1 = (0, 0, 1, 0, 0, 0)^{\mathrm{T}},$$
$$\mathbf{w}_2 = (0, 0, 0, 1, 0, 0)^{\mathrm{T}}, \tag{8.30}$$
$$\mathbf{w}_3 = (0, 0, 0, 0, 1, 0)^{\mathrm{T}}.$$

When considering friction in a planar contact, no relative motion is possible. Therefore, the planar contact with friction has zero degrees of freedom. The twist system is null, or empty. The wrench system is described by the following six vectors

$$\mathbf{w}_1 = (1, 0, 0, 0, 0, 0)^{\mathrm{T}},$$
$$\mathbf{w}_2 = (0, 1, 0, 0, 0, 0)^{\mathrm{T}},$$
$$\mathbf{w}_3 = (0, 0, 1, 0, 0, 0)^{\mathrm{T}},$$
$$\mathbf{w}_4 = (0, 0, 0, 1, 0, 0)^{\mathrm{T}}, \tag{8.31}$$
$$\mathbf{w}_5 = (0, 0, 0, 0, 1, 0)^{\mathrm{T}},$$
$$\mathbf{w}_6 = (0, 0, 0, 0, 0, 1)^{\mathrm{T}}.$$

In robotics we shall encounter another type of contact which will be called *soft finger*. The behavior of this contact is similar to the friction point contact, only that here the contact area is larger. In this way the contact resists moments acting about the contact normal. The soft finger has two degrees of freedom. As can be seen in Fig. 8.11, both are rotations $\mathbf{t}_1$ and $\mathbf{t}_2$.

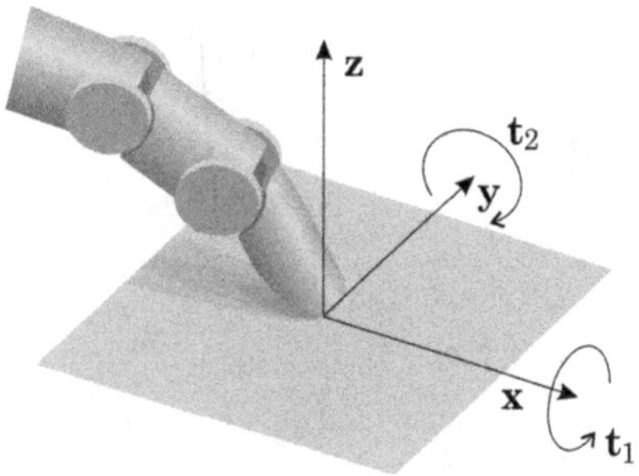

Fig. 8.11 Soft finger in twist system

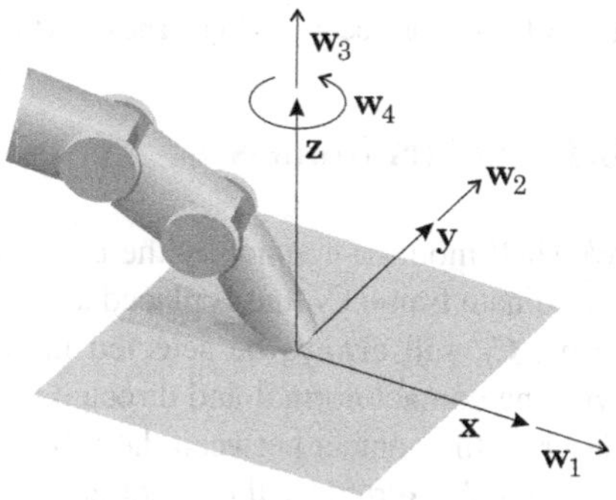

Fig. 8.12 Soft finger in wrench system

The soft finger has a two dimensional twist system. Both degrees of freedom are rotations in the tangential plane and can be described by the following equations

$$
\begin{aligned}
\mathbf{t}_1 &= (1, 0, 0, 0, 0, 0)^{\mathrm{T}}, \\
\mathbf{t}_2 &= (0, 1, 0, 0, 0, 0)^{\mathrm{T}}.
\end{aligned}
\tag{8.32}
$$

The dimension of the reciprocal wrench system of the soft finger is four. The wrenches are acting as forces along all three coordinate axes and as a moment around the vertical axis (Fig. 8.12)

$$
\begin{aligned}
\mathbf{w}_1 &= (1, 0, 0, 0, 0, 0)^{\mathrm{T}}, \\
\mathbf{w}_2 &= (0, 1, 0, 0, 0, 0)^{\mathrm{T}}, \\
\mathbf{w}_3 &= (0, 0, 1, 0, 0, 0)^{\mathrm{T}}, \\
\mathbf{w}_4 &= (0, 0, 0, 0, 0, 1)^{\mathrm{T}}.
\end{aligned}
\tag{8.33}
$$

We have presented the three basic types of contacts, point, line, and plane, in terms of their twist and wrench systems. In a similar way, other types of contacts with different geometrical constraints can be dealt with.

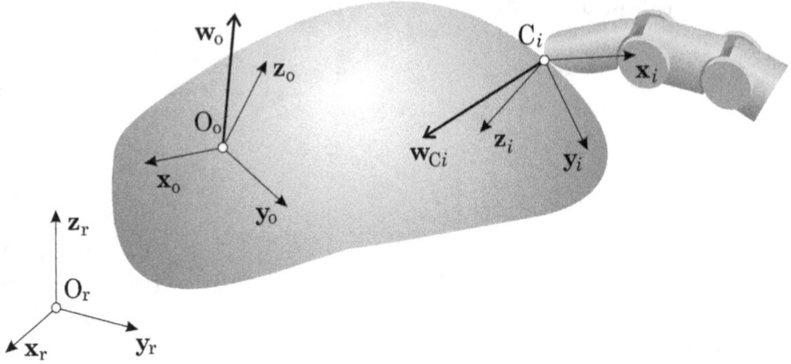

Fig. 8.13 Contact between finger and object

8.3 Contact Models

We shall model a contact by the use of a wrench $\mathbf{w}_{Ci}$ which acts at the origin of a coordinate frame C_i that is placed at the point of contact (Fig. 8.13). The coordinate frame C_i will always be selected in such a way, that the $\mathbf{z}$ axis will be aligned with the contact normal and directed towards the inside of the body. Later we shall describe the contact between the robot finger and the object as a mapping between the wrench exerted by the finger at the point of contact and the resultant wrench at a selected reference point of the object. As a reference point, typically the center of mass of the object is chosen. Another coordinate frame will be therefore attached to the center of mass of the object.

Usually a finger cannot exert forces or torques in an arbitrary direction. We shall make use of simple contacts which were described in the previous section. The simplest is the frictionless point contact. This is a contact where the finger can only exert a force normal to the surface of the object. We write the following wrench

$$\mathbf{w}_{Ci} = \mathbf{w}_1 F_{Ci} = \begin{bmatrix} 0 \\ 0 \\ 1 \\ 0 \\ 0 \\ 0 \end{bmatrix} F_{Ci}, \quad F_{Ci} \geq 0, \tag{8.34}$$

where F_{Ci} represents the magnitude of the force exerted by the finger in normal direction. The inequality, requiring only positive values of F_{Ci}, means that the selected contact can only push an object and cannot pull on it. Frictionless contacts do not occur in practical cases. Nevertheless, they represent useful models of contacts where the friction between the finger and the object is low or unknown. A grasp planned by considering contacts as frictionless will be more reliable than predicted, as friction in the contact will only enhance the grasp. With friction contacts, we must introduce a model of friction. We will use a simple Coulomb friction model.

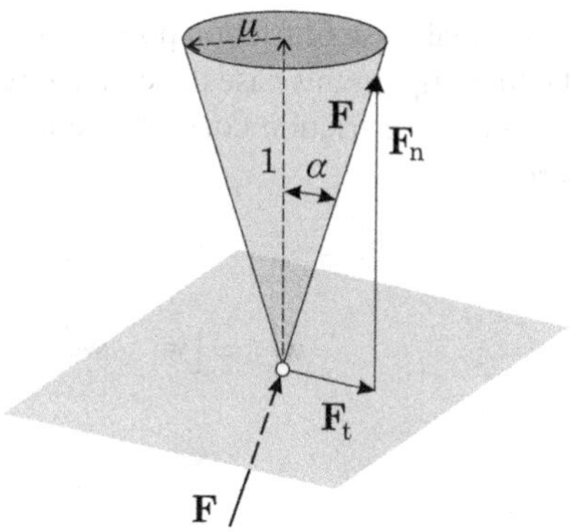

Fig. 8.14 Geometrical presentation of the Coulomb friction law

Table 8.1 Coefficients of friction for some materials

Steel on steel	0.58
Polyethylene on steel	0.3–0.35
Polyethylene on polyethylene	0.5
Wood on wood	0.25–0.5
Wood on metal	0.2–0.6

The Coulomb friction model is an empirical model, stating that the allowed tangential force is proportional to the normal force. The proportionality constant depends on the materials in contact. Let F_t be the magnitude of the tangential and F_n the magnitude of the normal force. Coulomb friction law says that slipping will start according to the following condition

$$|F_t| > \mu F_n. \tag{8.35}$$

In (8.35) μ is the coefficient of friction, where $\mu > 0$. This means that the range of tangential forces which can be applied in a contact is given by the following inequality

$$|F_t| \leq \mu F_n. \tag{8.36}$$

From the above inequality, it is evident that the force F_n must be positive. The equation describing the Coulomb friction can be presented geometrically. The set of forces which can be exerted at the contact must lie inside a cone, whose central axis is along the contact normal. This is called the friction cone (Fig. 8.14).

The angle of the cone with respect to the normal is

$$\alpha = \arctan \mu = \arctan \frac{F_t}{F_n}. \tag{8.37}$$

The coefficients of friction for some frequently encountered materials are given in Table 8.1. Typical values of μ are lower than 1, so that the opening angle of the friction cone is less than 45°.

We make use of friction point contact when friction exists between the object and the fingertip. In this case the contact exerts forces in all directions which are encompassed in the friction cone. The corresponding wrench is written in the following form

$$\mathbf{w}_{Ci} = \begin{bmatrix} \mathbf{w}_1 & \mathbf{w}_2 & \mathbf{w}_3 \end{bmatrix} \mathbf{F}_{Ci} = \begin{bmatrix} 1 & 0 & 0 \\ 0 & 1 & 0 \\ 0 & 0 & 1 \\ 0 & 0 & 0 \\ 0 & 0 & 0 \\ 0 & 0 & 0 \end{bmatrix} \mathbf{F}_{Ci}, \tag{8.38}$$

where

$$\begin{aligned} \mathbf{F}_{Ci} &= \begin{bmatrix} F_1 \\ F_2 \\ F_3 \end{bmatrix}, \\ \mu &> 0, \\ \sqrt{F_1^2 + F_2^2} &\le \mu F_3, \\ F_3 &\ge 0. \end{aligned} \tag{8.39}$$

A more realistic model is the soft finger contact. Apart from the forces inside the cone, also a moment about the contact normal is exerted. The moment is limited by the coefficient of friction γ. The wrench for the soft finger contact is

$$\mathbf{w}_{Ci} = \begin{bmatrix} \mathbf{w}_1 & \mathbf{w}_2 & \mathbf{w}_3 & \mathbf{w}_4 \end{bmatrix} \mathbf{F}_{Ci} = \begin{bmatrix} 1 & 0 & 0 & 0 \\ 0 & 1 & 0 & 0 \\ 0 & 0 & 1 & 0 \\ 0 & 0 & 0 & 0 \\ 0 & 0 & 0 & 0 \\ 0 & 0 & 0 & 1 \end{bmatrix} \mathbf{F}_{Ci}, \tag{8.40}$$

where

$$\begin{aligned} \mathbf{F}_{Ci} &= \begin{bmatrix} F_1 \\ F_2 \\ F_3 \\ F_4 \end{bmatrix}, \\ \mu &> 0, \\ \gamma &> 0, \\ \sqrt{F_1^2 + F_2^2} &\le \mu F_3, \\ F_3 &\ge 0, \\ |F_4| &\le \gamma F_3. \end{aligned} \tag{8.41}$$

In general we model a contact with a wrench basis $\mathbf{B}_{Ci}$ and the friction cone $\mathbf{F}_{Ci}$ as follows

$$\mathbf{w}_{Ci} = \mathbf{B}_{Ci}\mathbf{F}_{Ci}. \tag{8.42}$$

The wrench basis $\mathbf{B}_{Ci}$ has dimension $p \times m_i$. In all previous examples $p = 6$ was chosen. With planar contacts $p = 3$ can be selected. The dimension m_i represents the number of independent forces and moments which can be exerted by the contact.

Chapter 9
Robot Grasp

Abstract In this chapter it is our goal to determine the constraints required for a particular grasp to completely immobilize an object. Each contact between a finger and an object reduces the number of degrees of freedom of the object and at the same time allows external forces to be applied to the object. In this chapter we are not interested in single contacts between bodies; rather we are focused on the effects of multiple contacts on a grasped object. A mathematical description of a grasp with multiple fingers is introduced.

The effect of multiple contacts on the object can be studied either from the point of view of motion, or forces. Each contact constrains the object in such a way that the object can execute only a limited number of twists. When multiple contacts act on the object simultaneously, the resulting twist is represented by the intersection of the twists permitted by each individual contact. The resulting twist is a subset of those individual twists which have the same axis and pitch.

The description of a contact through motion is equivalent to the description by the use of forces. Let s_i represent the Plücker coordinates along or about which the body can execute the twist. As a result of the contact, there exist coordinates $\bar{s}_i$ reciprocal to s_i, along or about which a wrench is applied to the object.

A necessary and sufficient condition to constrain an object completely, requires that the intersection of the twists of all contacts equals zero

$$s_1 \cap s_2 \cap s_3 \cap \cdots \cap s_n = 0. \tag{9.1}$$

Conversely, an equivalent necessary and sufficient condition to constrain an object requires the union of the wrenches of all contacts must represent a six-dimensional space of all wrenches

$$\bar{s}_1 \cup \bar{s}_2 \cup \bar{s}_3 \cup \cdots \cup \bar{s}_n = \Re^6. \tag{9.2}$$

The first equation is satisfied only when also the second equation is satisfied and vice versa. When determining if an object is completely constrained or not, we will choose the equation which is more convenient for the problem under consideration.

J. Lenarčič et al., *Robot Mechanisms*,
Intelligent Systems, Control and Automation: Science and Engineering 60,
DOI 10.1007/978-94-007-4522-3_9, © Springer Science+Business Media Dordrecht 2013

In simple cases we can geometrically imagine the intersections and unions of the twists. Algebraic approaches will be applied in more complex cases.

Let us first consider a grasp with contacts having three degrees of freedom. Theoretically, two contacts with three degrees of freedom can completely constrain an object. In practice this is the case only when the twists of both contacts are reciprocal. This is not true for an arbitrary pair of basic contacts with three degrees of freedom.

Two point contacts with friction A point contact with friction is a three-dimensional system of twists. All unit twists are rotations. When constraining an object with two friction point contacts, the reciprocity will not be achieved. Each contact will have one common rotation around the line connecting both contact points. Therefore, two point contacts with friction cannot completely immobilize an object.

Two plane contacts without friction Also a planar frictionless contact is a three-dimensional system of twists. Two translations are parallel to the plane of contacts, while the rotation is about the contact normal. When constraining an object with two parallel planes, the object will slide between them and rotate about the common normal. When the planes are not parallel, the object will translate along the line of intersection of both planes.

Two point contacts with friction and one frictionless point contact We already know that two point contacts with friction allow rotation of an object about the common axis. When the third point contact is not directed along this common axis, an object can be immobilized. Two point contacts with friction and one frictionless point contact completely constrain an object when three contact points are not co-linear. In this way the intersection of the three twist systems is zero.

Two frictionless plane contacts and frictionless point contact When an object is constrained by two planar frictionless contacts, where the planes are not parallel, and another frictionless point contact is added, then the object is immobilized. The planar contacts permit translation along the line of intersection of both planes. Because of additional frictionless point contact, there is no intersection among the three twist systems.

Finally, let us state that all polyhedrons can be completely constrained with seven point contacts without friction. An object, which is not axis-symmetric, can be constrained with twelve frictionless point contacts. Axis-symmetric objects (e.g. sphere) cannot be constrained by the use frictionless point contacts.

9.1 Robot Grasp with Two Fingers

Before an object is successfully grasped with two fingers [42], first the phase of pushing and afterwards the phase of squeezing of an object occur. We will demonstrate that both operations are insensitive with respect to limited uncertainty of the initial pose of the object to be grasped.

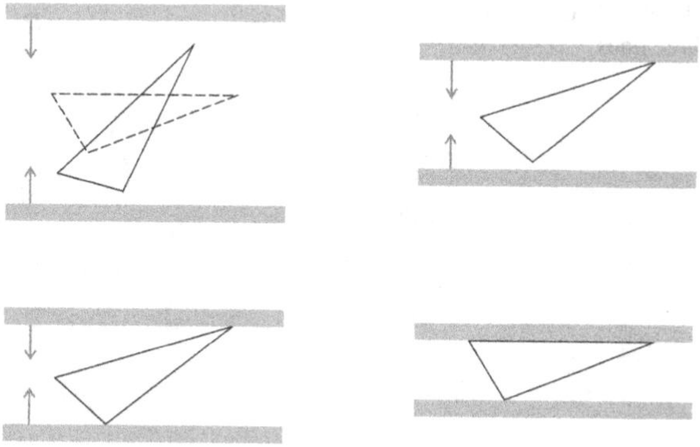

Fig. 9.1 Example of pushing and squeezing during robot grasping

Assume that we have a triangular object lying in a plane (Fig. 9.1). It is our aim to grasp this object by a robot gripper with two parallel fingers. The initial pose of the triangle is not completely determined, as shown in Fig. 9.1 above left. When the fingers come close to each other, it is very likely that one finger will contact the object before the other finger. At that moment the operation of pushing starts. The object is rotated about the point of contact between the finger and the object (Fig. 9.1 above right). The fingers travel closer to each other and soon also the second finger is in contact with the object. The squeezing phase is started at that instant (Fig. 9.1 below left). When the object is completely squeezed, the fingers cannot be further displaced. A stable grasp is achieved, shown in Fig. 9.1 below right. The task composed from pushing, squeezing, and grasping was successful even when a precise initial pose of the object was not known. The task described eliminates two (out of three) degrees of uncertainty in the pose of the triangular object. When the whole task is finished, the object is aligned with the gripper fingers. The gripper constrains the orientation of the object and also its position in the vertical direction. The positional uncertainty remains only in the horizontal direction.

Let us first examine the operation of pushing. Often it is more convenient to simply push an object from one point into another than first grasp the object, afterwards carry it into a new position and finally release it. Often we encounter the task of alignment of a larger number of objects. An example is in the alignment of playing cards. A pack of cards can be aligned more efficiently by pushing and squeezing then by displacing particular cards. Also, pharmacists, while counting tablets, make use of pushing them with the blade of a knife. Pushing is also efficient when displacing large and heavy objects. A heavy piece of furniture is usually pushed from one place to another in the room. The operation of pushing is also encountered with a fence placed above a belt conveyor. The objects are stopped and aligned at each such fence. It is important to stress that pushing is an efficient and very simple way of orienting the parts in a robot cell [9].

Several limitations will be adopted when studying robot pushing:

- a rigid object is pushed across a horizontal plane,
- a friction point contact occurs between the robot finger and the object,

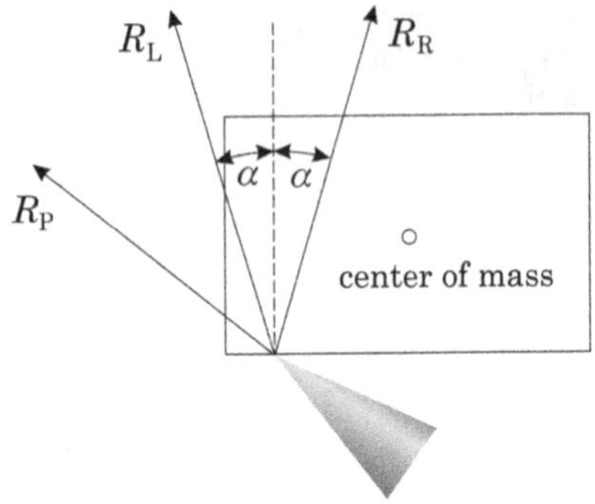

Fig. 9.2 Mason's rule of three lines

- the pushing forces are governed by the Coulomb friction law,
- the position of the center of mass of the object is known.

In the remaining text we shall familiarize ourselves with the basic mechanics of robotic pushing. We shall answer the question, in which direction will the object be rotated while being pushed. The answer is given by Mason's rule of three lines shown in Fig. 9.2:

- the line bordering on the right side of the projection of the friction cone R_R,
- the line bordering on the left side of the projection of the friction cone R_L, and
- the line representing the direction of the velocity of pushing R_P.

The three lines essentially "vote" for the direction of rotation. When at least two lines are e.g. on the left side of the center of mass, then the object will rotate in clockwise direction. When the result of the voting is undecided, then translational motion occurs.

Consider a few different pushing configurations. In all three examples, shown in Fig. 9.3, the object will rotate in a clockwise direction. In the first case the direction of rotation depends on the direction of pushing and the orientation of the finger. In the second example the object rotation only depends on the direction of pushing by the robot finger. In the last case the direction of rotation of the object is independent from either the orientation or the direction of motion of the finger. The last example does not appear to be intuitively logical. We can, however, convince ourselves about the correctness of the rule of the three lines, by simply pushing a glass ashtray over a smooth table with a pencil.

When both fingers of a robot gripper (Fig. 9.1) come in contact with the object, pushing is finished and the operation of squeezing is started [11]. While both fingers squeeze the object, the object is either further rotated or it is wedged between the fingers, so that further displacement of the fingers is not possible. In the example shown in Fig. 9.1, the triangular object rotates when the fingers are approaching closer together, as long as the edge of the object is aligned with the upper finger. The whole operation is finished by a stable grasp. In Fig. 9.4 we can see an example where the triangle is wedged between the fingers, as none of the edges is aligned with the fingers. This state is called wedging. It is our aim to plan such motions

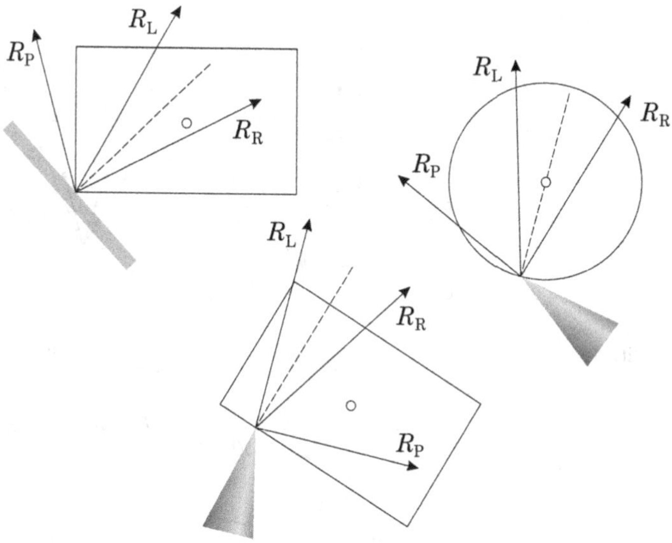

Fig. 9.3 Three examples of rotation of object during pushing

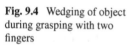

Fig. 9.4 Wedging of object during grasping with two fingers

of the fingers so that stable grasps are obtained and wedging is avoided. The orientation of the wedged object is not determined as the object can rotate about the axis connecting both points of contact. A wedged object is badly grasped and it can happen that during a fast motion of the robot, the object will slide out of the robot's grasp.

The problem of wedging is solved in a way similar to the Mason's rule of three lines. The object will be wedged between the fingers, when both fingers are in contact with the corners of the object and the imaginary line connecting both corners lies inside of both friction cones (Fig. 9.5). Wedging cannot occur when this condition is not fulfilled. Let as assume that the object lies on such a surface and that the friction between the object and the surface can be disregarded. The only external forces acting on the object occur because of the finger contacts. When an object is wedged, it is in a state of static equilibrium, meaning that all external forces and moments are balanced. This further means that the contact forces of both fingers must be of equal magnitude and directed one towards the other. This is only possible when the line connecting both corners lies inside the friction cones (Fig. 9.5

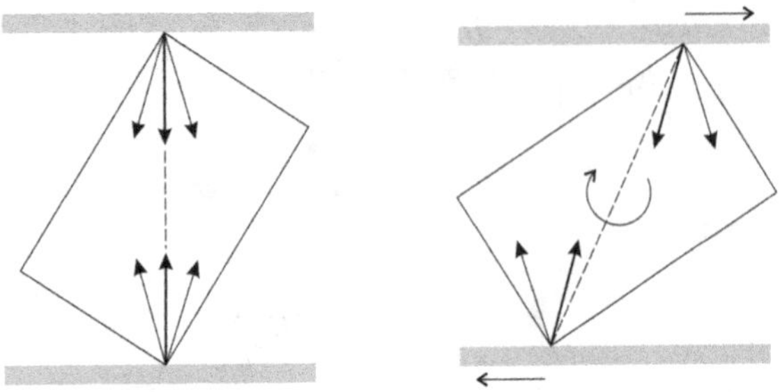

Fig. 9.5 Explanation of wedging rule

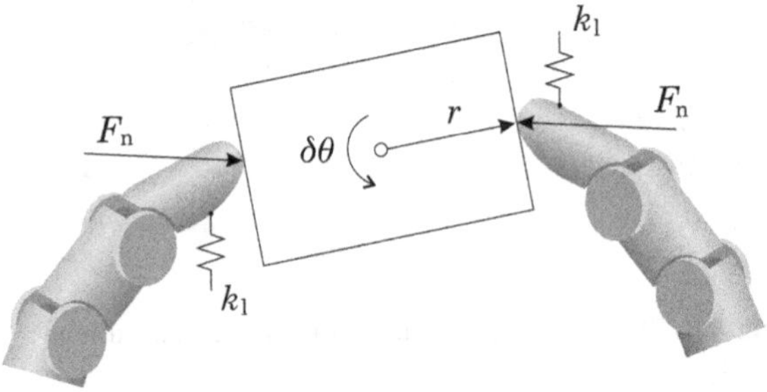

Fig. 9.6 Stability of robot grasp

left). When this is not the case, the finger forces lie on the border of the friction cones. As there is no more static equilibrium, the object slides and rotates between the fingers (Fig. 9.5 right).

Finally, let us examine the stability of a two-fingered grasp [16]. Stability of a robotic grasp is the ability of the gripper to reestablish a balance of the grasped object after the occurrence of an external disturbance.

When the block in Fig. 9.6 is rotated by an infinitesimal angle $\delta\theta$, the lateral elasticity of both fingers provokes a moment restoring the object into its initial pose

$$M_k = 2(k_l r \delta\theta) r. \tag{9.3}$$

As a consequence of the object's rotation, a moment resulting from the grasping forces F_n occurs

$$M_f = 2F_n r \delta\theta. \tag{9.4}$$

The object is influenced by the difference of the two moments

$$M_f - M_k = 2r(F_n - k_l r)\delta\theta. \tag{9.5}$$

A robot grasp is unstable when the difference in the moments is positive corresponding to a positive infinitesimal rotation of the object $\delta\theta$. An infinitesimally stable grasp is characterized by $F_n < k_l r$. With given dimensions of the object and given rigidity of the fingers this means, that stronger squeezing of the fingers results in a less stable grasp. Therefore, when planning robot grasps, it is not sensible

Table 9.1 Basic contacts

Contact	Friction	g_j
plane	yes	0
line	yes	1
soft finger	yes	2
point	yes	3
plane	no	3
line	no	4
point	no	5

to increase the grasping forces above the necessary values. This effect can be noticed when grasping along the edge of a coin between the thumb and index finger. When squeezing to strongly, the coin will collapse into a more stable pose where the fingers are in contact with the front and back faces of the coin.

9.2 Robot Grasp with Multiple Fingers

The development of robot grippers leads from simple grippers with two fingers towards universal grippers with multiple fingers, which are outfitted with various sensors. Such a universal gripper is more or less a copy of the human hand. The human hand has developed through long eras of evolution. The most important feature in the development of the human hand was the internal rotation of a thumb [64]. Due to this, the thumb opposes the other fingers. This feature is used in most of the grasps we apply in our daily lives. The situation is simpler with grippers having multiple fingers. The fingers are located on the palm in such a way that they represent the most efficient opposition with respect to each other. It is possible to demonstrate that a gripper with four fingers can grasp 99 % of the objects that can be manipulated by a gripper with five fingers. A gripper with three fingers can successfully grasp 90 % of all objects, while a gripper with two fingers only 40 % of the objects.

Between the fingers and the grasped object, there occur contacts (point, line, or plane) and friction. As demonstrated in previous section, we are dealing with basic types of contacts as shown in Table 9.1, where g_j represents the number of degrees of freedom of a particular contact. The number of degrees of freedom is given for a single contact between the finger and the object. A universal robot gripper has multiple fingers which are all connected to the same common segment. This segment is called the palm as with a human hand. When designing a gripper, we must first select the number of fingers and the number of joints of a particular finger. This choice depends on the desired mobility of the gripper and the desired connectivity between the gripper and the grasped object. The mobility (M) of an arbitrary kinematic system is defined as the number of independent parameters which must be specified at a given instant in order to give a complete description of the pose of each segment of the gripper. The connectivity (C) between two segments of a kinematic system

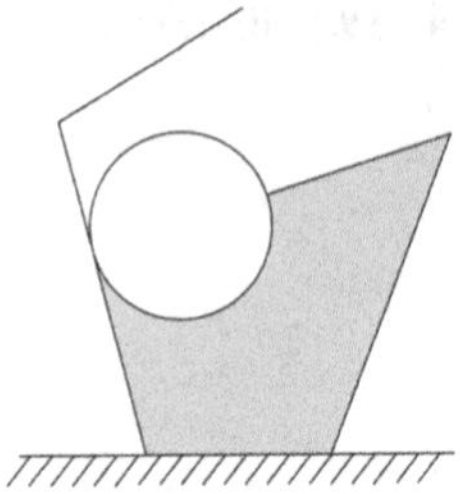

Fig. 9.7 Example of a loop when grasping with two fingers

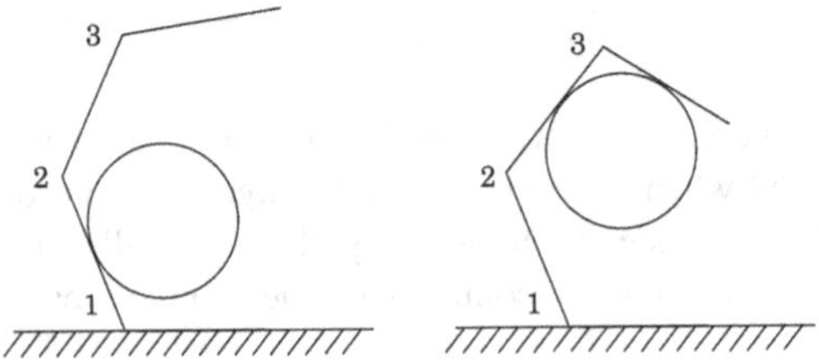

Fig. 9.8 Two examples of contact between finger and object

is defined as the number of independent parameters which must be specified at a given instant in order to completely describe the relative pose between the selected gripper segments. The mobility is important when the fingers are in motion. Good mobility means that the fingers are able to displace the object into various poses, in particular, in various orientations. The gripper connectivity is important when the fingers are closed. In this case the fingers should embrace the object as tightly as possible in order that the object will not escape from the gripper during fast robot movements.

The mobility of a robot gripper can be expressed by the following form of Grübler's formula

$$M \geq \sum_i f_i + \sum_j g_j - 6L. \tag{9.6}$$

In above equation M represents the mobility of the combined gripper-object system, f_i is the number of degrees of freedom of the i-th finger joint, g_j is the number of degrees of freedom of the j-th contact between the finger and the object, and L represents the number of loops in the gripper-object system. Figure 9.7 shows an example of a loop occurring when grasping with two fingers.

When determining the number of joints i, while calculating the mobility, we leave out all those joints which have no influence on the mobility of the grasped object. Consider an example of a contact with a finger having three joints as shown on the left side of Fig. 9.8. In this case we omit all the joints lying between the object and the finger tip. It is not difficult to realize that joints 2 and 3 do not effect the mobility of the object. The inequality in Grübler's formula arises from the fact that the constraints on the motion of a segment in a kinematic system are not always independent.

When the fingers are fixed during the grasp, while enclosing the object, we describe the mobility by the following equation

$$M' \geq \sum_j g_j - 6L. \tag{9.7}$$

Within robotic grasping we are also interested in the relative motion, or the lack of relative motion, between the object and the gripper. The connectivity is calculated from the mobility by taking away the degrees of freedom of the joints which do not contribute to increased connectivity in the gripper-object system. These are the joints lying between the palm and the object and are not in contact with the object. Such an example is shown on the right side of Fig. 9.8. In this case the first joint actuates a segment which is not in contact with the object and thus does not contribute to better connectivity of the object. If a joint has f_i degrees of freedom and the number of such joints is k, the connectivity is calculated as follows

$$C = M - \sum_{i=1}^{k} f_i. \tag{9.8}$$

During a fixed enclosed grasp the connectivity is equal to the mobility

$$C' = M'. \tag{9.9}$$

Consider a robot gripper with k fingers. Each finger can be related to the object in one of n different ways. With the finger having three joints we have eight possible ways (Fig. 9.9). The number of possible grasps for a gripper with multiple fingers is

$$\binom{n+k-1}{k} = \frac{n(n+1)\cdots(n+k-1)}{k!}. \tag{9.10}$$

Assume a robot gripper with three fingers ($k = 3$). Each finger has three joints. Suppose that only one type of contact can occur between the object and the gripper, e.g. a point contact without friction. We are dealing with $n = 8$ possible ways of grasping shown in Fig. 9.9. Also the case "0" should be considered as we have a gripper with three fingers. Not all three fingers have to be in contact with the object in order to accomplish a successful grasp. From (9.10), the total number of possible grasps is

$$\frac{8 \cdot 9 \cdot 10}{3!} = 120. \tag{9.11}$$

A dexterous robot gripper must be able to cause an arbitrary displacement of the object or exert arbitrary forces on the object. When the fingers are fixed and closed around the object, no unwanted displacements of the object should occur. Both conditions should be considered in grasp planning:

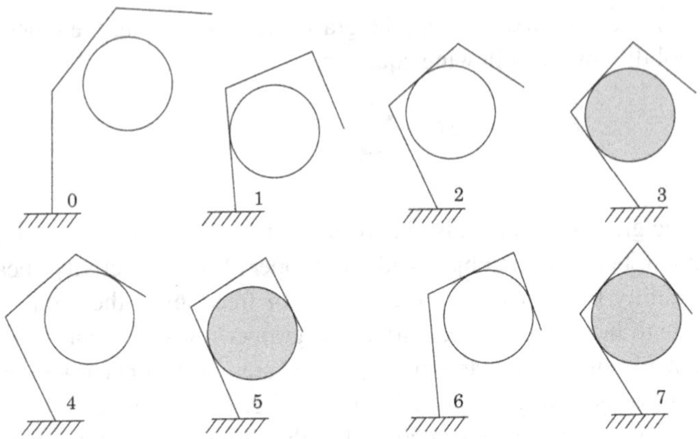

Fig. 9.9 Different ways of grasping by an individual finger with three joints

- a robot gripper must be able to exert arbitrary forces and arbitrary displacements of the object when the fingers are in motion;
- a robot gripper must firmly constrain the object when the finger joints are locked.

Let us describe both conditions in terms of connectivity. The first condition is saying that the connectivity must be $C = 6$, while the fingers are moving. In this way we can place the object grasped by the gripper into a desired position and orientation (in a limited workspace of gripper). The second condition is requiring that the connectivity with the finger joints locked is less than or equal to zero $C' \leq 0$.

In our further analysis of robot grasping we shall assume a point contact without friction between any of the individual fingers and the object. This is the contact with five degrees of freedom, $g_j = 5$. In this case we have 33 possible grasps out of the previous 120, which correspond to both conditions $C = 6$ and $C' \leq 0$. All 33 grasps are listed below. The numbers 1 through 7 represent different ways of grasping by an individual finger, as shown in Fig. 9.9. Each number belongs to one of the fingers of the gripper:

1. 7-7-0	12. 6-3-3	23. 7-7-4
2. 7-3-1	13. 7-3-3	24. 5-5-5
3. 7-5-1	14. 7-4-3	25. 6-5-5
4. 7-6-1	15. 5-5-3	26. 7-5-5
5. 7-7-1	16. 6-5-3	27. 6-6-5
6. 7-3-2	17. 7-5-3	28. 7-6-5
7. 7-5-2	18. 6-6-3	29. 7-7-5
8. 7-6-2	19. 7-6-3	30. 6-6-6
9. 7-7-2	20. 7-7-3	31. 7-6-6
10. 3-3-3	21. 7-5-4	32. 7-7-6
11. 5-3-3	22. 7-6-4	33. 7-7-7

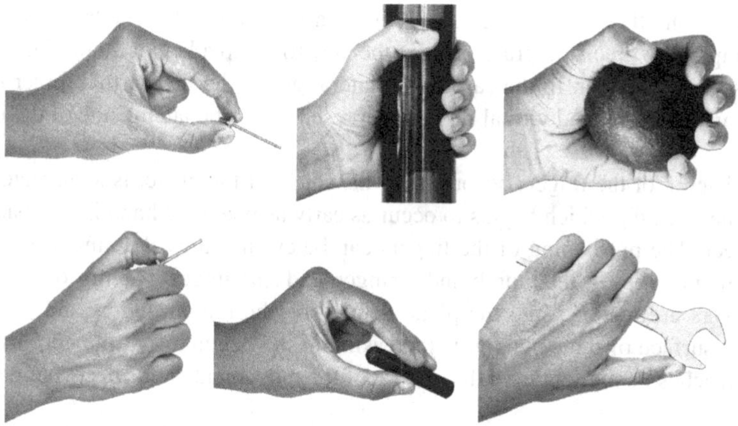

Fig. 9.10 The taxonomy of grasping: pinch, cylindrical, spherical, lateral, tripod, and hook grasp

Let us now calculate the mobility for the grasp 7-5-3. The mobility of the object is influenced by all three joints of the first and the second finger and only two joints of the third finger. We therefore have $i = 3+3+2 = 8$. The number of contacts between the individual finger and the object can be counted from Fig. 9.9, $j = 3+2+2 = 7$. Two loops occur within the first finger grasping, while there is only one loop with the second and the third finger. Another three loops arise between the palm, the three fingers and the object. Only two of them are independent. The number of independent loops is equal to $L = 2+1+1+2 = 6$. Each of the finger joints has only one degree of freedom $f_i = 1$. We have assumed point contact without friction, so $g_j = 5$. The mobility is calculated according to Grübler's formula (9.6) as $M = 8.1 + 7.5 - 6.6 = 7$. The connectivity is not influenced by the second joint (i.e. segment) of the second finger. We can write $C = M - 1 = 6$. For the finger joints locked, the mobility is calculated according (9.7), $M' = 7.5 - 6.6 = -1$. The connectivity is in this case equal to the mobility $C' = M' = -1$. The negative connectivity represents an excessive constraint.

Before concluding this section on grasping with multiple fingers, let us have a quick look at the characteristic properties of human grasping. The classification of human grasps is based on the shape of the object to be grasped [63]. Figure 9.10 shows several human grasps. The upper row, from left to right show the pinch, cylindrical and spherical grasps. The lower row shows the lateral, tripod, and hook grasps. Human grasps can be efficiently described by introducing a virtual finger [32]. A virtual finger can be either represented by a single finger or several fingers acting in the same direction. The palm can be also considered as a virtual finger. Usually with a particular grasp there occur two virtual fingers. Human grasps can be divided into power and precision grasps. Power grasps are characterized by large forces, high stability, and large contact surfaces of grasping. The cylindrical grasp is an example of a power grasp. It is used when grasping different tools, a glass, or a bottle. With the cylindrical grasp all five fingers represent the virtual finger, while the palm is the opposing virtual finger. On the contrary, precision grasps are characterized by low forces, small contact surfaces, and high grasping dexterity. The lateral grasp is an example of a precision grasp. This is the way to grasp a key or a thin flat object, such as piece of paper. In the case of the lateral grasp, the first virtual finger is the thumb, while the second virtual finger is represented by the lateral side of the index finger.

Preshaping of the fingers according to the shape of the object is a characteristic of human grasping which begins to occur as early as when the hand is approaching the object. The preshaping of the fingers can be evaluated by defining a pentagon connecting the tips of the thumb, index finger, and ring finger. The tips of the other two fingers are mapped into the plane defined by the thumb, index, and ring finger. The surface of the pentagon is increasing in the beginning of the approaching phase, reaches its maximum in the middle of the movement and is decreasing afterwards [81].

An interesting parameter for the evaluation of the approaching phase is the angle between the pentagon normal and the object normal. The time history of the angle has a saddle shape and can be divided into three phases: fast turn of the wrist, transport phase, where the angle remains almost constant, and final preshaping of the fingers according to the shape of the object. The velocity profile has a triangular shape during the approaching phase. The rising of the velocity in the first part of the approaching phase is greater in magnitude than the decreasing of the velocity is in the second part of the approaching phase.

9.3 Grasp Matrix

Three coordinate frames will be used in our further mathematical description of a grasp with multiple fingers. The origin of the frame C_i is at the contact point between the finger and the grasped object. The origin of the frame $\mathrm{O_o}$ is the center of mass of the object. The motion of the grasped object is expressed relative to reference frame $\mathrm{O_r}$, which means that all vectors shown in Fig. 9.11 are expressed relative to the reference coordinate frame $\mathrm{O_r}$.

We shall first write the equations describing the relationship between the translational velocities of the origins of the frames $\mathrm{O_o}$ and C_i and their angular velocities, with respect to the reference frame $\mathrm{O_r}$

$$
\begin{aligned}
\dot{\mathbf{p}}_i &= \dot{\mathbf{p}}_\mathrm{o} + \boldsymbol{\omega}_\mathrm{o} \times \mathbf{p}_{\mathrm{o},i}, \\
\boldsymbol{\omega}_i &= \boldsymbol{\omega}_\mathrm{o}.
\end{aligned}
\tag{9.12}
$$

The cross product can be replaced by the multiplication with the skew-symmetric matrix $\mathbf{P}_{\mathrm{o},i}$, composed from the components of the vector $\mathbf{p}_{\mathrm{o},i}$, according to the definition (1.30)

$$
\mathbf{P}_{\mathrm{o},i} = \mathbf{p}_{\mathrm{o},i} \otimes \mathbf{I}. \tag{9.13}
$$

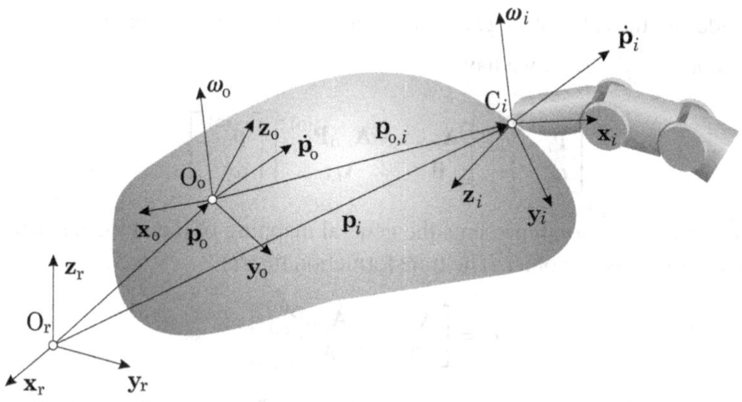

Fig. 9.11 Contact between the finger and the object

We obtain

$$\begin{bmatrix} \dot{\mathbf{p}}_i \\ \boldsymbol{\omega}_i \end{bmatrix} = \begin{bmatrix} \mathbf{I} & -\mathbf{P}_{o,i} \\ \mathbf{0} & \mathbf{I} \end{bmatrix} \begin{bmatrix} \dot{\mathbf{p}}_o \\ \boldsymbol{\omega}_o \end{bmatrix}. \tag{9.14}$$

In the above matrix equation all vectors are referred to frame O_r. Usually the vectors are expressed in local coordinate frames, so that we write

$$\begin{aligned} \mathbf{p}_{o,i} &= \mathbf{p}_{o,i}^{(r)} = \mathbf{A}_{r,o}\mathbf{p}_{o,i}^{(o)}, \\ \dot{\mathbf{p}}_o &= \dot{\mathbf{p}}_o^{(r)} = \mathbf{A}_{r,o}\dot{\mathbf{p}}_o^{(o)}, \\ \boldsymbol{\omega}_o &= \boldsymbol{\omega}_o^{(r)} = \mathbf{A}_{r,o}\boldsymbol{\omega}_o^{(o)} \end{aligned} \tag{9.15}$$

and

$$\begin{aligned} \dot{\mathbf{p}}_i &= \dot{\mathbf{p}}_i^{(r)} = \mathbf{A}_{r,i}\dot{\mathbf{p}}_i^{(i)} = \mathbf{A}_{r,o}\mathbf{A}_{o,i}\dot{\mathbf{p}}_i^{(i)}, \\ \boldsymbol{\omega}_i &= \boldsymbol{\omega}_i^{(r)} = \mathbf{A}_{r,i}\boldsymbol{\omega}_i^{(i)} = \mathbf{A}_{r,o}\mathbf{A}_{o,i}\boldsymbol{\omega}_i^{(i)}. \end{aligned} \tag{9.16}$$

Here, $\mathbf{A}_{r,o}$ and $\mathbf{A}_{o,i}$ are rotation matrices describing the transformations between the frames O_r, O_o and C_i. They can be calculated as explained in Sect. 1.3.1. By inserting the above expressions into (9.14), we obtain

$$\mathbf{A}_{r,o}\mathbf{A}_{o,i} \begin{bmatrix} \dot{\mathbf{p}}_i^{(i)} \\ \boldsymbol{\omega}_i^{(i)} \end{bmatrix} = \begin{bmatrix} \mathbf{I} & -\mathbf{P}_{o,i}^{(r)} \\ \mathbf{0} & \mathbf{I} \end{bmatrix} \mathbf{A}_{r,o} \begin{bmatrix} \dot{\mathbf{p}}_o^{(o)} \\ \boldsymbol{\omega}_o^{(o)} \end{bmatrix}. \tag{9.17}$$

Multiplication with $\mathbf{A}_{o,i}^T\mathbf{A}_{r,o}^T$ brings

$$\begin{bmatrix} \dot{\mathbf{p}}_i^{(i)} \\ \boldsymbol{\omega}_i^{(i)} \end{bmatrix} = \begin{bmatrix} \mathbf{A}_{o,i}^T & -\mathbf{A}_{o,i}^T(\mathbf{A}_{r,o}^T\mathbf{P}_{o,i}^{(r)}\mathbf{A}_{r,o}) \\ \mathbf{0} & \mathbf{A}_{o,i}^T \end{bmatrix} \begin{bmatrix} \dot{\mathbf{p}}_o^{(o)} \\ \boldsymbol{\omega}_o^{(o)} \end{bmatrix}. \tag{9.18}$$

By considering the rules in (1.27), it is not difficult to demonstrate that $\mathbf{A}_{\mathrm{r,o}}^{\mathrm{T}}\mathbf{P}_{\mathrm{o},i}^{(\mathrm{r})}\mathbf{A}_{\mathrm{r,o}} = \mathbf{P}_{\mathrm{o},i}^{(\mathrm{o})}$. Since $\mathbf{A}_{\mathrm{o},i}^{\mathrm{T}} = \mathbf{A}_{i,\mathrm{o}}$, we have

$$\begin{bmatrix} \dot{\mathbf{p}}_i^{(i)} \\ \boldsymbol{\omega}_i^{(i)} \end{bmatrix} = \begin{bmatrix} \mathbf{A}_{i,\mathrm{o}} & -\mathbf{A}_{i,\mathrm{o}}\mathbf{P}_{\mathrm{o},i}^{(\mathrm{o})} \\ \mathbf{0} & \mathbf{A}_{i,\mathrm{o}} \end{bmatrix} \begin{bmatrix} \dot{\mathbf{p}}_{\mathrm{o}}^{(\mathrm{o})} \\ \boldsymbol{\omega}_{\mathrm{o}}^{(\mathrm{o})} \end{bmatrix}. \tag{9.19}$$

This matrix operation represents the general mapping of the velocities from one coordinate frame into another. The transformation matrix

$$\mathcal{A}_i = \begin{bmatrix} \mathbf{A}_{i,\mathrm{o}} & -\mathbf{A}_{i,\mathrm{o}}\mathbf{P}_{\mathrm{o},i}^{(\mathrm{o})} \\ \mathbf{0} & \mathbf{A}_{i,\mathrm{o}} \end{bmatrix} \tag{9.20}$$

is referred to as the adjoint transformation matrix. The forces and moments which will occur during the contact are related with the transposed adjoint matrix

$$\mathcal{A}_i^{\mathrm{T}} = \begin{bmatrix} \mathbf{A}_{\mathrm{o},i} & \mathbf{0} \\ \mathbf{P}_{\mathrm{o},i}^{(\mathrm{o})}\mathbf{A}_{\mathrm{o},i} & \mathbf{A}_{\mathrm{o},i} \end{bmatrix}. \tag{9.21}$$

In this way the following relation exists between the wrench in the center of mass of the object $\mathbf{w}_{\mathrm{o}}$ and the wrench in the contact point $\mathbf{w}_{\mathrm{C}i}$ (Fig. 8.13)

$$\mathbf{w}_{\mathrm{o}} = \mathcal{A}_i^{\mathrm{T}}\mathbf{w}_{\mathrm{C}i}. \tag{9.22}$$

This equation can be rewritten using the wrench basis $\mathbf{B}_{\mathrm{C}i}$ and friction cone $\mathbf{F}_{\mathrm{C}i}$, introduced in Sect. 8.3 which discussed contact models

$$\mathbf{w}_{\mathrm{o}} = \mathcal{A}_i^{\mathrm{T}}\mathbf{B}_{\mathrm{C}i}\mathbf{F}_{\mathrm{C}i}. \tag{9.23}$$

The matrix $\mathcal{A}_i^{\mathrm{T}}$ maps the contact wrenches into the object wrenches. Let us define the contact matrix $\mathbf{G}_i$ as the product of $\mathcal{A}_i^{\mathrm{T}}$ and $\mathbf{B}_{\mathrm{C}i}$

$$\mathbf{G}_i = \mathcal{A}_i^{\mathrm{T}}\mathbf{B}_{\mathrm{C}i}. \tag{9.24}$$

When there are k fingers in contact with the object, the total wrench acting on the body is given by the sum of the wrenches produced by each of the contacting fingers. The map between the contact forces of the fingers and the total resultant force on the object is called the grasp matrix. Because of the linearity of the contact matrices, the wrenches can be summed

$$\mathbf{w}_{\mathrm{o}} = \mathbf{G}_1\mathbf{F}_{\mathrm{C}1} + \cdots + \mathbf{G}_k\mathbf{F}_{\mathrm{C}k} = \begin{bmatrix} \mathbf{G}_1 & \ldots & \mathbf{G}_k \end{bmatrix} \begin{bmatrix} \mathbf{F}_{\mathrm{C}1} \\ \vdots \\ \mathbf{F}_{\mathrm{C}k} \end{bmatrix}. \tag{9.25}$$

The grasp matrix $\mathbf{G}$ has the following form

$$\mathbf{G} = \begin{bmatrix} \mathcal{A}_1^{\mathrm{T}}\mathbf{B}_{\mathrm{C}1} & \ldots & \mathcal{A}_k^{\mathrm{T}}\mathbf{B}_{\mathrm{C}k} \end{bmatrix}. \tag{9.26}$$

The total wrench, acting on the object, can be written as follows

$$\mathbf{w}_o = \mathbf{G}\mathbf{F}_C, \tag{9.27}$$

where

$$\mathbf{F}_C = \begin{bmatrix} \mathbf{F}_{C1} & \dots & \mathbf{F}_{Ck} \end{bmatrix}^T. \tag{9.28}$$

First consider the contact matrix describing the point contact without friction

$$\mathbf{w}_o = \begin{bmatrix} \mathbf{A}_{o,i} & \mathbf{0} \\ \mathbf{P}_{o,i}^{(o)}\mathbf{A}_{o,i} & \mathbf{A}_{o,i} \end{bmatrix} \begin{bmatrix} 0 \\ 0 \\ 1 \\ 0 \\ 0 \\ 0 \end{bmatrix} F_{Ci}. \tag{9.29}$$

The matrix $\mathbf{A}_{o,i}$ can be expressed in terms of the direction cosines of the angles between the axes of the frames C_i and O_o

$$\mathbf{A}_{o,i} = \begin{bmatrix} \mathbf{l}_{Ci} & \mathbf{m}_{Ci} & \mathbf{n}_{Ci} \end{bmatrix}, \tag{9.30}$$

where e.g. the unit vector $\mathbf{n}_{Ci}$ means the direction of the normal to the object surface ($\mathbf{z}$ axis runs along the normal) in the object coordinate frame O_o. After multiplication we have the following equation for each contact

$$\mathbf{w}_o = \begin{bmatrix} \mathbf{n}_{Ci} \\ \mathbf{p}_{o,i} \times \mathbf{n}_{Ci} \end{bmatrix} F_{Ci}. \tag{9.31}$$

When grasping with multiple fingers we have

$$\mathbf{w}_o = \begin{bmatrix} \mathbf{n}_{C1} & \dots & \mathbf{n}_{Ck} \\ \mathbf{p}_{o,1} \times \mathbf{n}_{C1} & \dots & \mathbf{p}_{o,k} \times \mathbf{n}_{Ck} \end{bmatrix} \begin{bmatrix} F_{C1} \\ \vdots \\ F_{Ck} \end{bmatrix}. \tag{9.32}$$

The polyhedrons can be completely constrained by seven point contacts without friction. Let us first find the grasp matrix $\mathbf{G}$ for the case, where a cube with the side length 1 is constrained with seven point contacts without friction. The object coordinate frame is placed onto one of the corners of the cube, as shown in (9.12).

From Fig. 9.12 we identify the vectors $\mathbf{p}_{o,i}$ and $\mathbf{n}_{Ci}$, which are entered into Table 9.2. With regard to (9.32), the grasp matrix $\mathbf{G}$ has the following form

$$\mathbf{G} = \begin{bmatrix} 0 & 0 & -1 & 0 & 0 & -1 & 1 \\ 0 & 1 & 0 & 0 & -1 & 0 & 0 \\ -1 & 0 & 0 & 1 & 0 & 0 & 0 \\ 0 & -1 & 0 & 1 & 0 & 0 & 0 \\ 1 & 0 & -1 & 0 & 0 & -1 & 0.5 \\ 0 & 1 & 0 & 0 & 0 & 1 & -1 \end{bmatrix}. \tag{9.33}$$

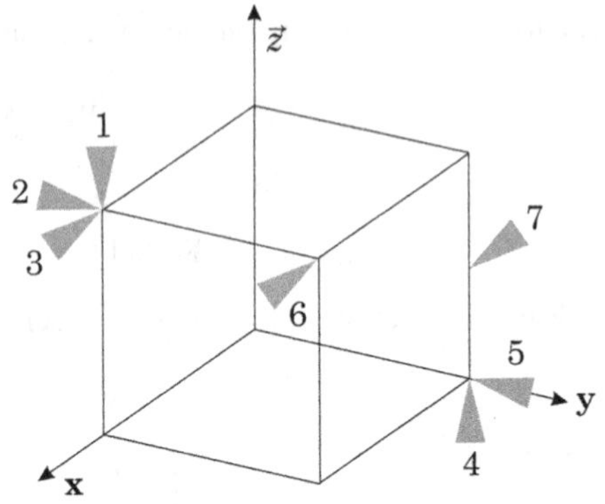

Fig. 9.12 Cube constrained by seven point contacts without friction

Table 9.2 Vectors $\mathbf{p}_{o,i}$ and $\mathbf{n}_{Ci}$ for the cube from Fig. 9.12

i	1	2	3	4	5	6	7
$\mathbf{p}_{o,i}$	1	1	1	0	0	1	0
	0	0	0	1	1	1	1
	1	1	1	0	0	1	0.5
$\mathbf{n}_{C,i}$	0	0	−1	0	0	−1	1
	0	1	0	0	−1	0	0
	−1	0	0	1	0	0	0

The Plücker coordinates, which are the columns of the grasp matrix, can be inferred directly from the figure, so the grasp matrix can be written directly. The line moments must be referenced to the origin of the coordinate frame. Attention must be paid to the correct direction of the line moments. By taking away only one point contact, the cube is no longer completely constrained.

Consider another example where a box is grasped by two soft fingers as shown in Fig. 9.13. First we shall describe the orientation and the position of the first finger contact $\mathbf{C}_1$ with respect to the frame placed at the center of mass of the box. For the first finger we have

$$\mathbf{A}_{0,1} = \begin{bmatrix} 0 & 1 & 0 \\ 0 & 0 & 1 \\ 1 & 0 & 0 \end{bmatrix}, \qquad \mathbf{p}_{0,1}^{(0)} = \begin{bmatrix} 0 \\ -r \\ 0 \end{bmatrix}, \tag{9.34}$$

and for the second finger

$$\mathbf{A}_{0,2} = \begin{bmatrix} 1 & 0 & 0 \\ 0 & 0 & -1 \\ 0 & 1 & 0 \end{bmatrix}, \qquad \mathbf{p}_{0,2}^{(0)} = \begin{bmatrix} 0 \\ r \\ 0 \end{bmatrix}. \tag{9.35}$$

The matrix $\mathbf{B}_{Ci}$, describing the soft finger contact, has the form given by (8.40), which we developed in the previous chapter. By using (9.24) the contact matrix for

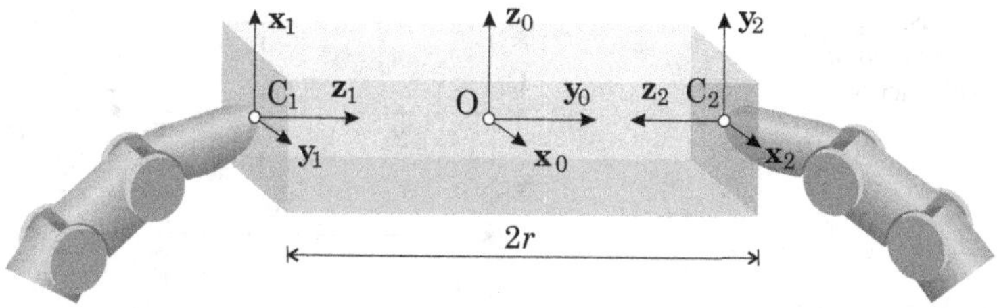

Fig. 9.13 Box grasped by two soft fingers

the first finger is obtained

$$\mathbf{G}_1 = \begin{bmatrix} 0 & 1 & 0 & 0 \\ 0 & 0 & 1 & 0 \\ 1 & 0 & 0 & 0 \\ -r & 0 & 0 & 0 \\ 0 & 0 & 0 & 1 \\ 0 & r & 0 & 0 \end{bmatrix}. \tag{9.36}$$

In an analogous manner the contact matrix $\mathbf{G}_2$ for the second finger is calculated. Both matrices are combined into the grasp matrix $\mathbf{G}$ as follows

$$\mathbf{G} = \begin{bmatrix} 0 & 1 & 0 & 0 & 1 & 0 & 0 & 0 \\ 0 & 0 & 1 & 0 & 0 & 0 & -1 & 0 \\ 1 & 0 & 0 & 0 & 0 & 1 & 0 & 0 \\ -r & 0 & 0 & 0 & 0 & r & 0 & 0 \\ 0 & 0 & 0 & 1 & 0 & 0 & 0 & -1 \\ 0 & r & 0 & 0 & -r & 0 & 0 & 0 \end{bmatrix}. \tag{9.37}$$

Finally, consider the description of grasping in the plane. Assume that the contact forces act in the plane, which is perpendicular to the $\mathbf{z}$ axis, while the moment occurs about the $\mathbf{z}$ axis. In (9.22) we shall omit the force along the $\mathbf{z}$ axis and the moments around the axes $\mathbf{x}$ and $\mathbf{y}$. In this way the wrenches $\mathbf{w}_o$ and $\mathbf{w}_{Ci}$ have only two components of force and only a single moment component. The rotation matrix $\mathbf{A}_{o,i}$, describing the rotation around the $\mathbf{z}$ axis, has dimension 2×2, as already seen in Sect. 1.8.3. When inserting $p_{o,iz} = 0$ in (9.22) and considering (8.10), the following simplified relation between the contact wrench and the object wrench in the plane is obtained

$$\begin{bmatrix} F_x^{(o)} \\ F_y^{(o)} \\ M_z^{(o)} \end{bmatrix} = \begin{bmatrix} \mathbf{A}_{o,i} & \begin{matrix} 0 \\ 0 \end{matrix} \\ \begin{bmatrix} -p_{o,iy} & p_{o,ix} \end{bmatrix} \mathbf{A}_{o,i} & 1 \end{bmatrix} \begin{bmatrix} F_x^{(i)} \\ F_y^{(i)} \\ M_z^{(i)} \end{bmatrix}. \tag{9.38}$$

Consider the example of planar grasping of a rectangle. Both fingers are grasping using point contacts with friction as shown in Fig. 9.14. When describing planar

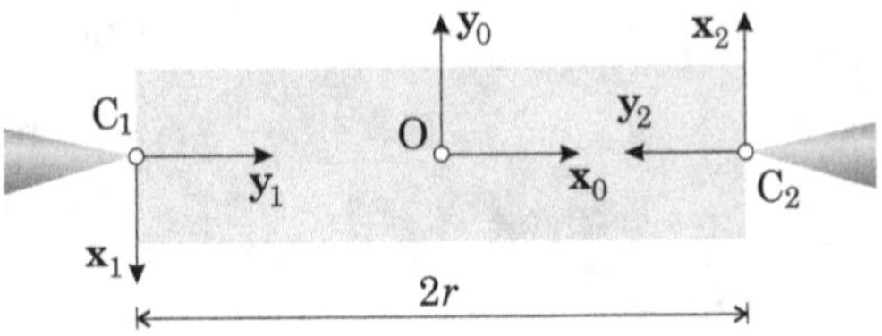

Fig. 9.14 Grasping a rectangle by two point contacts with friction

grasping, the **y** axis is aligned with the contact normal. The position and orientation of the first finger can be read from Fig. 9.14

$$\mathbf{A}_{0,1} = \begin{bmatrix} 0 & 1 \\ -1 & 0 \end{bmatrix}, \qquad \mathbf{p}_{0,1}^{(0)} = \begin{bmatrix} -r \\ 0 \end{bmatrix} \tag{9.39}$$

and for the second finger

$$\mathbf{A}_{0,2} = \begin{bmatrix} 0 & -1 \\ 1 & 0 \end{bmatrix}, \qquad \mathbf{p}_{0,2}^{(0)} = \begin{bmatrix} r \\ 0 \end{bmatrix}. \tag{9.40}$$

The contact matrix of the finger $\mathbf{C}_1$ is obtained as a product of the transposed adjoint matrix (9.38) and the matrix $\mathbf{B}_{\mathrm{C}1}$ describing the planar point contact with friction

$$\mathbf{G}_1 = \mathcal{A}_1^{\mathrm{T}} \mathbf{B}_{\mathrm{C}1} = \begin{bmatrix} 0 & 1 & 0 \\ -1 & 0 & 0 \\ r & 0 & 1 \end{bmatrix} \begin{bmatrix} 1 & 0 \\ 0 & 1 \\ 0 & 0 \end{bmatrix} = \begin{bmatrix} 0 & 1 \\ -1 & 0 \\ r & 0 \end{bmatrix}. \tag{9.41}$$

The grasp matrix for both fingers has the following form

$$\mathbf{G} = \begin{bmatrix} 0 & 1 & 0 & -1 \\ -1 & 0 & 1 & 0 \\ r & 0 & r & 0 \end{bmatrix}. \tag{9.42}$$

Finally, consider the application of the grasp matrix while developing a system for evaluation and training of human hand dexterity [43]. When assessing human hand dexterity, the subject grasps an object with the tips of the fingers and changes its orientation. It is unpractical to measure the finger movements by the use of optical methods. First, it is difficult to place markers on the knuckles of the fingers and second, the markers are often hidden during the hand motion. The measurements were performed in isometric conditions, where the subject only exerts the finger forces while the fingers are fixed. The change of the orientation of the object is executed in the virtual environment based on the measured finger forces. The subject is sitting in front of the screen displaying the virtual scene.

The isometric device designed to simultaneously measure the forces and torques applied by the thumb, index, and middle finger is shown in Fig. 9.15. The isometric device consists of three, three-dimensional force-torque sensors, which are positioned close to the fingertips. The sensors are mounted on the aluminum assembly, which provides firm support for the sensors during the measurement. The measurement range of the sensors is ± 150 N for the lateral forces and ± 300 N for the axial

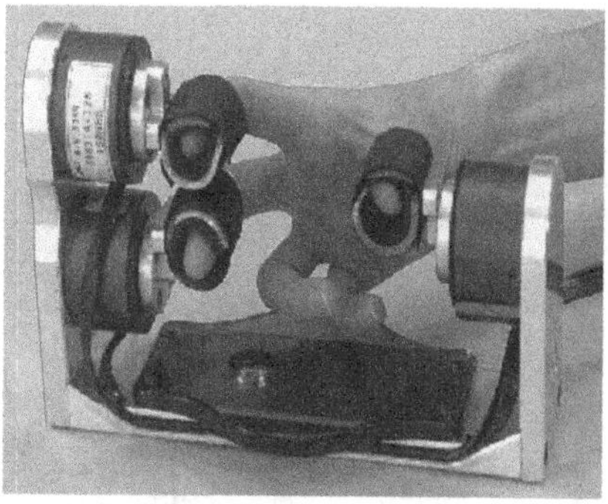

Fig. 9.15 Isometric finger device designed to simultaneously measure forces and torques applied by the thumb, index, and middle finger

force with a torque range of ± 8 Nm. During the measurement the hand is positioned between the thumb and the index and middle finger sensors. Finger supports are used to position the fingers in the correct configuration while providing transfer of forces and torques to the sensors. The finger supports are made of acrylic which provides a rigid connection between the fingertip and the sensor. The shape of the finger support was designed considering anthropometric and ergonomic factors. The fingers are fastened to the support using Velcro straps. The device can be applied to either left or right hand measurement by rotating the orientation of the sensor platform.

A mathematical model was developed with the aim of transforming finger forces measured in the real environment into displacements of the grasped object in the virtual environment. To describe the three-fingered grasping, the grasp matrix $\mathbf{G}$ was used, mapping the finger forces and torques $\mathbf{F}_{Ci}$ into the resultant wrench $\mathbf{w}_{\text{O}}$. The mapping was performed by use of (9.27). The model of grasping a virtual object is shown in Fig. 9.16. In our model of grasping, we assumed that the location of the fingers, when in contact with the object, is fixed relative to its center of mass. Two contact types were implemented in our model: point contact with friction and soft finger contact. A Coulomb friction model was employed in which the normal force was defined to be positive and the lateral forces proportional to the applied normal force.

For realistic interaction with an object in a virtual environment, a mathematical model of the body dynamics is needed. The model describes dynamic behavior of the object as influenced by the fingertip forces and external forces or torques. In the model, the object's motion is constrained by stiffness and friction (virtual springs and dampers) in all six degrees of freedom. Dynamic behavior of the object is controlled by adjusting the stiffness and friction parameters. With high stiffness of the virtual springs and sufficient friction, the speed of movement of the object can be directly proportional to the grasping force. The dynamic model of the object incorporates the object's mass, moment of inertia, basic shape (e.g. sphere, box), and location of its mass center. To describe the virtual object's dynamics in the local coordinates, we used the Newton-Euler equations written in matrix form as follows

$$\mathbf{M}\ddot{\mathbf{x}} + \mathbf{C}\dot{\mathbf{x}} + \mathbf{N}\mathbf{x} = \mathbf{w}_{\text{O}}. \tag{9.43}$$

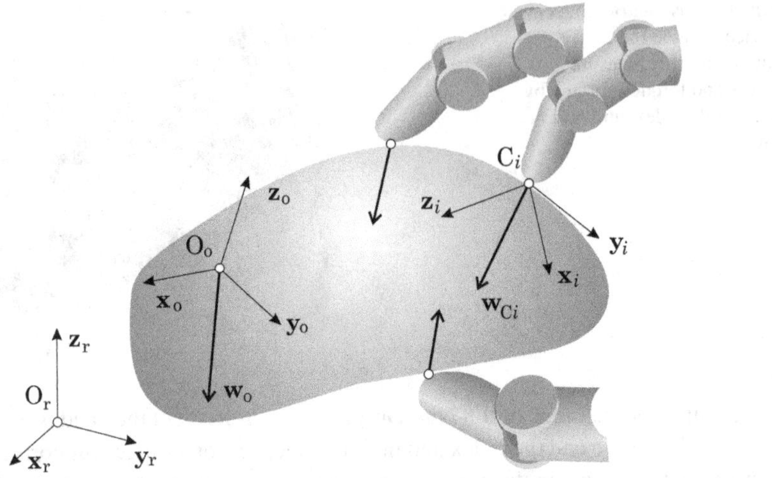

Fig. 9.16 Model of grasping a virtual object with three fingers

Fig. 9.17 Grasping a box in virtual environment

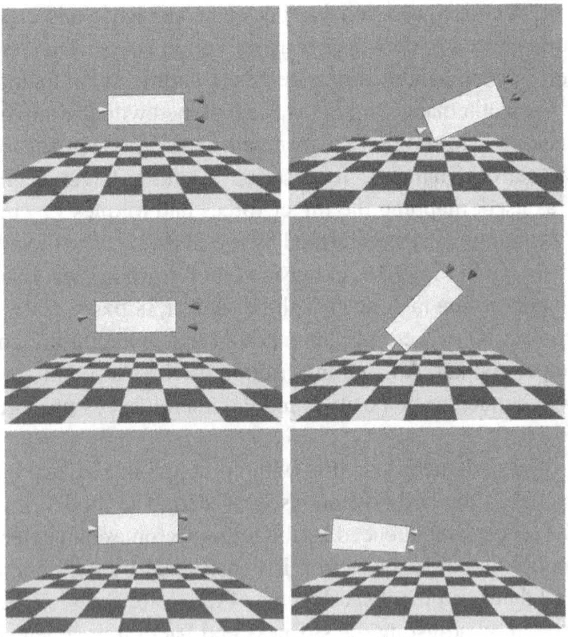

In the above equation the vector **x** describes the object's pose (three parameters for position and roll-pitch-yaw parameters for orientation), the matrix **M** is the inertia matrix consisting of object's mass and mass moments of inertia, **C** is a diagonal matrix of friction coefficients, and **N** is a diagonal matrix of stiffness coefficients. The total wrench on the object $\mathbf{w}_O$, which is the consequence of the three fingers acting on the virtual object, was derived from (9.27). In our environment, gravity was excluded from the model. The acceleration vector is expressed in the following form

$$\ddot{\mathbf{x}} = \mathbf{M}^{-1}(\mathbf{w}_o - \mathbf{C}\dot{\mathbf{x}} - \mathbf{N}\mathbf{x}), \tag{9.44}$$

while the position and orientation of the object is obtained by integration

$$\mathbf{x} = \iint \ddot{\mathbf{x}} dt = \iint \mathbf{M}^{-1}(\mathbf{w}_o - \mathbf{C}\dot{\mathbf{x}} - \mathbf{N}\mathbf{x}) dt. \tag{9.45}$$

The equation describes the dynamic behavior of the virtual object in space and time resulting from the total wrench on the object, its physical properties, and given environmental variables.

Changing of the orientation of a box by a three-finger grasp in the virtual environment is shown in Fig. 9.17. The virtual symbolic fingers are shown as geometric cones. Their position on the box is fixed. The virtual finger comes into contact with the object when its force exceeds a preselected threshold. When the threshold is exceeded, the color of the virtual finger is changed. When only one finger is in contact with the object, pushing occurs in the direction of the force measured at this particular finger. The virtual object is grasped and rotated with two or three fingers acting in opposition. The approach described was used to evaluate the grasp dexterity of stroke victims [43].

Chapter 10
Kinematic Model of the Human Hand

Abstract In the last chapter we present the direct and inverse kinematic model of the thumb and fingers of a human hand. The model of the hand is developed by assembling the palm, the fingers, and the thumb and expressing their poses in the reference coordinate frame of the wrist. The introduced kinematic model was obtained based on a series of optical measurements of the human hand. The introduced model enables us to analyze the motion of the human hand, depending on the length and width of the palm.

The skeleton of the hand is shown in Fig. 10.1. It consists of twenty seven bones: fourteen finger bones, five palm bones, and eight wrist bones. The finger bones are interconnected through the proximal and distal interphalangeal joints, while they are connected to the palm bones through the metacarpophalangeal joints.

The anatomical structure of the thumb differs from that of the other fingers. The metacarpal bones are interconnected with muscles, tendons, and soft tissue. The relative displacements among them are small and difficult to measure. In contrast with fingers, the metacarpal bone of the thumb is completely mobile. Its motion is provided by the carpometacarpal joint connecting the palm bone to one of the wrist bones (*os trapezium*). The surfaces of both parts of the carpometacarpal joint have a saddle shape and are loosely interconnected. Such anatomical structure enables the thumb to execute, apart from flexion-extension and abduction-adduction movements, also passive internal rotation. This internal rotation provides opposition to the rest of the fingers of the human hand. The thumb has only two segments: proximal and distal. The proximal segment is connected to the palm bone by the metacarpophalangeal joint, while the interphalangeal joint links both segments together.

The main advantage of the kinematic models of the finger and the thumb presented in this chapter, in comparison to the models existing in the literature, is the adaptability to conform to the size of the hand of a particular subject. The kinematic parameters are varied with respect to the length a and the width b of the hand. Here, the length of the hand is measured as the distance between the distal wrist flexion fold and the tip of the middle finger. The width of the hand is defined as

J. Lenarčič et al., *Robot Mechanisms*,
Intelligent Systems, Control and Automation: Science and Engineering 60,
DOI 10.1007/978-94-007-4522-3_10, © Springer Science+Business Media Dordrecht 2013

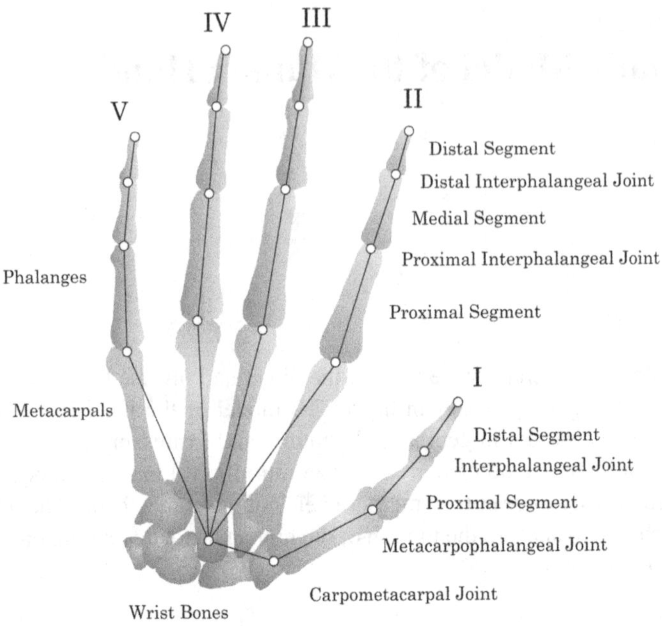

Fig. 10.1 Anatomical properties of human hand consisted from wrist, palm, and finger segments

the width of the palm in the region of the metacarpophalangeal joints, when the fingers are placed parallel to each other and completely extended. The length and the width of the hand of the subject participating in this study was $a = 204$ mm (8.03 in) and $b = 90$ mm (3.54 in). The lengths of particular bone segments and the positions of the most proximal joints (metacarpophalangeal joints for the fingers and carpometacarpal joint for the thumb) were calculated by the use of the statistical anthropometrical parameters. The models presented here correspond to the right hand.

The pose of the thumb, palm, and fingers will be described with respect to the reference coordinate frame of the hand. It is attached to a point of the wrist bone (*os capitatum*) lying on the axis of the radial-ulnar deviation of the wrist. The **z** axis of the reference frame is aligned with the tip of the middle finger when the middle finger is completely extended and the abduction-adduction angle in its metacarpophalangeal joint is zero. The **y** axis is perpendicular to the palm and directed away from the dorsum of the hand. The **x** axis completes the right-handed coordinate frame. It is level with the palm and directed towards the little finger.

10.1 Kinematic Model of the Finger

The kinematic model of a finger is developed using the method of Vector Parameters introduced in the second chapter. The rotation axes are in Fig. 10.2 marked

Fig. 10.2 Kinematic finger model

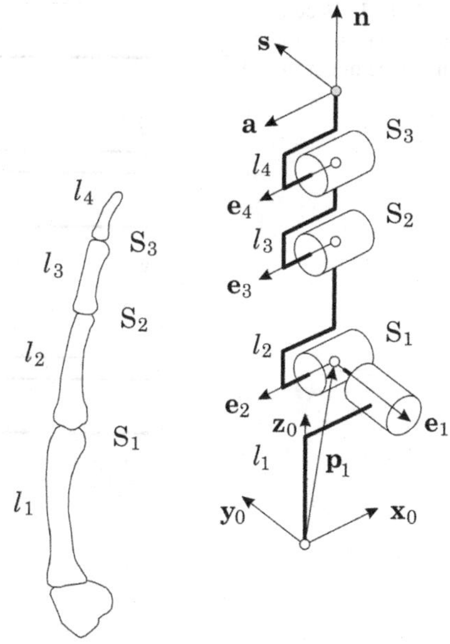

by the Cartesian vectors $\mathbf{e}_1, \ldots, \mathbf{e}_4$, while the corresponding rotations are denoted as $\theta_1, \ldots, \theta_4$. In this way every finger has four degrees of freedom. The metacarpophalangeal joint S_1 was replaced by the universal joint (Fig. 2.4). The first degree of freedom, being perpendicular to the palm, belongs to the abduction-adduction of the metacarpophalangeal joint. The second degree of freedom describes the flexion-extension of the metacarpophalangeal joint. The proximal interphalangeal joint S_2 and the distal interphalangeal joint S_3 were replaced by the rotational joints. All finger joints flexion-extension axes are parallel, while the rotation axes of the metacarpophalangeal joint S_1 intersect perpendicularly. The lengths of the proximal, medial, and distal segments are denoted with parameters l_2, l_3, and l_4. The length of the palm l_1 is included in the components of vector $\mathbf{p}_1$. These are the components x_{S1}, y_{S1}, and z_{S1}, describing the position of the metacarpophalangeal joint in the hand reference frame. The kinematic model of the finger can be described by the vector parameters collected in Table 10.1 [85].

The direct kinematic finger model can be written as the following product of the homogeneous matrices

$$\mathbf{H} = \begin{bmatrix} \mathbf{n}^{(0)} & \mathbf{s}^{(0)} & \mathbf{a}^{(0)} & \mathbf{p}^{(0)} \\ 0 & 0 & 0 & 1 \end{bmatrix} = \mathbf{H}_{0,1}\mathbf{H}_{1,2}\mathbf{H}_{2,3}\mathbf{H}_{3,4}\mathbf{H}_{4,5}. \tag{10.1}$$

The vectors **n**, **s**, **a** are the axes of the coordinate frame attached to the tip of the finger. The vector **p** represents the position of the finger tip in the reference frame $\mathbf{x}_0, \mathbf{y}_0, \mathbf{z}_0$. The above transformation matrices are obtained by the use of expression

Table 10.1 Vector parameters describing kinematic finger model

i	1	2	3	4
θ_i	θ_1	θ_2	θ_3	θ_4
d_i	0	0	0	0

i	1	2	3	4
$\mathbf{e}_i^{(i-1)}$	0	−1	−1	−1
	−1	0	0	0
	0	0	0	0

i	1	2	3	4	5
$\mathbf{b}_{i-1,i}^{(i-1)}$	x_{S1}	0	0	0	0
	y_{S1}	0	0	0	0
	z_{S1}	0	l_2	l_3	l_4

(2.12) in the following form

$$\mathbf{H}_{0,1} = \begin{bmatrix} c_1 & 0 & -s_1 & x_{S1} \\ 0 & 1 & 0 & y_{S1} \\ s_1 & 0 & c_1 & z_{S1} \\ 0 & 0 & 0 & 1 \end{bmatrix},$$
$$\mathbf{H}_{1,2} = \begin{bmatrix} 1 & 0 & 0 & 0 \\ 0 & c_2 & s_2 & 0 \\ 0 & -s_2 & c_2 & 0 \\ 0 & 0 & 0 & 1 \end{bmatrix},$$
$$\mathbf{H}_{2,3} = \begin{bmatrix} 1 & 0 & 0 & 0 \\ 0 & c_3 & s_3 & 0 \\ 0 & -s_3 & c_3 & l_2 \\ 0 & 0 & 0 & 1 \end{bmatrix}, \tag{10.2}$$
$$\mathbf{H}_{3,4} = \begin{bmatrix} 1 & 0 & 0 & 0 \\ 0 & c_4 & s_4 & 0 \\ 0 & -s_4 & c_4 & l_3 \\ 0 & 0 & 0 & 1 \end{bmatrix},$$
$$\mathbf{H}_{4,5} = \begin{bmatrix} 0 & 0 & -1 & 0 \\ 0 & 1 & 0 & 0 \\ 1 & 0 & 0 & l_4 \\ 0 & 0 & 0 & 1 \end{bmatrix}.$$

Table 10.2 Anthropometric factors for calculation of kinematic finger parameters

i	II	III	IV	V	
l_2	0.245	0.266	0.244	0.204	$\times a$
l_3	0.143	0.170	0.165	0.117	$\times a$
l_4	0.097	0.108	0.107	0.093	$\times a$
x_{S1}	−0.251	0	0.206	0.402	$\times b$
z_{S1}	0.447	0.446	0.409	0.368	$\times a$

The anthropometric factors are collected in Table 10.2. They must be multiplied by the length of the hand a, in order to calculate the lengths of the segments l_2, l_3, and l_4 for each particular finger. The positions of the metacarpophalangeal joints x_{S1} and z_{S1} are obtained by multiplying the anthropometric factors with the hand width b and the hand length a. The Roman numerals II to V denote the fingers from the index to the little finger. As the anthropometric data for the hand are only given in the plane, the values of the parameters y_{S1} for the finger metacarpophalangeal joints equal zero.

10.2 Inverse Kinematics of the Finger

The kinematic model of the human finger used here is a serial mechanism with four degrees of freedom. The joint coordinates $\theta_1, \ldots, \theta_4$ represent abduction-adduction and flexion-extension in the metacarpophalangeal joint S_1, flexion-extension in the proximal interphalangeal joint S_2 and flexion-extension in the distal interphalangeal joint S_3. The kinematic model of the finger, viewing onto the frontal plane, is shown on the left side of Fig. 10.3. The right side of the same figure presents the view onto the sagittal plane, when the joint variables are rotated in a positive direction. The vector $\mathbf{p}$ represents the position of the finger tip, while the position of the metacarpophalangeal joint S_1 is given by the vector $\mathbf{p}_1$ with the components x_{S1}, y_{S1}, and z_{S1}. The vector $\mathbf{r}_3$ connects the position of the metacarpophalangeal joint S_1 to the distal interphalangeal joint S_3.

Given the position of the finger tip, defined by the vector $\mathbf{p}$, and the pointing direction of the distal segment, defined by the unit vector $\mathbf{n}$, the inverse kinematics problem can be solved algebraically. We have

$$\mathbf{r}_3^{(0)} = \mathbf{p}^{(0)} - l_4\mathbf{n}^{(0)} - \mathbf{p}_1^{(0)}. \tag{10.3}$$

Define the length of the vector $\mathbf{r}_3$ and the lengths of the projections of the vectors $\mathbf{r}_3$ and $\mathbf{n}$ to the plane perpendicular to the axis $\mathbf{y_0}$ in the following way

$$r_3 = \sqrt{r_{3\mathrm{x}}^2 + r_{3\mathrm{y}}^2 + r_{3\mathrm{z}}^2}, \qquad r_{3\mathrm{xz}} = \sqrt{r_{3\mathrm{x}}^2 + r_{3\mathrm{z}}^2}, \qquad n_{\mathrm{xz}} = \sqrt{n_{\mathrm{x}}^2 + n_{\mathrm{z}}^2}.$$

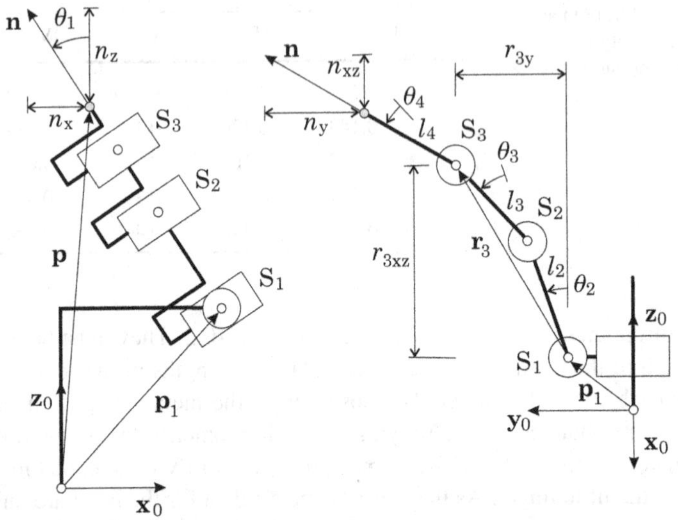

Fig. 10.3 Kinematic finger model presented in frontal plane (*left*) and sagittal plane (*right*)

The solution to the inverse kinematics problem for the human finger have already been developed, in Sect. 3.3, while studying the 3R mechanism and they do not need to be redeveloped here in detail. First, we analyze the inclined triangle determined by the joint centers S_1, S_2, and S_3. In this triangle the side l_3 lies opposite to the point S_1, side r_3 opposite to S_2, and l_2 opposite to the point S_3. The angle θ_3, lying at the point S_2, is obtained by the law of cosines

$$\theta_3 = \pi - \arccos\left(\frac{l_2^2 + l_3^2 - r_3^2}{2l_2l_3}\right). \tag{10.4}$$

The angle θ_2 is found by calculating the angle lying in the triangle at the point S_1

$$\theta_2 = \arctan\left(\frac{r_{3y}}{r_{3xz}}\right) - \arccos\left(\frac{r_3^2 + l_2^2 - l_3^2}{2r_3l_2}\right), \tag{10.5}$$

likewise, from Fig. 10.3 the following relation is developed

$$\theta_4 = \arctan\left(\frac{n_y}{n_{xz}}\right) - (\theta_2 + \theta_3). \tag{10.6}$$

In this manner, all four joint angles $\theta_1, \ldots, \theta_4$ are determined. When solving inverse trigonometric equations we must be aware of the multiple solutions. This is not a great problem for human fingers, where the expected values of the joints θ_2, θ_3, and θ_4 are constrained by anatomical ranges of motion.

Because of the specific biomechanical structure of the finger, where the tendons are running over several joints, the joint variables at certain joints are coupled. Based on the measurements, the following relation between the angles in proximal and dis-

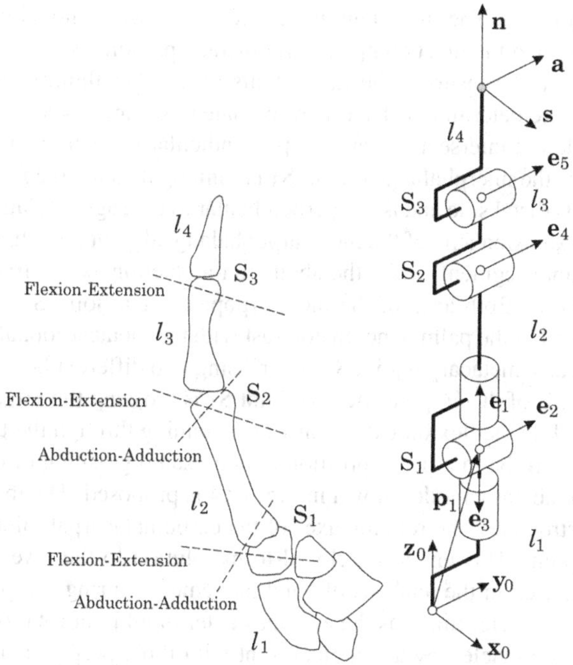

Fig. 10.4 The rotational axes in human thumb and its kinematic model

tal interphalangeal joints can be introduced into the inverse kinematic finger model

$$\theta_4 = c\theta_3, \tag{10.7}$$

where c is a coefficient which is constant during movement. Its values are 0.32, 0.36, 0.16, and 0.25 for the index, middle, ring, and little finger respectively. Such a relation between the joint variables of the kinematic finger model simplifies the calculation of the inverse kinematics. With this relation, it is only necessary to know the position of the tip of the finger, the orientation of the distal segment is a consequence of the fingertip position. The above equations of the inverse finger kinematics can be reorganized in such a way, that the position of the distal interphalangeal joint S_3 is used as input data, which is advantageous when measuring the hand movements during grasping.

10.3 Kinematic Model of the Thumb

Because of its different role in grasping objects, the movements of the thumb are more complex when compared to the fingers. The main difference is the number of degrees of freedom of the carpometacarpal joint S_1. The movements of this joint are of crucial importance for grasping objects. The descriptions of the number and the layout of the thumb joint axes together with their positions differ in the literature. Researchers are more or less in agreement regarding the simplest interphalangeal joint S_3. The important differences occur when modeling the metacarpophalangeal S_2 and the most complex carpometacarpal joint S_1.

By measuring the movements of cadaver thumbs, the five main rotation axes were determined. Their approximate poses are shown in Fig. 10.4. The thumb axes do not intersect and are not perpendicular to each other. The flexion-extension axis of the interphalangeal joint S_3 is running through the proximal finger segment from its distal side and is not perpendicular to the sagittal plane of the thumb. The flexion-extension axis of the metacarpophalangeal joint S_2 is fixed with respect to the proximal segment, while the abduction-adduction axis is fixed with respect to the palm bone. Both axes of the metacarpophalangeal joint S_2 are going through the distal end of the palm bone. In contrast with the metacarpophalangeal joint, the axes of the carpometacarpal joint S_1 are crossing two different bones. The abduction-adduction axis of the carpometacarpal joint S_1 is crossing the proximal end of the palm bone, while the flexion-extension axis is running through the trapezium bone.

To be in a better position to understand grasping of different objects, the thumb kinematic model shown in Fig. 10.4 is proposed. The model has a simple kinematic structure. The rotation axes of the carpometacarpal joint S_1 are united into a single point. This enables a closed-form solution to the inverse kinematics, which is required in the analysis of hand movements during grasping. The center of the joint S_1 is determined as the average center of all joint rotations. The degrees of freedom are modeled by a spherical joint with three perpendicular rotation axes. The rotation about the first axis represents the abduction-adduction of the carpometacarpal joint S_1, while the rotation about the second axis belongs to flexion-extension. The degree of freedom about the axis aligned with the segment represents the internal rotation in the carpometacarpal joint S_1. This enables opposition of the thumb with respect to the other fingers of the hand.

The metacarpophalangeal joint S_2 is modeled as a rotational joint describing only the flexion-extension movement. The abduction-adduction was neglected, as the joint mobility in this direction is very small. In the case when the third rotation in the joint S_1 equals zero, the flexion-extension axes in the joints S_1 and S_2 are parallel. The simplest thumb movement occurs in the interphalangeal joint which is described by the rotational joint S_3. This joint is considerably more mobile than the metacarpophalangeal joint S_2 and has a very important role in dexterous manipulation. It was assumed that the joint axes S_2 and S_3 are parallel.

The rotation axes in Fig. 10.4 are denoted by the vectors $\mathbf{e}_1 \dots \mathbf{e}_5$. The directions of the rotation axes are selected in such a way that the flexion angles about $\mathbf{e}_2$, $\mathbf{e}_4$, and $\mathbf{e}_5$ are positive. With positive internal rotation about the axis $\mathbf{e}_3$ the thumb rotates towards the palm. The parameters l_2, l_3, and l_4 represent the lengths of the palm bone, and proximal and distal segments respectively. The length l_1 describes the distance from the carpometacarpal joint S_1 to the origin of the wrist coordinate frame. The coefficients necessary to calculate the kinematic parameters from the external hand dimensions a and b are listed in Table 10.3. The parameters x_{S1}, y_{S1}, and z_{S1} determine the position of the carpometacarpal joint with respect to the reference frame in the wrist. The parameters l_2, l_3, l_4, and z_{S1} are obtained by multiplying the anthropometric factors with the length of the hand a, while the parameter x_{S1} is

Table 10.3 Factors for calculation of thumb kinematic parameters

l_2	l_3	l_4	z_{S1}	x_{S1}	y_{S1}
$0.251 \times a$	$0.196 \times a$	$0.158 \times a$	$0.073 \times a$	$-0.196 \times b$	10 mm

Table 10.4 Vector parameters describing kinematic model of the thumb

i	1	2	3	4	5
θ_i	θ_1	θ_2	θ_3	θ_4	θ_5
d_i	0	0	0	0	0

i	1	2	3	4	5
$\mathbf{e}_i^{(i-1)}$	0	0	0	0	0
	0	1	0	1	1
	1	0	-1	0	0

i	1	2	3	4	5	6
$\mathbf{b}_{i-1,i}^{(i-1)}$	x_{S1}	0	0	0	0	0
	y_{S1}	0	0	0	0	0
	z_{S1}	0	0	l_2	l_3	l_3

given by multiplying the factors with the hand width b. The carpometacarpal joint of the thumb does not lie in the same plane as the metacarpophalangeal finger joints. As the anthropometric hand data are given only in the plane, the value $y_{S1} = 10$ mm was assumed.

The vector parameters describing the kinematic thumb model from Fig. 10.4 are listed in Table 10.4.

The direct kinematics of the thumb is given by the product of the transformation matrices

$$\mathbf{H} = \begin{bmatrix} \mathbf{n}^{(0)} & \mathbf{s}^{(0)} & \mathbf{a}^{(0)} & \mathbf{p}^{(0)} \\ 0 & 0 & 0 & 1 \end{bmatrix} = \mathbf{H}_{0,1}\mathbf{H}_{1,2}\mathbf{H}_{2,3}\mathbf{H}_{3,4}\mathbf{H}_{4,5}\mathbf{H}_{5,6}. \quad (10.8)$$

The vectors $\mathbf{n}$, $\mathbf{s}$, $\mathbf{a}$ belong to the axes of the coordinate frame attached to the finger tip, while the vector $\mathbf{p}$ represents the position of the finger tip in the reference frame $\mathbf{x}_0$, $\mathbf{y}_0$, $\mathbf{z}_0$. The transformation matrices are calculated according (2.12) as follows

$$
\mathbf{H}_{0,1} = \begin{bmatrix} c_1 & -s_1 & 0 & x_{S1} \\ s_1 & c_1 & 0 & y_{S1} \\ 0 & 0 & 1 & z_{S1} \\ 0 & 0 & 0 & 1 \end{bmatrix},
$$

$$
\mathbf{H}_{1,2} = \begin{bmatrix} c_2 & 0 & s_2 & 0 \\ 0 & 1 & 0 & 0 \\ -s_2 & 0 & c_2 & 0 \\ 0 & 0 & 0 & 1 \end{bmatrix},
$$

$$
\mathbf{H}_{2,3} = \begin{bmatrix} c_3 & s_3 & 0 & 0 \\ -s_3 & c_3 & 0 & 0 \\ 0 & 0 & 1 & 0 \\ 0 & 0 & 0 & 1 \end{bmatrix},
\qquad (10.9)
$$

$$
\mathbf{H}_{3,4} = \begin{bmatrix} c_4 & 0 & s_4 & 0 \\ 0 & 1 & 0 & 0 \\ -s_4 & 0 & c_4 & l_2 \\ 0 & 0 & 0 & 1 \end{bmatrix},
$$

$$
\mathbf{H}_{4,5} = \begin{bmatrix} c_5 & 0 & s_5 & 0 \\ 0 & 1 & 0 & 0 \\ -s_5 & 0 & c_5 & l_3 \\ 0 & 0 & 0 & 1 \end{bmatrix},
$$

$$
\mathbf{H}_{5,6} = \begin{bmatrix} 0 & 1 & 0 & 0 \\ 0 & 0 & 1 & 0 \\ 1 & 0 & 0 & l_4 \\ 0 & 0 & 0 & 1 \end{bmatrix}.
$$

10.4 Inverse Kinematics of the Thumb

Figure 10.5 shows the kinematic model of the thumb in a pose where the values of all joint variables $\theta_1, \ldots, \theta_5$ are positive. On the left hand side the mechanism is shown in the reference coordinate frame and on the right hand side in the sagittal plane. The solution to the inverse kinematics problem of the thumb can be solved algebraically when the position of the tip of the finger $\mathbf{p}$ and the orientation of the distal segment of the thumb (i.e. the orientation of the frame $\mathbf{n}$, $\mathbf{s}$, $\mathbf{a}$) are given. Vector $\mathbf{r}$ connects the center of the carpometacarpal joint S_1 with the finger tip, while vector $\mathbf{r}_3$ runs from the center of the joint S_1 to the interphalangeal joint S_3 (Fig. 10.5). From Fig. 10.5 the following relations can be written

$$
\begin{aligned}
\mathbf{r}^{(0)} &= \mathbf{p}^{(0)} - \mathbf{p}_1^{(0)}, \\
\mathbf{r}_3^{(0)} &= \mathbf{p}^{(0)} - l_4\mathbf{n}^{(0)} - \mathbf{p}_1^{(0)}.
\end{aligned}
\qquad (10.10)
$$

The lengths of both vectors are computed as

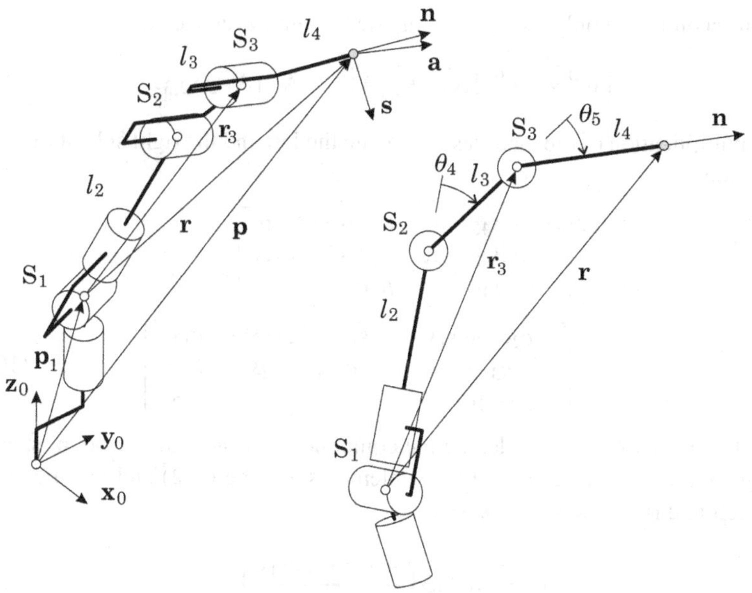

Fig. 10.5 Kinematic thumb model in reference frame (*left*) and in sagittal plane (*right*)

$$r = \sqrt{r_x^2 + r_y^2 + r_z^2}, \qquad r_3 = \sqrt{r_{3x}^2 + r_{3y}^2 + r_{3z}^2}. \tag{10.11}$$

The angles θ_4 and θ_5, rotating about the axes perpendicular to the sagittal plane through the joints S_2 and S_3, are calculated by the law of cosines applied to the inclined triangles determined by the sides r, r_3, and l_4 and r_3, l_2, and l_3 respectively (Fig. 10.5). The result is

$$\theta_4 = \pi - \arccos\left(\frac{l_2^2 + l_3^2 - r_3^2}{2l_2 l_3}\right), \tag{10.12}$$

and

$$\theta_5 = \pi - \arccos\left(\frac{r_3^2 + l_4^2 - r^2}{2r_3 l_4}\right) - \alpha, \tag{10.13}$$

where

$$\alpha = \arccos\left(\frac{l_3^2 + r_3^2 - l_2^2}{2l_3 r_3}\right).$$

The values of the angles θ_4 and θ_5 are then used to calculate the angles θ_1, θ_2, and θ_3 in the carpometacarpal joint S_1. This is accomplished by post-multiplying the transformation matrix $\mathbf{H}$ from (10.8) by the inverse transformation matrices $\mathbf{H}_{3,4}^{-1}$, $\mathbf{H}_{4,5}^{-1}$, and $\mathbf{H}_{5,6}^{-1}$, which gives

$$\mathbf{H}\left(\mathbf{H}_{5,6}^{-1}\mathbf{H}_{4,5}^{-1}\mathbf{H}_{3,4}^{-1}\right) = \mathbf{H}_{0,1}\mathbf{H}_{1,2}\mathbf{H}_{2,3}. \tag{10.14}$$

It is sufficient to use only the rotation matrices from the above equation

$$\left[\mathbf{n}^{(0)}\mathbf{s}^{(0)}\mathbf{a}^{(0)}\right]\mathbf{A}_{5,6}^{\mathrm{T}}\mathbf{A}_{4,5}^{\mathrm{T}}\mathbf{A}_{3,4}^{\mathrm{T}} = \mathbf{A}_{0,1}\mathbf{A}_{1,2}\mathbf{A}_{2,3}. \tag{10.15}$$

After multiplication of the matrices we equate the left and the right side of the above expression

$$\begin{bmatrix} s_{\mathrm{x}}c_{45}+n_{\mathrm{x}}s_{45} & a_{\mathrm{x}} & n_{\mathrm{x}}c_{45}-s_{\mathrm{x}}s_{45} \\ s_{\mathrm{y}}c_{45}+n_{\mathrm{y}}s_{45} & a_{\mathrm{y}} & n_{\mathrm{y}}c_{45}-s_{\mathrm{y}}s_{45} \\ s_{\mathrm{z}}c_{45}+n_{\mathrm{z}}s_{45} & a_{\mathrm{z}} & n_{\mathrm{z}}c_{45}-s_{\mathrm{z}}s_{45} \end{bmatrix} = \begin{bmatrix} c_1c_2c_3+s_1s_3 & -s_1c_3+c_1c_2s_3 & c_1s_2 \\ s_1c_2c_3-c_1s_3 & c_1c_3+s_1c_2s_3 & s_1s_2 \\ -s_2c_3 & -s_2s_3 & c_2 \end{bmatrix}. \tag{10.16}$$

Here, the indices x, y, and z denote the components of the unit vectors **n**, **s**, and **a**. Taking the ratio of the $(2,3)$ and $(1,3)$ elements and the $(3,2)$ and $(3,1)$ elements of the left and right sides of (10.16) gives

$$\begin{aligned} \theta_1 &= \arctan_2\left(\frac{n_{\mathrm{y}}c_{45}-s_{\mathrm{y}}s_{45}}{n_{\mathrm{x}}c_{45}-s_{\mathrm{x}}s_{45}}\right), \\ \theta_3 &= \arctan_2\left(\frac{-a_{\mathrm{z}}}{-s_{\mathrm{z}}c_{45}-n_{\mathrm{z}}s_{45}}\right). \end{aligned} \tag{10.17}$$

The solution for θ_2 comes from the $(3,3)$ element

$$\theta_2 = \arccos(n_{\mathrm{z}}c_{45}-s_{\mathrm{z}}s_{45}). \tag{10.18}$$

While calculating the inverse kinematics of the thumb we may encounter considerable problems because of the multiple solutions which arise from the inverse trigonometric functions. We must take into consideration whether a combination of the calculated values for the joint angles is a mathematically correct result. Even mathematically correct results can represent solutions which are not within the biomechanical constraints of the human thumb and must be therefore discarded when studying human hand movements.

10.5 Thumb and Fingers Pose with Respect to Palm

Figure 10.6 shows the pose of the thumb and the fingers in the reference frame of the wrist, according to the parameters from Tables 10.2 and 10.3. The upper portion of the figure shows the mechanisms of the fingers and thumb in the reference pose as described in kinematic models from Figs. 10.2 and 10.4, using the parameters given in Tables 10.1 and 10.4.

All joints and finger segments, when in the reference pose, are lying in a plane defined by the vectors $\mathbf{x}_0, \mathbf{z}_0$. The thumb however is lying in a parallel plane which is

Fig. 10.6 Thumb and fingers in initial (*above*) and arbitrary pose (*below*)

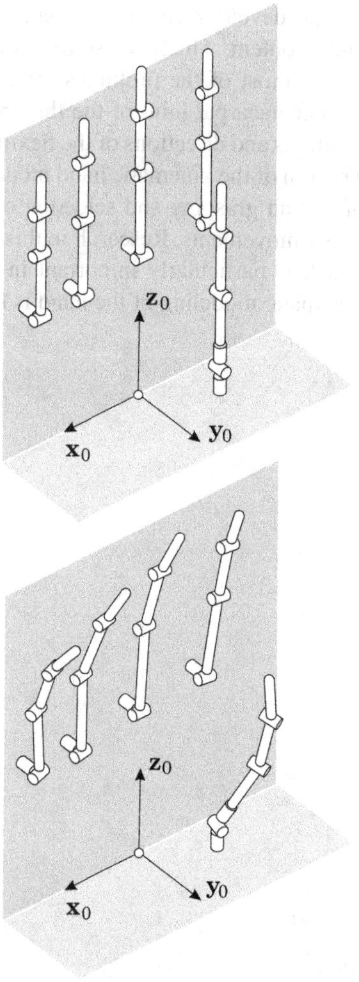

shifted in the positive direction of vector $\mathbf{y}_0$. The flexion-extension axes in metacarpophalangeal and proximal and distal interphalangeal joints of the fingers are parallel to the vector $\mathbf{x}_0$. The abduction-adduction axes in metacarpophalangeal finger joints are parallel to vector $\mathbf{y}_0$. In the reference pose of the thumb, the axes of abduction-adduction and internal rotation of the carpometacarpal joint are parallel to vector $\mathbf{z}_0$, while the flexion-extension axis is aligned with vector $\mathbf{y}_0$. Also, in the reference pose, the flexion-extension axes in the metacarpophalangeal and the interphalangeal joint are parallel to vector $\mathbf{y}_0$. Such a pose of the hand occurs when the hand is lying on a level surface with the fingers completely extended. The lower portion of Fig. 10.6 shows the pose of the fingers and the thumb when the joint variables are nonzero. The values of the joint variables were selected in such a way that the pose of the mechanism resembles that of a human hand.

The development of a kinematic model of the human hand is a rather complex problem, which is also true for other parts of human locomotor systems. With hands, most of the problems arise when determining the kinematic model of the carpometacarpal joint of the thumb. In the literature we find scarce data about the positions and directions of the flexion-extension and abduction-adduction joint axes. The aim of the kinematic hand model described in this chapter is mainly the analysis of human grasping and selection of the variables appropriate for the evaluation of finger movements. Research in this area can help us understand human movements, which is particularly important in sports and medicine. Of utmost importance in kinematic modeling of the hand is the development of humanoid robots.

References

1. J. Angeles, *Rational Kinematics* (Springer, New York, 1988)
2. J. Angeles, Int. J. Robot. Res. **11**(3), 196 (1992)
3. J. Angeles, *Fundamentals of Robotic Mechanical Systems*, 3rd edn. (Springer, New York, 2007)
4. J. Babič, J. Lenarčič, in *On Advances in Robot Kinematics*, ed. by J. Lenarčič, C. Galletti (Kluwer Academic Publishers, Dordrecht, 2004), pp. 155–162
5. J. Babič, L. Bokman, D. Omrčen, J. Lenarčič, F. Park, J. Mech. Robot. **1**, 1 (2009)
6. T. Bajd, M. Mihelj, J. Lenarčič, A. Stanovnik, M. Munih, *Robotics* (Springer, Dordrecht, 2010)
7. D.R. Baker, C.W. Wampler, Int. J. Robot. Res. **7**(2), 3 (1988)
8. R.S. Ball, *A Treatise on the Theory of Screws* (Cambridge University Press, Cambridge, 1900)
9. Z. Balorda, T. Bajd, IEEE Trans. Robot. Autom. **10**(4), 535 (1994)
10. O. Bottema, B. Roth, *Theoretical Kinematics* (Dover, New York, 1979/1990)
11. R.C. Brost, Int. J. Robot. Res. **7**(1), 3 (1988)
12. M. Ceccarelli, A. Vinciguerra, Int. J. Robot. Res. **14**(2), 152 (1995)
13. G.S. Cirikjian, J.W. Burdick, IEEE Trans. Robot. Autom. **10**(3), 343 (1994)
14. R. Clavel, in *Proceedings 18th International Symposium on Industrial Robots*, Sydney, Australia (1988), pp. 91–100
15. J.J. Craig, *Introduction to Robotics: Mechanics and Control*, 3rd edn. (Pearson/Prentice-Hall, Upper Saddle River, 2005)
16. M.R. Cutkosky, *Robot Grasping and Fine Manipulation* (Kluwer Academic, Boston, 1985)
17. J. Denavit, R.S. Hartenberg, J. Appl. Mech. **22**(2), 215 (1955)
18. R. Di Gregorio, V. Parenti-Castelli, in *Advances in Robot Kinematics: Analysis and Control*, ed. by J. Lenarčič, M.L. Husty (Kluwer Academic, Dordrecht, 1998), pp. 49–58
19. F. DiCaprio, M.M. Stanišić, J. Mech. Des. **116**(1), 17 (1994)
20. P. Dietmaier, in *Advances in Robot Kinematics: Analysis and Control* (Kluwer Academic, Dordrecht, 1998), pp. 7–16
21. J. Duffy, *Analysis of Mechanisms and Robot Manipulators* (Arnold, London, 1980)
22. R. Featherstone, Int. J. Robot. Res. **2**(2), 35 (1983)
23. R. Featherstone, *Robot Dynamics Algorithms* (Kluwer Academic, Dordrecht, 1987)
24. J. Furusho, S. Onishi, in *Proceedings 15th International Conference on Industrial Robots*, Tokyo (1985), pp. 1051–1058
25. C. Gosselin, J. Angeles, J. Mech. Transm. Autom. Des. **110**, 35 (1988)
26. V.E. Gough, S.G. Whitehall, in *Proceedings 9th International Technical Congress F.I.S.I.T.A.*, Institution of Mechanical Engineers (1962), pp. 117–135
27. J.M. Hollerbach, K.C. Suh, IEEE J. Robot. Autom. **3**(3), 308 (1987)

J. Lenarčič et al., *Robot Mechanisms*,
Intelligent Systems, Control and Automation: Science and Engineering 60,
DOI 10.1007/978-94-007-4522-3, © Springer Science+Business Media Dordrecht 2013

28. M. Honegger, A. Codourey, E. Burdet, in *Proceedings IEEE Robotics and Automation Conference*, Albuquerque (1997)
29. K.H. Hunt, *Kinematic Geometry of Mechanisms* (Clarendon Press, Oxford, 1978)
30. K.H. Hunt, J. Mech. Transm. Autom. Des. **105**, 705 (1983)
31. M.L. Husty, Mech. Mach. Theory **31**(4), 365 (1996)
32. T. Iberall, Int. J. Robot. Res. **16**, 285 (1997)
33. C. Innocenti, V. Parenti-Castelli, Mech. Mach. Theory **28**(4), 553 (1993)
34. J. Kieffer, IEEE Trans. Robot. Autom. **10**(1), 1 (1994)
35. J. Kieffer, J. Lenarčič, in *Proceedings 3rd International Symposium on Advances in Robot Kinematics*, Ferrara, Italy (1992), pp. 65–72
36. C.A. Klein, C.H. Huang, IEEE Trans. Syst. Man Cybern. **13**(3), 245 (1983)
37. C.A. Klein, C. Chu-Jenq, S. Ahmed, IEEE Trans. Robot. Autom. **11**(1), 50 (1995)
38. V.C. Klema, A.J. Laub, IEEE Trans. Autom. Control **25**(2), 164 (1980)
39. P. Kovacs, G. Hommel, in *Advances in Robot Kinematics*, Ferrara, Italy (1992), pp. 88–95
40. P. Kovacs, G. Hommel, in *Computational Kinematics*, ed. by J. Angeles, G. Hommel, P. Kovacs (Kluwer Academic, Dordrecht, 1993), pp. 27–39
41. A. Kumar, K.J. Waldron, J. Mech. Des. **103**, 665 (1981)
42. G. Kurillo, T. Bajd, R. Kamnik, J. Autom. Control **12**(1), 38 (2002)
43. G. Kurillo, M. Mihelj, M. Munih, T. Bajd, Presence **16**, 239 (2007)
44. J. Lenarčič, Robotica **1**, 205 (1983)
45. J. Lenarčič, Robotica **3**, 21 (1985)
46. J. Lenarčič, in *International Encyclopedia of Robotics*, ed. by R. Dorf (Wiley, New York, 1988)
47. J. Lenarčič, Lab. Robot. Autom. **6**(6), 293 (1994)
48. J. Lenarčič, Lab. Robot. Autom. **8**, 11 (1996)
49. J. Lenarčič, in *Proceedings IEEE International Conference on Robotics and Automation*, Leuven, Belgium (1998), pp. 3235–3240
50. J. Lenarčič, Robot. Auton. Syst. **30**, 231 (2000)
51. J. Lenarčič, CIT, J. Comput. Inf. Technol. **10**(2), 125 (2002)
52. J. Lenarčič, M.M. Stanišić, IEEE Transactions on Robotics and Automation **19** (2003)
53. J. Lenarčič, U. Stanič, U. Oblak, Robot. Comput.-Integr. Manuf. **5**(2/3), 235 (1989)
54. J. Lenarčič, U. Stanič, P. Oblak, in *Proceedings 23rd International Symposium on Industrial Robots*, Barcelona, Spain (1992), pp. 277–282
55. J. Lenarčič, M.M. Stanišić, V. Parenti-Castelli, in *Proceedings IEEE Robotics and Automation Conference*, San Francisco (2000), pp. 27–32
56. A.A. Maciejewski, C.A. Klein, Int. J. Robot. Res. **4**(3), 109 (1985)
57. M.T. Mason, J.K. Salisbury, *Robot Hands and the Mechanics of Manipulation* (MIT Press, Cambridge, 1985)
58. J.M. McCarthy, *An Introduction to Theoretical Kinematics* (MIT Press, Cambridge, 1990)
59. J.M. McCarthy, *Geometric Design of Linkages* (Springer, New York, 2000)
60. J.P. Merlet, *Parallel Robots*, 2nd edn. (Springer, Dordrecht, 2006)
61. V. Milenkovic, in *Robots 11, RI/SME* (1987), pp. 13.29–13.42
62. Y. Nakamura, H. Hanafusa, T. Yoshikawa, Int. J. Robot. Res. **6**(2), 3 (1987)
63. J. Napier, J. Bone Jt. Surg. **38-B**, 902 (1956)
64. J. Napier, *Hands* (Princeton University Press, Princeton, 1980)
65. D.N. Nenchev, J. Robot. Syst. **6**(6), 769 (1989)
66. V. Parenti-Castelli, R. Di Gregorio, in *Advances in Robot Kinematics*, ed. by J. Lenarčič, M.M. Stanišić (Kluwer Academic, Dordrecht, 2000), pp. 333–334
67. R. Paul, *Robot Manipulators: Mathematics, Programming and Control* (MIT Press, Cambridge, 1981)
68. R.P. Paul, C.N. Stevenson, Int. J. Robot. Res. **2**(1), 31 (1983)
69. D.L. Pieper, B. Roth, in *Proceedings 2nd International Congress of the Theory of Machines and Mechanisms*, Zakopane, Poland (1969), pp. 159–169
70. S. Remis, M.M. Stanišić, IEEE Trans. Robot. Autom. **9**(6), 816 (1993)

71. R.G. Roberts, A.A. Maciejewski, IEEE Trans. Robot. Autom. **12**(4), 543 (1996)
72. M. Rosheim, *Robot Evolution, The Development of Anthrobotics* (Wiley, New York, 1994)
73. B. Roth, in *Computational Kinematics*, ed. by J. Angeles, G. Hommel, P. Kovacs (Kluwer Academic, Dordrecht, 1993), pp. 3–14
74. B. Roth, in *Advances in Robot Kinematics and Computational Geometry*, ed. by J. Lenarčič, B. Ravani (Kluwer Academic, Dordrecht, 1994), pp. 7–16
75. A. Ružič, in *Proceedings 4th Workshop on Robotics in Alpe-Adria Region*, Pörtschach, Austria (1995), pp. 59–62
76. S.J. Ryu, J. Kim, J.C. Hwang, C.B. Park, H.S. Cho, K. Lee, Y.H. Lee, U. Cornel, F.C. Park, J.W. Kim, J. Manuf. Sci. Eng. **8**, 681 (1998)
77. L. Sciavicco, B. Siciliano, *Modeling and Control of Robot Manipulators*, 2nd edn. (Springer, London, 2000)
78. M.M. Stanišić, O. Duta, IEEE Trans. Robot. Autom. **6**(5), 562 (1990)
79. Y. Stepanenko, M. Vukobratović, Math. Biosci. **28**, 137 (1976)
80. D. Stewart, Proc. Inst. Mech. Eng. **180**, 371 (1965)
81. T. Šupuk, T. Kodek, T. Bajd, Engineering &. Physics **27**, 790 (2005)
82. J. Trevelyan, P. Kovesi, M. Ong, D. Elford, Int. J. Robot. Res. **4**(4), 71 (1986)
83. L.W. Tsai, in *Recent Advances in Robot Kinematics*, ed. by J. Lenarčič, V. Parenti-Castelli (Kluwer Academic, Dordrecht, 1996), pp. 401–409
84. L.W. Tsai, *Robot Analysis, The Mechanics of Serial and Parallel Manipulators* (Wiley, New York, 1999)
85. M. Veber, T. Kodek, T. Bajd, M. Munih, Meccanica **42**, 451 (2007)
86. M. Vukobratović, M. Kirćanski, *Kinematics and Trajectory Synthesis of Manipulation Robots* (Springer, Berlin, 1986)
87. C.W. Wampler, A.P. Morgan, A.J. Sommesse, J. Mech. Des. **112**, 59 (1990)
88. P. Wenger, D. Chablat, in *Advances in Robot Kinematics*, ed. by J. Lenarčič, M.M. Stanišić (Kluwer Academic, Dordrecht, 2000), pp. 305–314
89. P. Wenger, J. El Omri, in *Advances in Robot Kinematics and Computational Geometry*, ed. by J. Lenarčič, B. Ravani (Kluwer Academic, Dordrecht, 1994), pp. 29–38
90. D.E. Whitney, IEEE Trans. Man-Mach. Syst. **MMS-10**, 47 (1969)
91. K. Wohlhart, Mech. Mach. Theory **29**, 581 (1994)
92. T. Yoshikawa, Int. J. Robot. Res. **4**(2), 3 (1985)
93. V.M. Zatsiorsky, *Kinematics of Human Motion* (Human Kinetics, Champaign, 1998)

Index

J. Lenarčič et al., *Robot Mechanisms*,
Intelligent Systems, Control and Automation: Science and Engineering 60,
DOI 10.1007/978-94-007-4522-3, © Springer Science+Business Media Dordrecht 2013